VOLUME 1: BAYOU TERREBONNE

HARD SCRABBLE to HALLELUJAH
LEGACIES
of TERREBONNE PARISH, LOUISIANA

By **CHRISTOPHER E. CENAC**, SR., M.D., F.A.C.S.
with **CLAIRE DOMANGUE JOLLER**
Foreword by **CARL A. BRASSEAUX**, PH.D. and **DONALD W. DAVIS**

Volume 1: Bayou Terrebonne

HARD SCRABBLE *to* HALLELUJAH
LEGACIES *of*
TERREBONNE PARISH, LOUISIANA

By **CHRISTOPHER E. CENAC**, SR., M.D., F.A.C.S.
with **CLAIRE DOMANGUE JOLLER**
Foreword by **CARL A. BRASSEAUX**, PH.D. *and* **DONALD W. DAVIS**

This contribution has been supported with funding provided by the Louisiana Sea Grant College Program (LSG) under NOAA Award # NA14OAR4170099. Additional support is from the Louisiana Sea Grant Foundation. The funding support of LSG and NOAA is gratefully acknowledged, along with the matching support by LSU.

Logo created by Louisiana Sea Grant College Program.

HARD SCRABBLE TO HALLELUJAH:
Legacies of Terrebonne Parish, Louisiana
Volume 1: Bayou Terrebonne
Christopher Everette Cenac, Sr., M.D., F.A.C.S.

Copyright © 2016 by JPC, LLC
All Rights Reserved
Printed in the United States of America
First Edition - 2016

Library of Congress Control Number: 2016954356

ISBN: 9780989759410

Book and cover design:
Scott Carroll Designs, Inc., New Orleans, Louisiana

Book index created by:
Portier Gorman Publications, Thibodaux, Louisiana

Distributed by University Press of Mississippi

Cover: Painting of Hard Scrabble plantation house by Godfrey J. Olivier (1928-2005)

Front and back end sheets: 1846 La Tourrette Map

George G. Rodrigue paintings found on pages 29, 34, 96, 162, 194, 218, 292, 312, 346, 425, 466, 492, 497, 500
George Rodrigue Foundation of the Arts
747 Magazine Street, New Orleans, Louisiana

JPC, LLC
3661 Bayou Black Drive
Houma, LA 70360

THE COAST OF LOUISIANA IS WASHING AWAY. THE CULTURE AND HISTORY OF FRENCH SPEAKING RURAL SOUTHERN LOUISIANA IS *RUNNING* AWAY.

– Christopher E. Cenac, Sr., M.D., F.A.C.S.

BOOKS BY CHRISTOPHER EVERETTE CENAC, SR., M.D., F.A.C.S.

Eyes of an Eagle: Jean-Pierre Cenac, Patriarch:
An Illustrated History of Early Houma-Terrebonne

Livestock Brands and Marks: An Unexpected Bayou Country History:
1822-1946 Pioneer Families: Terrebonne Parish, Louisiana

Carl A. Brasseaux and Donald W. Davis, series editors

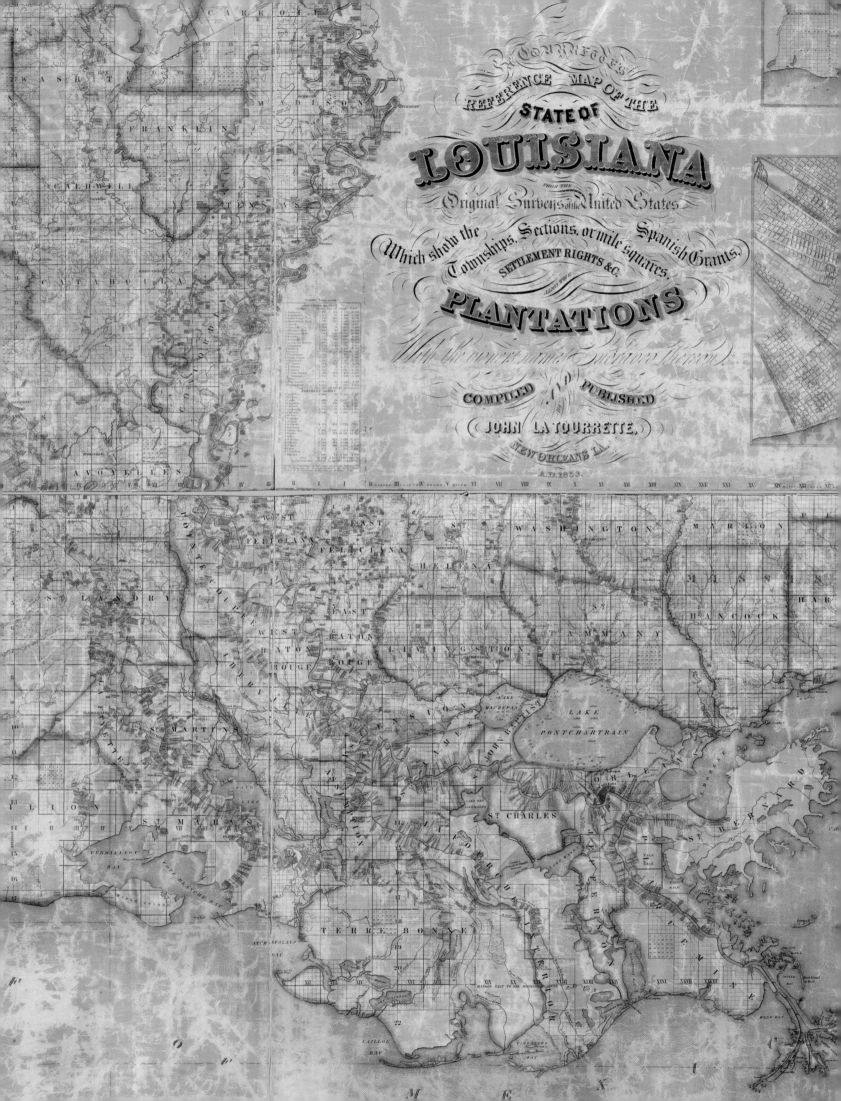

ACKNOWLEDGEMENTS

I thank the following for their assistance in providing information, photos, maps, graphics or other services for this publication: Faye Parker, Peggy G. Darsey, and Ramona A. Griffin of the Cenac medical office staff; Scott LaPée Carroll, Erin Callais, and Dean Cavalier of Scott Carroll Designs, Inc.; Vincent P. and Ying M. Kreamer for photographic services; Clifton P. Theriot, Archivist of the Ellender Memorial Library at Nicholls State University and author of *Lafourche Parish (Images of America)*; Carolyn Portier Gorman; Patty A. Whitney of the Bayou History Center; Carl M. Bennett, Jr.; Chester F. Morrison; Debra S. Benoit, Director of Research and Sponsored Programs, Nicholls State University, and member of the Terrebonne Parish School Board; Terrebonne Parish Schools Superintendent Philip E. Martin and School Board employees Becky L. Breaux and Melissa A. Hagen; Robert S. Brantley, author of *Henry Howard: Louisiana's Architect*; Garth K. Swanson; Laura A. Browning of LBC Consulting LLC and author of *Faith, Family, Friends: 150 Years of Sacred Heart and Montegut*; Debi L. Lauret of The J.M. Burguières Co. Ltd.; Terrebonne Parish Assessor Loney J. Grabert and Cary R. "Buddy" Hebert of the Assessor's office; Michael Gene Burke; Leonard J. Chauvin, CPE; Logan H. Babin, Jr. and L.H. Babin III of Logan Babin Real Estate; Terral J. Martin, Jr., PLS; Arthur A. DeFraites, Jr., PE, PLS; Keneth L. Rembert; William Clifford Smith, PE, PLS and Kenneth W. Smith, PE, PLS of T. Baker Smith; Grant J. Dupre (the previous eleven, for providing maps and other graphics); Dan H. Davis; Rachel E. Cherry, former manager of Southdown Plantation and author of *Forgotten Houma (Images of America)* and *Southdown Plantation: The House that Sugar Built*; Terrebonne Parish Clerk of Court Theresa A. Robichaux; Jean J. Fugatt, Skye G. Bardeleben, Mary B. Champagne, and Renee L. Ledet of the Terrebonne Parish Clerk of Court's Office; Dr. Florent Hardy, Jr., Louisiana State Archivist; Al J. Levron, Terrebonne Parish Consolidated Government; Mary Lou Eichhorn of The Historic New Orleans Collection; Tara Z. Laver and Jessica Lacher-Feldman of the Louisiana State University Hill Memorial Library, Special Collections; Donna McGee Onebane, Ph.D., author of *The House that Sugarcane Built: The Louisiana Burguières*; Charles J. Christ and Dexter A. Babin of the Regional Military Museum in Houma; Wayne M. Fernandez, Wendy Wolfe Rodrigue, Jacques George and Mallory Page Rodrigue of the George Rodrigue Foundation of the Arts; Judith A. Soniat, Carlos B. Crockett and Mary Cosper LeBoeuf of the Terrebonne Parish Public Library; former manager of Waubun Terry P. Guidroz, Jody A. Davis, and Dr. Thomas E. Powell III and Glenn Walker of the Carolina Biological Supply Company; Lillian Joseph; Randolph A. Bazet, Jr.; Elmo P. Bergeron, Jr.; Sheri Lee Guidry Bergeron; Richard J. Bourgeois and Angela M. Cheramie; Lester C. Bourgeois, Sr.; Lester C. Bourgeois, Jr.; L. Philip Caillouet, Ph.D., FHIMSS; Dolly Domangue Duplantis; Coralie and Wallace J. Ellender, Jr.; Glyn V. Farber; Wanda L. and Wilson J. Gaidry III; Montella Doescher and David A. Guidry; Dr. Jamie Ellis-John Hutchinson; N. Dean Landreneau; Dwayne Lyons and Warren Lyons, Sr. of Residence Baptist Church; Claire Moreau Mahalick; Keith J. and Susan P. Manning; Murphy H. Savoie; Albert P. Naquin, Traditional Chief of the Isle à Jean Charles Band of Biloxi-Chitimacha-Choctaw; Lynn F. and William T. Nolan; Barbara K. and William A. Ostheimer; Clara A. Redmond; Dorothy "Dot" Rogers; Elizabeth M. Scurto; James M. Sothern; Martha Richardson South; Virginia R. and Michael X. St. Martin, Sr., Celeste St. Martin Wedgeworth, Charlean St. Martin Dickson; Russell W. Talbot; Janelle M. Moen; James R. Carrere, Sr.; Cyrus J. Theriot, Jr. and Adruel B. Luke of the Harry Bourg Corp.; Brian W. Larose; Gail H. and Garland Anthony Trahan; Lori R. and Dr. Herman E. Walker, Jr.; Carolyn Walker Mabile; Tina H. and Dr. Craig M. Walker; Elward P. Whitney; Linwood P. Whitney; Debra J. Fischman; Claude J. Bourg; Gary D. Lipham; Brian Cheramie; Leryes J. Usie; Albert P. Ellender, DDS; Arlen B. Cenac, Sr. and Jacqueline Guidry Cenac; Susan O. and Douglas P. Patterson; Melissa N. and Jerry P. Thibodaux; Veranese E. Douglas, Geraldine L. Lagarde; Kevin J. Allemand of the Diocese of Houma-Thibodaux; George J. Jaubert; Francis Deoma Callahan; David B. Kelley, Ph.D. of Coastal Environments, Inc.; Joseph J. Bergeron, Jr. of TPGS; Rudy R. Aucoin, deceased; Sharon A. Alford of the Houma Area Convention and Visitors Bureau; Patrick H. Yancey and Stanley E. Yancey; David J. Morgan of The C.A.R.T.E. Museum; Angela Marie Fonseca Trahan; Farmand J. Matherne, Jr.; Tegwyn Murphy and Joseph John Weigand, Jr.; David D. Plater, author of *The Butlers of Iberville Parish, Louisiana;* Charles Kevin Champagne; Bert A. Guiberteau; Richard Paul "Dickie" Brown; Richard Anthony Arcement; Kirby A. Verret, former Tribal Chairman of the United Houma Nation; Judge Edward J. Gaidry, Sr.; Judge Timothy C. Ellender, Sr.; Rev. Michael A. Bergeron, Jeanette F. Schexnayder, Lawrence C. Chatagnier; Gayle B. Cope; Rosalie M. and Gerald J. Voisin; Beverly P. and Prosper J. Toups Jr.; Nancy C. and Edward L. Diefenthal; Bethany C. and Eric A. Paulsen; Cindy T. Cenac; Emil W. Joller; and a very special thanks to A.B. Cenac, Jr. for his contributions to complete this work, and the Arlen B. Cenac, Jr. Foundation.

Opposite page: 1853 John La Tourrette map

Blue Dog print by George G. Rodrigue, 2003

DEDICATION

In memory of my friend

George G. Rodrigue

(March 13, 1944 - December 14, 2013)

Christopher E. Cenac, Sr., M.D., F.A.C.S.

TABLE OF CONTENTS

Foreword...13
Chapter One......................................14
Chapter Two.....................................22
Land Measurement Terms......................30
Ownership Sources..............................32
Sugar Crops in Terrebonne Parish.............33
Malbrough Settlement..........................36
Sargeant -Armitage.............................39
   Bourgeois-Thibodaux -Barras House........44
   Bourgeois Carriage House...................46
Johnson Ridge...................................48
Ducros-Julia.....................................53
Schriever..71
Waubun-Magnolia Grove.......................77
St. George..85
Balsamine..93
Hobertville.......................................95
St. Bridget..98
   Andrew Price School........................102
Isle of Cuba....................................105
   Levy Town.....................................108
Evergreen.......................................116
Hedgeford......................................123
Gray..128
Bernard's Open Kettle Syrup & LaCuite.....130
Beattie-Batey...................................132
Halfway..137
Ayo...143
Orange Grove-Pilié............................146
Bayou Cane....................................148
   Nicholas Leret Claim........................148
   Filican Duplantis............................150
   Williamsburg.................................152
   Bayou Cane School.........................153
Gabriel J. Montegut II Residence.............157
Wade Claim.....................................161
Barataria Canal................................164
   Houma Cypress Co., Ltd.....................170
   Houma Brick & Box Co., Inc.................173
Joseph Haché Claim - City of Houma.........175

Houma Additions..............................183
   Newtown 01/25/1870
   Celestin Addition to Newtown 01/05/1899
   Deweyville-05/19/1899
   Oscar Daspit Addition to Newtown 08/01/1899
   Honduras Addition-04/1923
   Crescent Park Addition-04/03/1924
   Breaux-Morrison Addition-08/19/1932
   Daspit-Breaux Addition-05/27/1937
   Clark Place Addition-02/04/1938
   Houma Colored School
   The Alley
   Silver City
   Good Templars Hall
   Odessa's Place
Burguières, Smith, St. Martin Houses.........187
Magnolia Cemetery-
   Terrebonne Memorial Park..............196
Homestead.....................................197
Joseph A. Gagné House......................202
Daigleville......................................210
   Right Bank
     Boardville-01/10/1882
     J. H. Hellier-03/31/1902
     Elizabeth Place -04/03/1935
     Alidore J. Mahler Addition-03/07/1938
     Houma Heights-06/06/1938
     Bellview Place-06/09/1938
     Theogene J. Engeron Subdivision-04/24/1939
     Roselawn Subdivision-04/06/1940
     Harry J. Bourg Subdivision-08/23/1945
     Authement Subdivision-11/28/1948
     G. P. Boquet Subdivision-03/14/1949
     Boquet Subdivision Addendum # 1-04/28/1951
   Left Bank
     Connely Row-01/03/1906
     Connely Subdivision-03/15/1926
     Residence Subdivision-07/29/1930
     St. Michel Subdivision-04/23/1942
     Lebouef Subdivision-06/28/1949
     George Pitre Lane
Residence......................................221
Mechanicsville-Barrowtown..................234

RobertaGrove..236
   South Terrebonne Subdivision-04/14/1951
   Barrow Subdivision-10/30/1952
   Roberta Grove Subdivision-11/01/1952
   Oleander Subdivision-11/12/1952
   Cole Subdivision-07/18/1954
U.S. Naval Air Station (LTA) Houma 242
Prisoner of War Camps......................260
Myrtle Grove 274
Presqu'ile 280
Frontlawn 284
Edmund Fanguy...............................286
Oakwood (Semple & Shields).................288
Pecan Grove 291
Bourg...295
   Company Canal
   Canal Belanger
   Newport
   Hubert Belanger Property
   Marianne Marie Nerisse Iris-Spanish Grant
   Jean Baptiste Guidry
   Matherne Dairy
   Lyes J. Bourg Sawmill
Bourg Agricultural School.....................311
LeCompte Property-Billiot Claim 314
Rural Retreat..................................316
St. Agnes.......................................320
Klondyke.......................................322
Hope Farm.....................................324
Bayou Pointe-aux-Chênes 328
   Easter Lilies................................328
   Ile à Jean Charles..........................334
   FaLa Village...............................342
Deroche Brothers Syrup Mill 348
Aragon..352
St. Peter's Baptist Church 359
Pointe Farm....................................361
Lower Terrebonne Refinery 370
Montegut-le Terrebonne
   Sandersville/Crochetville..................387
Angela..395
Magenta..396
Eliza..401

Sunbeam..402
Live Oak 404
Dugas Cemetery...............................406
Hard Scrabble - "Caillou Field" 408
Humble Canal 415
Argene..416
Red Star..418
Orange Grove..................................422
Methodist Mission Chapel 424
Pointe-au-Barré................................427
   Lower Montegut Indian School 428
   St. Peter Catholic Mission Chapel 430
Eloise...432
Pecan Tree 435
   Texas Company Yard-Lower Montegut.....436
Lapeyrouse Canal & Store 438
Boyne Boat Works 441
Rhodes Brothers 442
Madison Canal.................................444
Bush Canal.....................................447
End of the Road 449
Sea Breeze 454
Afterword......................................462
About the Authors.............................464
Index...468
Photo Credits 493

GEORGE G. RODRIGUE PAINTINGS

Macque Chou................................... 29
Low Tide.. 34
John Courrege's Pirogue 96
Cajun Bride of Oak Alley 162
Cora's Restaurant 194
Good Wine, Good Friends....................218
Spinning Cotton in Erath..................... 292
Farmer's Market................................312
Aioli Dinner 346
The First Cajuns............................... 425
High Water at Whiskey Bay 466
Evangeline 492
The Class of Marie Courrege 497
Doctor on the Bayou 500

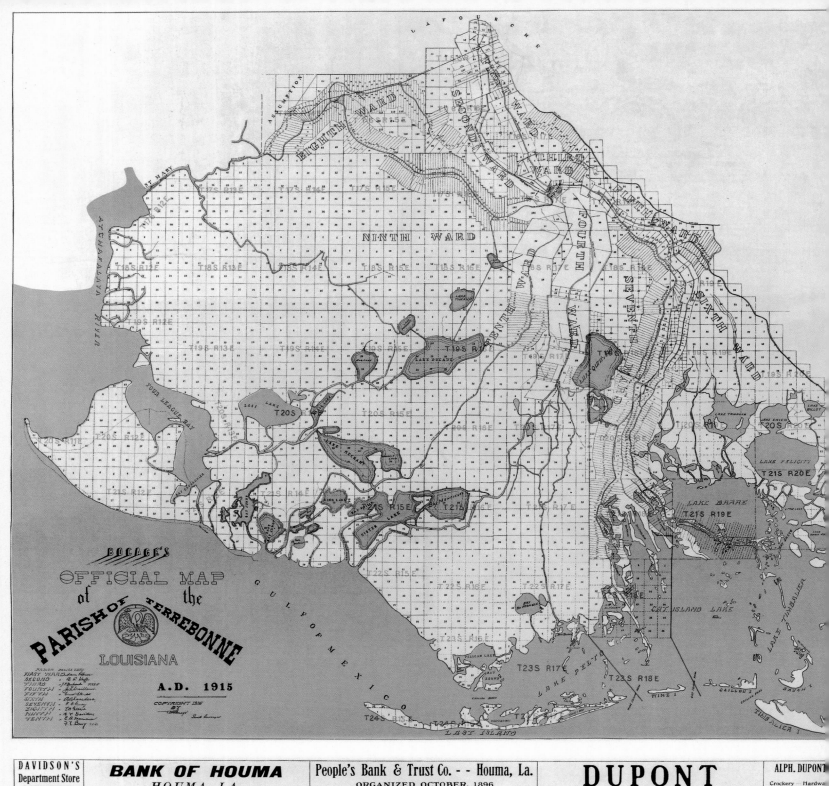

FOREWORD

In order to comprehend the macro, competent historians and historical geographers must first thoroughly understand the micro, and some academic pundits have likened good historical writing to the creation of an interpretive mosaic through the assembly of factual tesserae. The more skillfully and logically arranged the tiles, the more sharply defined the image. However, these traditional interpretive standards are rarely met by local and regional historians, whose works are routinely compilations of at best loosely connected minutiae. Scrupulously avoiding interpretation of the data, these writers abandon their readers to formulate the big picture—if they can somehow negotiate their way through such works' confounding factual morasses.

In a refreshing departure from this dismal modern historiographical norm, Christopher Cenac and Claire Domangue Joller's *Hard Scrabble to Hallelujah, Legacies, Terrebonne Parish, Louisiana* provides a laudable contemporary model for local and regional historical writing. Cenac, a Houma native and renowned orthopedic surgeon, and Joller, a local freelance writer and award-winning columnist, in their third collaborative book, carefully examine the region's unique geographic personality and cultural landscape through the prism of Terrebonne Parish's distinctive historical experience. Their introductory theses provide readers a crucial backdrop and interpretive framework for the topical essays that constitute the book's main corpus. Scores of highly detailed and opulently illustrated micro-histories of individual property holdings—*Hard Scrabble to Hallelujah*—track the parish's historic settlement patterns, as farmsteads radiated southward from the Houma area's plantation belt to ever more economically tenuous footholds in the coastal marsh. These sketches are complemented by numerous essays recounting the development of now largely forgotten communities, businesses, religious institutions, transportation milestones, and miscellaneous landmarks, as well as dimly remembered historical events. These brief narratives, in turn, provide invaluable information regarding the contributions of myriad racial and ethnic groups that collectively created and shaped Terrebonne's highly complex cultural landscape.

This rich textual tapestry is enhanced immeasurably by Cenac's and Joller's insider perspectives. As lifelong residents of Terrebonne Parish, they recognize, record, and interpret subtle historical, geographical, and demographic nuances that are routinely lost on outsiders. As "insiders," they also recognize that their tome probably constitutes the last, best opportunity to document thoroughly the portion of North America most likely to literally vanish by the dawn of the twenty-second century. In fact, *Hard Scrabble to Hallelujah* could not appear at a more propitious time. Imminent environmental change on a biblical scale—driven by subsidence, erosion, and ocean-level rise—threatens much of Terrebonne Parish with inundation, and, if current projections prove accurate, many, perhaps most, of the communities Cenac and Joller document will lie beneath the Gulf's restless waves within three generations. The memory of these sunken districts will live on, however, in the pages of this excellent historical encyclopedia/gazetteer.

Donald W. Davis
Carl A. Brasseaux, Ph.D.

Opposite page: Charles William Bocage's
Official Map of the Parish of Terrebonne 1915

CHAPTER ONE

The very origin of the landscape of Terrebonne Parish in south central Louisiana gave rise to the lucrative sugar industry that peaked locally in the nineteenth century and still endures in lesser eminence today. Sugar cane production in the region dates back to before the time the Louisiana legislature established Terrebonne as a parish (county) in 1822.

Built as delta land of the unfettered Mississippi River over the course of unknown centuries, Terrebonne Parish's fertile land was fed by siltation from Bayous Lafourche and Terrebonne, both distributaries of the Mississippi. On the western limits of the parish, the same scenario occurred through delta-building action of the Atchafalaya River, which is itself a powerful distributary of the Mississippi.

The arable land thus built by the 1700s was particularly fruitful along the bayous of various lengths that transect the parish, from twenty to fifty miles long. The most important of these is Bayou Terrebonne, 50 miles in distance from its headwaters at Bayou Lafourche to Sea Breeze at the Gulf of Mexico. Fertile land rested along these bayous which were hugged by high headlands that tapered downward to swamps and wetlands on both banks. Each bank generally had a "strip of high ground from a quarter to one mile in width"[1] parallel to the bayou.

While this geography could have been limiting in the relatively narrow widths of acreage friendly to cultivation along the bayous, the landscape more than made up for that limitation by existence of the natural waterways that allowed shipping of produce by flatboat, and, farther upstream, by steamboats. This was imperative at the beginning of sugar cane cultivation, especially, since adequate roadways were nonexistent in the coastal parish during early settlement. Bayous also served as natural drainage for arable lands along their banks.

However, one source wrote that the head of Bayou Terrebonne, "as were those of the other area bayous, Blue, Petit Black, Chacahoula, was silted over long before the coming of

Potato farm Bayou Black 1920

The Sugar Harvest, A.R. Waud
Harper's Weekly, 1875

the white men."[2] Just as the Mississippi fed the Lafourche and the Lafourche fed the Terrebonne in the long-ago geological past, Bayou Terrebonne was the source of all the major bayou waterways in Terrebonne Parish. Those that had their direct source from Bayou Terrebonne are Bayou Black (Little and Big), Bayou Grand Caillou, Bayou Little Caillou, Bayou Cane, and Bayou Pointe-aux-Chênes. Bayou Black, in turn, gave rise to Bayou Buffalo (Dularge) and Bayou Chacahoula. Bayou Blue had its source from Bayou Lafourche and paralleled Bayou Terrebonne for a distance, but it is not a distributary of the Terrebonne.

First recorded inhabitants of the area were the Native American Houmas tribe which had drifted after 1784 into what was to become Terrebonne[3]. A few hardy French families, "principally from the older colonies of Louisiana," inhabited the lower reaches of the parish by the late 1700s[4], establishing their homesteads not far from the Gulf of Mexico coast. Possessors of land grants who had received them for service in the Revolutionary War when Louisiana was a Spanish colony, and some for other reasons, also settled there. Others, many of them Acadians from Nova Scotia, traversed neighboring Lafourche and St. Mary parishes to settle along other Terrebonne bayous in what was then known as the Lafourche Interior.

Historian Alcee Fortier listed the first settlers of Terrebonne Parish as Royal Marsh on "Black bayou," the "Boudreaus" on

CHAPTER ONE | 15

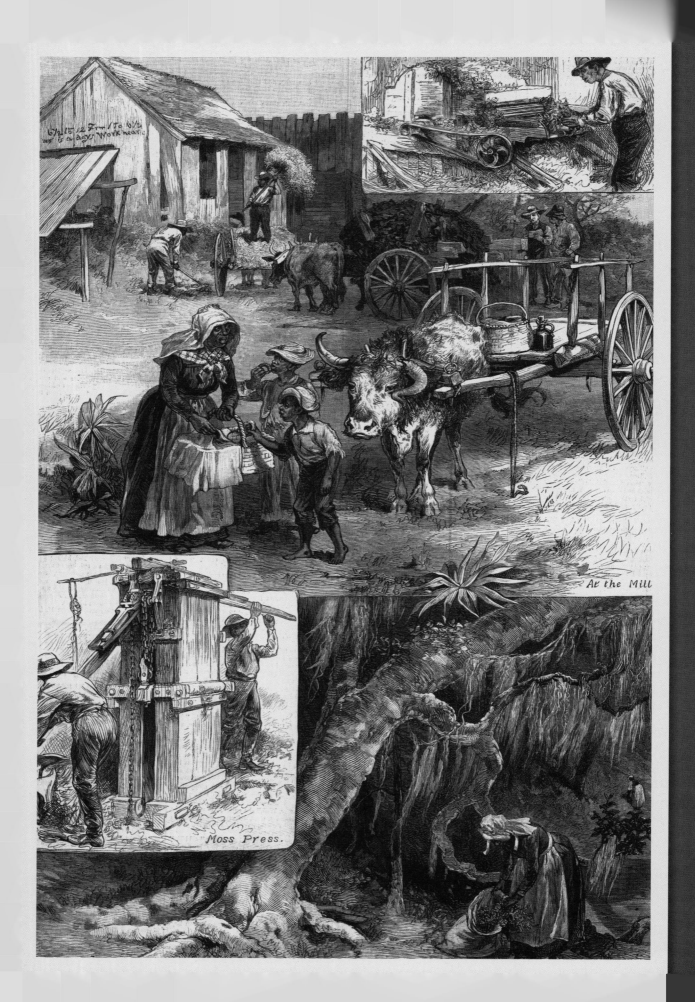

Little Caillou and the Terrebonne, the Belanger family along lower Terrebonne, Prevost, who "started a plantation on Grand Caillou," the "Shuvin [Chauvin] family on Little Caillou, the "Marlboroughs" in the northern part of the parish, and other sections' settlers Curtis Rockwood, the D'Arbonnes, LeBoeufs, Trahans, Bergerons, R.H. and James B. Grinage near Houma.[5]

Those people were overwhelmingly Acadians (Cajuns) who had been expelled from Nova Scotia for not swearing allegiance to the English monarch, and some Creoles (Louisiana-born descendants of European ancestry), with a sprinkling of "Americans, Spaniards and Germans."[5]

The great majority of inhabitants until the second quarter of the 1800s were subsistence farmers who grew cotton, corn, rice, peas and fruits of all kinds. Adequacy, not bounty, seems to have been the status quo in Terrebonne from its early settlement in the last days of the 1700s. It took almost two decades after New Orleans planter Etienne de Boré achieved granulation of sugar in 1795 for the parish to begin cultivation of the white gold with which the local area was so identified for more than a century, and which has a healthy presence even today.

At one time in the bayou-webbed civil parish, more than 100 plantations, many of them each worked by only one man and his family, hugged the water transportation corridors fanning out from the parish seat in all directions. Most were of modest acreage on properties abutting each other from bayou headlands.

But as the parish grew, the countryside was made majestic by the unbroken view of verdant fields in an area renowned for its status in the state's Sugar Bowl. The sheer expanse of waving greenery can be imagined from the fact that in its agricultural heyday, 1830s-1920s, the civil parish was then the largest of the state with its 2,080 square miles, larger than the entire state of Delaware.

Terrebonne's vast reaches of pristine arable land later became a magnet for more materially ambitious planters from points north and east who settled in Terrebonne to amass sweeping estates. Some locals, and many newcomers, developed sugar estates and large, even grand, homes. The architecture of their homes ranged from Greek Revival to Queen Anne to Louisiana Raised Cottage, to Eastlake, to Colonial Revival, to early Victorian styles, adorning the bayou landscapes among their fields.

Among findings of the 1850 U.S. Census is that Terrebonne Parish was the site of 550 dwellings that year, mostly families. Population was totaled for persons who lived along the various bayou corridors of the parish. Bayou Terrebonne had 1149 people along its banks; Bayou Black, 906; Bayou Little Caillou, 717; Bayou

Opposite page: The Moss Industry in the South, Harper's Weekly, 1882

Above, Negro cabin Terrebonne Parish c. 1900
Below, Southdown Engine #5 c. 1920

Picking moss 1937

Left, Lirette Field blowout 1908

Grand Caillou, 109; Bayou Dularge, 77; Bayou Pointe-aux-Chênes, 27; Bayou Bleu, 22; Bayou Grand Coteau, 18.

By far, Terrebonne farmers (302) and laborers (223) exceeded the numbers of other occupations, according to 1850 U.S. Census figures. Not surprisingly, overseers, carpenters, and coopers (44, 42, and 36, respectively) were the next most numerous occupational group, since those jobs were vital to plantation operation.

After the Civil War, estates were broken up and sold off, either by owners or the Freedmen's Bureau. Many stately homes were abandoned, and owners moved away. The sugar farmers who survived were, for the most part, producers with small estates whose families had traditionally worked the land themselves, not relying on slave labor. Mosaic sugar cane disease finished off many planters in the early part of the 20th century.

It is important to know that only a few substantive reminders of Terrebonne's once-eminent status in national sugar production remain, and whereas at one time sugar houses were common on almost every plantation, not even one sugar house still exists in the parish. This, in spite of the fact that sugar cane remains the dominant local crop.

By 1901, twenty-three gas wells had been drilled in Terrebonne Parish, thereby marking the beginning of the shift from white gold's dominance to that of black gold in the local economy. Oil and gas production reached its zenith locally in the mid-1900s and beyond.

Today's residents who drive past Honduras School on Grand Caillou Road, Greenwood School at Gibson, and St. Bridget Catholic Church in Schriever may not be aware that those places

are among vestiges of "high sugar" days, surviving only through names that once signified different land owners' holdings.

Communities identified on signage or by locals as Hallelujah, Peterville, Levy Town, and other outposts from "town" (Houma), are the descendants of communities formed for the most part during slavery, or after emancipation, by former plantation slave laborers.

Other settlements are of long standing among Cajuns and other locals who inhabited spots along stretches of bayous as early as the 1700s—among them Dulac, Pointe-aux-Chênes, Cocodrie, and Daspit.

Many of those communities have survived. Not so the plantations. Local residents who have not had the advantage of Grandpa or Grandma's pointing out the former locations of Mandalay, Aragon, Roberta Grove, Honduras, Waterproof, Oak Forest, Concord, Windermere, Poverty Flat or Eureka will be interested to read about these and other places of agriculturally historic significance. Every effort has been made to locate in words, maps, and through other means the named sugar estates that now live only in street, subdivision, and school designations. There were, however, a considerable number of many other, unnamed, farms of which only some owners' names live in official records.

This book attempts to fulfill the wish of many student and other researchers who have had no one volume to turn to when seeking local data about the years when sugar was king.

Concerted efforts at accuracy and completeness preceded the actual writing of this book, with research spanning international, national, state, and local sources. This prompted research also into other subjects that are interesting or important to the parish's history.

The reader will find here adjunct information on Terrebonne prisoner of war camps of the 1940s, on the U.S. Naval Air Station (LTA) upon which grounds the local airport grew, and on bygone syrup mills, dairies, schools and post offices. These topics are included to give a credible resource about Terrebonne subjects for which information would probably not otherwise be found in one written source.

SOURCES
1. J. Carlyle Sitterson, *Sugar Country: The Cane Sugar Industry in the South, 1753-1950*, University of Kentucky Press, 1953
2. Bayou Terrebonne History from *Historical Scenes of Thibodaux* book, Nicholls State University Archives
3. John Swanton, *Bureau of American Ethnology, Indian Tribes*
4. Marguerite Watkins, *History of Terrebonne Parish to 1861*, thesis submitted to the graduate faculty of Louisiana State Department of History, 1939
5. Alcee Fortier, *Sketches of Parishes, Towns, Events, Institutions, and Persons: Terrebonne Parish*, 1914

Le Danois Gas Well, Lirette Oil Field, 1901

Moonlight and Magnolias *by Mort Künstler 1997*

U.S. Naval Air Station (Houma) LTA c. 1943

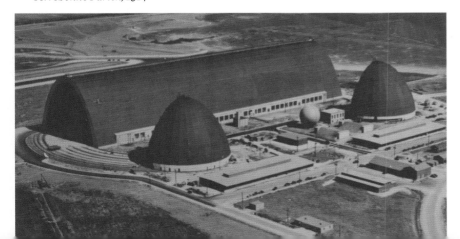

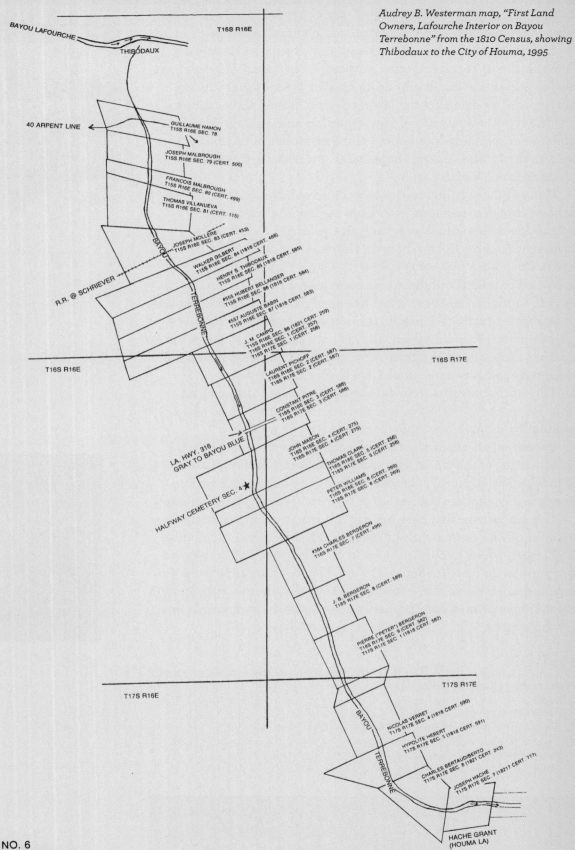

Audrey B. Westerman map, "First Land Owners, Lafourche Interior on Bayou Terrebonne" from the 1810 Census, showing Thibodaux to the City of Houma, 1995

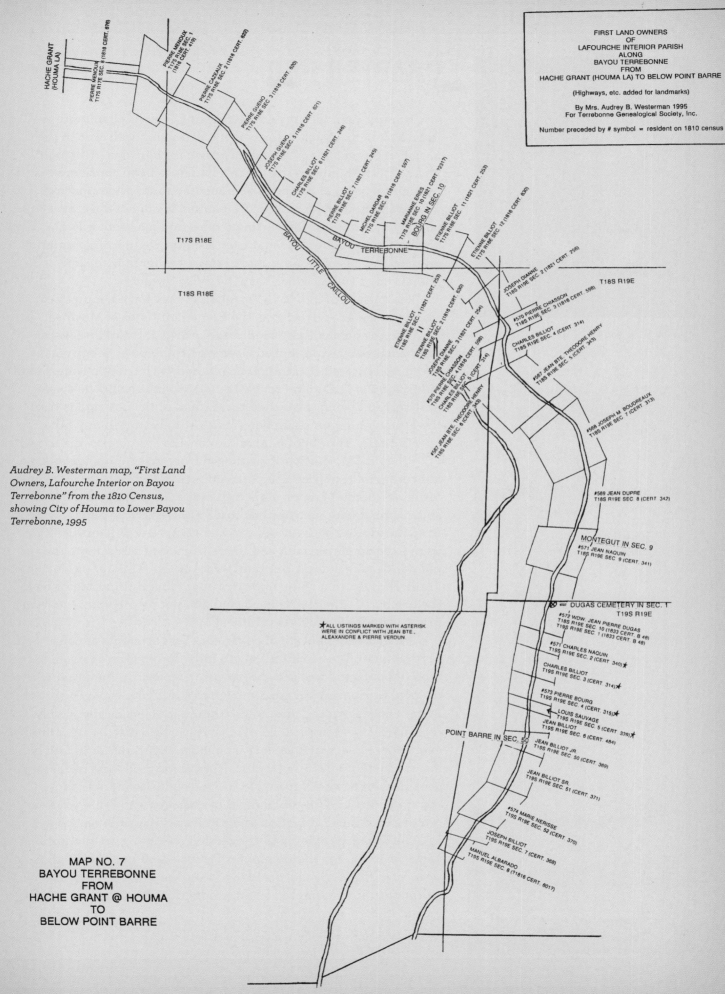

Audrey B. Westerman map, "First Land Owners, Lafourche Interior on Bayou Terrebonne" from the 1810 Census, showing City of Houma to Lower Bayou Terrebonne, 1995

CHAPTER ONE | 21

CHAPTER TWO

It might surprise current Terrebonne Parish inhabitants to learn that before sugar cane was the dominant cash crop in the parish, its forerunner was indigo grown for blue dye. Although much of Terrebonne in the 1820s could have been considered frontier territory, it has been documented that indigo was probably among the first major cash crops planted locally in the early 1800s, notably by James Bowie of Alamo fame and his brother Rezin. The brothers grew indigo on what later became Southdown plantation on the outskirts of Houma, and perhaps earlier at Live Oak plantation at Dulac, which they then also owned.

James Bowie (1796-1836)

The culture of indigo in Terrebonne may have begun even earlier, since as early as 1733 the French Minister of Marine wrote to the colony that "sugar cane growing, from the several failures already experienced, seemed ill suited to the province and urged instead that indigo be planted."[1] This led to widespread cultivation of indigo in French Louisiana, and the Bowies were probably following the agricultural trend of the day rather than the French minister's mandate. It is lost to history if any other specific early Terrebonne inhabitants were indigo growers as well.

When indigo cultivation suffered from insect damage and later from falling price competition from the same product from India, the official French stance changed. That is not to say that some enterprising entities had totally bowed to the official word from the mother country. As early as 1742, New Orleans Jesuits were experimenting with seed cane from Santo Domingo. Indigo planter Claude Joseph Dubreuil wrote to the Minister of Marine in France in 1754 that "I am endeavoring to found sugar factories in the colony. . . . I believe this culture will not be prejudicial to that of the islands [French West Indies]. . . ."[1]

100-pound Cuban sugar sack, Sasco Sugar Co., N.Y. and 100-pound Honduras sugar sack, Azucar Rio Lindo Compania

Widespread planting of sugar cane and attempts at sugar production took until after 1795 to fully flower. One obstacle to the crop in Louisiana was the shorter growing season than that of the tropical Caribbean Islands, which had a full-year cycle of growth. Sugar cane is a grass, and in the tropics, it does not have to be replanted. In subtropical Louisiana, the crop has to be replanted every three years, and each season must be harvested in late fall in anticipation of winter freezes. Also, sugar mills and necessary accoutrements were expensive, as were salaries for skilled sugar makers and for the number of enslaved and other workers requisite to cane cultivation and sugar-making success.[1]

Sugar Levee, New Orleans, La., c. 1890s

The greatest impetus to conversion to sugar cane plantations came when in 1793 indigo planter Etienne de Boré, whose lands were near New Orleans, decided to take the great risk of cane cultivation and attempted sugar production. When he achieved granulation of sugar from his cane in 1795, he was hailed as the savior of Louisiana, and his experiment prompted numbers of landowners to follow his lead.

22 | HARD SCRABBLE TO HALLELUJAH

Those numbers increased in most areas of south Louisiana. But author J. Carlyle Sitterson wrote of the early-1800s Terrebonne-Lafourche area, "There were few aristocrats, self-styled or otherwise, and life moved slowly, undisturbed either by the thought that sugar wealth was to be had from the land or the fear that small farms would be swallowed up by the expanding sugar plantations. As one early sugar planter of Bayou Lafourche recalled, the people lived economically and out of debt. Although the farmers were 'neither very energetic, or enterprizing [sic], . . . they were contented and lived a free, and happy life.'"[1]

But in 1810, Henry Schuyler Thibodaux from Lafourche began to accumulate lands that would become one of the earliest extensive sugar plantations in Terrebonne Parish. Thibodaux founded his home place in what is now Schriever in the northern reaches of the parish. He named his large farm St. Brigitte, French for the namesake saint of his wife Brigitte Belanger Thibodaux of Terrebonne. When the parish was legally established in 1822, he became known as the Father of Terrebonne.

Native Louisiana entrepreneurs were not the only businessmen to be attracted to sugar cane cultivation and land speculation in the 1820s. Following the Louisiana Purchase in 1803, Anglo-Americans had begun to filter into the state; many of them set their sights on Terrebonne Parish in the 1820s.

In 1828 three Anglo-Americans founded early sugar cane plantations in Terrebonne Parish. In that year, William John Minor, James Dinsmore, and John A. Quitman bought land from the Bowie brothers in two separate locations. Quitman, who owned Monmouth plantation and other cotton lands in Mississippi, bought two tracts in Dulac which he made into the vast Live Oak plantation. Minor lived at Concord, the residence of the early Spanish governors near Natchez, which became the Minor family home. He and Dinsmore purchased Bowie lands near Houma, present-day Southdown. Their initial purchase grew over the years to approximately 22,000 acres.

Relatively soon thereafter, numbers of wealthy Americans from Mississippi, Tennessee, Virginia, the Carolinas, Maryland, New York, and Pennsylvania, and other places, began to purchase Terrebonne lands to be made into working sugar plantations. Indeed, "in the first quarter of the nineteenth century, the number of Anglo-Saxon sugar planters was greater in Terrebonne Parish than in any other south central Louisiana parish. . . .A few of the property owners along Bayou Black by the mid-1840's, in addition to W.J. Minor, were: R.R. Barrow, Thomas Butler, Tobias Gibson, Richard Ellis, E. Ogden and William A. Shaffer. . . .By 1858 there were 59 American planters to 20 French planters. In 1831, there had been 11 French and 10 American sugar planters. Virtually all of these men, including Minor, usually began by acquiring land from the small farmers in the area and gradually assembled sugar plantations of 1,000 acres and larger."[2]

However, ownership of such huge estates was not all mint juleps and tea parties. Before such large acreages could begin to

Coastal Planter hailing a New Orleans Steamer, Harper's Weekly, 1889

Tobias Gibson (1800-1872)

William J. Minor (1808-1869)

William A. Shaffer (1796-1887)

Robert Ruffin Barrow, Sr. (1798-1875)

CHAPTER TWO | 23

THE SUGAR INDUSTRY OF LOUISIANA.—

1. View in Sugar District of Louisiana. 2. Stripping and Cutting. 3. Bringing in the Cane.

The Sugar Industry of Louisiana, Harper's Weekly, 1863

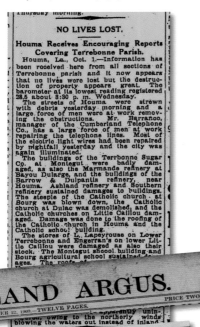

1915 Hurricane Damage, The Times Picayune, *October 2, 1915*

produce profits, the land and buildings required construction and improvements by both skilled and unskilled labor. There were "brick makers and bricklayers, carpenters, fence builders, drivers, coopers, and other laborers to train or hire. Ditching of the land required large numbers of men to ditch, trench, clean the waste lands and hew down the forests."[1] Not only for actual planting and harvesting of cane, but also in production of sugar from the cut cane, labor was intensive and sometimes dangerous for the people planters relied upon to produce their prized commodity. Both black slave labor and white labor were utilized to do many of these jobs.

By 1843, the granulation part of the process became safer and less arduous through the invention of Norbert Rillieux, New Orleans-born son of a French engineer and a former slave. Rillieux received his higher education in Paris, where he invented the prototype model for a sugar evaporation system to replace the previous process that required handling of boiling hot liquids in open kettles.

But no invention could obliterate other challenges to plantations in Terrebonne and other locations. As with farmers of every type, sugar plantation owners gambled on weather suitable for their cane to flourish. That gamble sometimes failed when freezing weather appeared, impairing or destroying cane that had not yet been harvested. Severe storms and hurricanes could decimate entire stands of cane. "A severe storm on Bayou Black late in September, 1860, did considerable damage to the sugar crops in Terrebonne and the adjoining parish. About thirty sugarhouses were damaged, some entirely blown down," according to author J. Carlyle Sitterson.[1]

Storms severely injured the sugar crop in only five antebellum years, (1812, 1824, 1832, and 1860).[3] This does not include the August 1856 "Last Island" hurricane which also struck south Louisiana, causing loss of life and ruined properties in Terrebonne. Strong storms in the three consecutive years of 1886, 1887, and 1888 destroyed not only crops, but also homes, farm animals, and other possessions, especially of the Terrebonne Gulf coast. Two damaging hurricanes followed each other in September and October of 1893. An uncharacteristic snowfall of February 1895 had disastrous effects on another crop that had become important locally in the late 1800s, that of oranges and other citrus. The hurricanes of 1909, 1915, and 1926 took huge tolls on both crops and structures in Terrebonne Parish.[4]

With its location between the Mississippi and Atchafalaya rivers, neither of which were effectively leveed to withstand overtopping and crevasses until much later, pre-Civil War Terrebonne was subject to flooding. Tobias Gibson of the Tigerville, or Gibson, community wrote in 1851, "I am now busily engaged in preparing to protect this place [Live Oak plantation] effectually from liability to Mississippi waters. The Levees on the River have become so uncertain that our only safety consists in protecting each for himself—I am about to construct a levee in

Left, 1909 Terrebonne Parish storm news in The Rock Island Argus, *MN, September 22, 1909*

the rear and at the same time put up a large draining machine."[5] Individual effort was more effective on the levees at the rear of the plantation, where it was necessary to protect the fields from the rising swamp water.[1]

It goes without saying that all the travails and uncertainties paid off abundantly when the weather was good and the rivers calm. As early as 1844 the top producer (James Cage of Woodlawn plantation on Grand Caillou) in Terrebonne made 965 hogsheads of sugar (each hogshead containing approximately 1,000 pounds). The total parish crop amounted to 12,661 hogsheads.

The high water mark for sugar production in Terrebonne was in 1861-62, when planters yielded a total of 28,839 hogsheads. Top producers for that year were D.S. and A.G. Cage of Woodlawn, with a crop that made 1,284 hogsheads.

These statistics were compiled by P.A. Champomier (followed later by Louis and Alcee Bouchereau) for each year as *Statement of the Sugar Crop Made in Louisiana* for the state and individual parishes. Bouchereau's reports were entitled *Statement of the Sugar and Rice Crops Made in Louisiana* and *The Louisiana Sugar Report*. Their reports recorded the names of planters (owners and lessees) and producers. The reports faithfully listed planters' names regardless of whether or not they actually produced sugar that year, but the total number of producers is deducible from the sugar reports.

Among Champomier's reports available for research, the year with the highest number of Terrebonne producers was 145 for the 1872-73 season. That number is a telling indicator of what happened to large plantations during and after the Civil War. According to Louis Bouchereau, of the 2,400 sugar producers in Louisiana on the eve of the war, fewer than 200 farms still cultivated cane in 1864. The value of the sugar industry in the state had fallen from $200,000,000 in 1861 to no more than $30,000,000 in 1865. Bouchereau's reports documented the fact that less than 40 percent of the formerly enslaved pre-Civil War work force toiled in the cane fields under free labor and "many

Right, Storm article, Parts 1 & 2
Times Democrat *1888*

Below, Snow in Houma February 15, 1895

Plantation police checking passes, Harper's Weekly c. 1895

fine plantations" were now lying waste and idle for the want of the labor to work them.

This often led to large plantations being broken down into smaller holdings—thus the high number of 145 producers for Terrebonne in 1872-73. Small plantation or farm owners did not give up on sugar cane planting; four years recorded as few as one hogshead produced by five, six, eight, and nine individual producer-planters. An upswing in production occurred statewide by the late 1800s. But by 1916-17, Terrebonne sugar producers' numbers had fallen to 13 when the effects of the sugar cane mosaic disease began to affect the local crop.

The chart on page 33 details records for Terrebonne Parish for 21 years of Terrebonne's significant contributions to Louisiana's Sugar Bowl from 1844 to 1916.

SOURCES

1. J. Carlyle Sitterson, *Sugar Country: The Cane Sugar Industry in the South, 1753-1950*, University of Kentucky Press, 1953
2. W. J. Minor & Family Papers
3. P. A. Champomier, *Statement of the Sugar Crop Made in Louisiana, 1856-1857*
4. *Eyes of An Eagle: Jean-Pierre Cenac, Patriarch,* by Christopher Everette Cenac, Sr., M.D., F.A.C.S., with Claire Domangue Joller
5. T. Gibson to Randal Gibson, May 6, 1851, Weeks Collection, Louisiana State University Library

Examples of the sugar reports, Champomier 1859-60, Bouchereau 1874-75, Bouchereau 1868-69

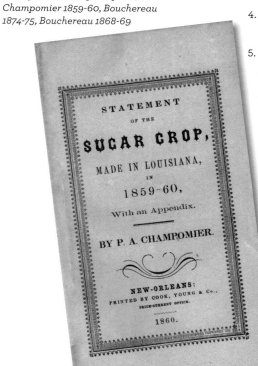
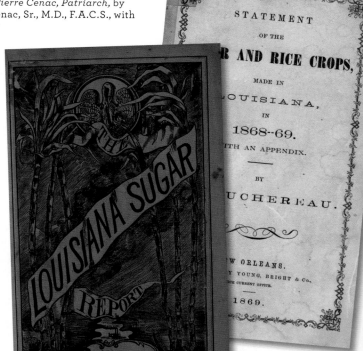

Macque Chou, 1986, Oil on canvas, 36 x 24 inches,
George G. Rodrigue

CHAPTER TWO | 29

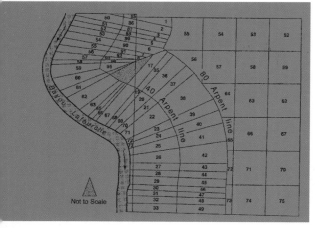

Map of the Lafourche Country

LAND MEASUREMENT TERMS

During French ownership of the Louisiana Territory, the colonial government granted land along rivers and other waterways using the arpent as a unit of measure in the French long-lot system. An arpent is an old French and Canadian linear measure of 191.83 feet. In the French system, the back edges of the plots are more or less even, forming a 40 arpent line. The importance of waterways' natural levees cultivation and development played a role in the plot divisions.[1] Irrigation, transportation, and the use of animals to perform field work by limiting the number of their turns were considered when developing this format. The surveyor was the *arpenteur*.[2]

The French system was utilized from the mid-1750s until replaced by the Spanish after they gained control of Louisiana with the Treaty of Fontainebleau in 1762. The Spanish abandoned the practice of aligning the back boundary, but they did recognize and respect all extant surveys. They granted sitios, (ranchos) and estates (vacheries), large parcels of land, to encourage the ranching industry, particularly in southwest Louisiana.[2] A sitio often measured one or one-half league square (a league equal to 2.63 miles), with a complete sitio consisting of over 4,400 acres.[1] The units utilized were *varas* (rods) and *ligas* (leagues). This survey system left a "Texan Air" in the cowboy country of southwest Louisiana,[2] *La Savanne*.

The British system utilized "metes and bounds" in the regions of North America in which they settled. This system documented parcels of land by the streams, ridges, boulders, trees, and/or roads, which met to form the boundary of the property. This system worked well until trees died, roads or streams moved, and/or boulders became displaced. "Witness trees" (marked with

French long lots

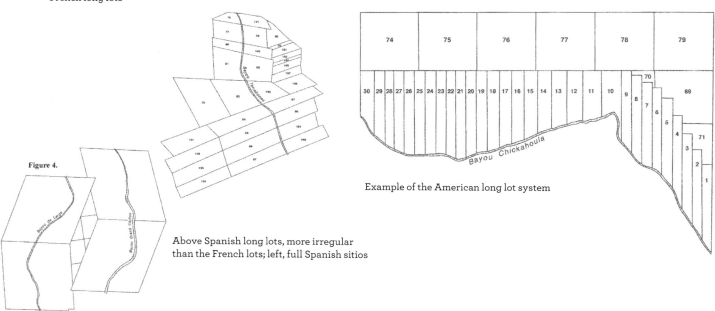

Above Spanish long lots, more irregular than the French lots; left, full Spanish sitios

Example of the American long lot system

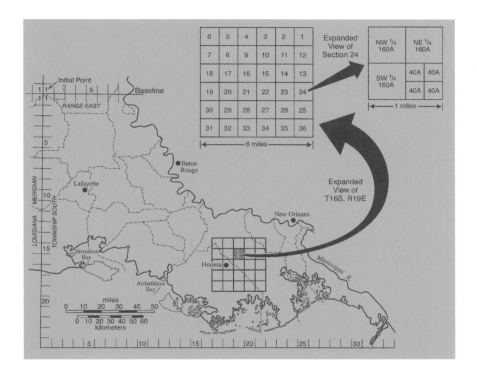

All photos Spanish Sitios diagrams from The Lafourche Country, Vol. III

Basic elements of the Township and Range survey system for the Southeastern District. Ideally, a township square is subdivided into 36 one-square-mile sections.

brands or by other methods) became a tool of the surveyor. This system did not function in the long term and became an obstacle in documenting land ownership in the "wealth creation" cycle of the future.[3]

After Louisiana became part of the United States in 1812, the township-and-range system overlaid the older systems, and those who had been granted land claims in earlier times were guaranteed valid ownership as previously surveyed.[1]

The Americans, namely Thomas Jefferson, developed an entirely different and revolutionary system with his Land Ordinance of 1785 to sell federal property acquired through the Enabling Act.[2] This act required that when a territory agreed to enter the Union, all unclaimed land entered the public domain.[1] Those lands considered "inappropriate lands"[3] were appropriated as property of the United States and made available for sale as land patents, surveyed according to the Township and Range system. The Public Land Survey (PLS) "township-and-range" system was introduced in 1807.[2] This modified long lot system used the acre as a unit of measure (an acre was long ago the amount that could be plowed by a yoke of oxen in a day). The acre was standardized to equal 43,560 feet with 208.71 feet linear measure square.[1]

Louisiana is distinct among the 50 states in that it retains the imprint of all four powers of the North American stage on its geography.[2]

SOURCES
1. *The Lafourche Country, Volume III: Annals and Onwardness*, editors John P. Doucet and Stephen S. Michot, Lafourche Heritage Society, 2010
2. "Geographer's Space," Richard Campanella in *Louisiana Cultural Vistas*, Volume 27 Number 1, Spring 2016, The Magazine of the Louisiana Endowment for the Humanities
3. Richard B. Crowell, *Chenier Plain*, Friesens, Inc., Altona, Manitoba, 2015

Property map of the northern Bayou Lafourche surrounding Donaldsonville, showing the variety of long lot survey patterns

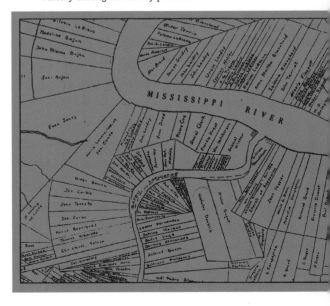

LAND MEASUREMENT TERMS | **31**

OWNERSHIP SOURCES

For every property purchase or transfer described in this volume, sources of information are either Terrebonne Parish Consolidated Government records from the Clerk of Court's Office or the Assessor's Office, or as indicated in the annual compilation *Statement of the Sugar Crop Made in Louisiana*, first by P.A. Champomier and later by L. Bouchereau, printed in New Orleans. The annual statistics were later retitled *Statement of the Sugar and Rice Crops Made in Louisiana* and *The Louisiana Sugar Report*. Other sources of ownership information are from the *Standard History of New Orleans, Louisiana,* edited by Henry Rightor and published by the Lewis Publishing Company of Chicago in 1900, as well as personal interviews with current owners of the properties. *The Historical Encyclopedia of Louisiana* published by the Louisiana Historical Bureau in 1940 was used to glean information about plantation owners, and the Historic New Orleans Collection provided information about property transfers as well as about owners themselves. Maps which were also essential in determining ownership parameters were provided by Michael Gene Burke; Leonard J. Chauvin, PE, PLS; Carey Francis "Buddy" Hebert; Grant J. Dupre; Clifford T. Smith, PE, PLS; Keneth L. Rembert, PLS; Terral J. Martin, Jr., PLS; and Arthur A. DeFraites, PE, PLS.

Note: Although every effort has been made to document ownership particulars accurately, this book should be used as a guide only, and not as an official legal description.

TERREBONNE PARISH SUGAR CROP REPORT 1844-1917

YEAR	NO. PRODUCERS	SUGAR HOUSES	TOTAL CROP HHDS	TOTAL CROP POUNDS	TOP PRODUCER IN HHDS OR POUNDS
1844	42		12,661		965 James Cage, Gr. Caillou
1845-46	16		12,080		714 Widow L. Tanner, Schriever
1846-47		for three			675 Wm. Bisland and J. Watson
1847-48	47 (1846-48)	years			893 James H. Cage, Grand Caillou
1858-59	79		22,815		970 H. Cage, Woodlawn, G.C.
1860-61*	82		16, 222		765 D.S. & A.G. Cage, Woodlawn
					765 A.A. Williams Ardoyne, B. Black
1861-62*	87		28,839		1,284 D.S. & A.G. Cage, Woodlawn
1868-69	62	57	8,924		585 T.M. Cage & Co., Woodlawn
1869-70	76	74	6,893		425 T.M. Cage & Co., Woodlawn
---48 producers in single or double digit numbers of hhds. produced this year---					
1870-71	107	91	11,389		540 E. Guidry, Little Caillou
1872-73	145	105	11,268		816 Hare & Baker, Blackwater
---100 planters in single or double digit numbers of hhds. produced this year---					
1873-74	116	106	6,583		672 by G.D. Craigin, Dulac and Live Oak Plantations combined
---91 planters in single or double digit numbers of hhds. produced this year---					
1876-77	110	104	12,331		606 T.M. Gage & Bro., Woodlawn
1877-78	92	103	10,620		600 Wm. A. Shaffer, Crescent Farm
---five sugar houses are noted as having been destroyed---					
1884-85	90	84	18,452		1,610 H. C. Minor, Southdown
--35 planters in single or double digit numbers of hhds. produced this year---					
1886-87	72	68	11,618		855 H.C. Minor, Southdown
1904-05	22	50		76,536,250	16,150,000 lbs., Lower Terrebonne Refinery & Manufacturing. Co.
---86 plantations named---					
1905-06	20	47		84,329,713	13,775,00 lbs., Lower Terrebonne Refinery & Manufacturing. Co.
1906-07	20	52		53,377,936	7,371,000 lbs., H.C. Minor Estate Partnership Southdown
1907-08	19	51		64,827,741	9,623,340 lbs., Lower Terrebonne Refinery & Manufacturing. Co.
---86 plantations named---					
1916-17	13	46		47,961,850	11,904,432 lbs., Terrebonne Sugar Company Central Factory at Montegut
---87 plantations named, 24 planters denoted only by names---					

*1861 is recognized as the high water mark in Louisiana Sugar crop production

OWNERSHIP SOURCES | 33

Low Tide, 2009, Oil on canvas, 15 x 30 inches, George G. Rodrigue

MALBROUGH SETTLEMENT
CÔTE DE MALBROUGH

Joseph and François Malbrough were among the pioneer settlers of the Lafourche Interior before the creation of Terrebonne Parish, which encompassed part of their land grant upon the parish's founding. After Louisiana became a state, the Malbroughs registered their long-standing claim with the U.S. government on March 13, 1816. It was confirmed on May 11, 1820.[1]

The Malbroughs' claim extended from Bayou Lafourche to the railroad crossing at Terrebonne Station, reaching across the boundary between Lafourche and Terrebonne parishes as they now exist.

Côte de Malbrough (literally, Malbrough Coast, from the French) was the most northerly settlement in Terrebonne Parish, and incorporated an extensive expanse up to Bayou Lafourche.

A survey plat from 1856[2] shows Bayou Terrebonne "curving onto Joseph Malbrough's lands around present City Hall, crossing Canal Boulevard in the area of the present post office, where it took a northeasterly curve to the present area of Patriot and West Fourth" streets [in Thibodaux, Lafourche Parish].[3]

According to a "Bayou Terrebonne History" from the book *Historical Scenes of Thibodaux*, "The western boundary of the Malbrough tract was Jackson Street, later known as *le grand chemin du Bayou* Terrebonne, the Bayou Terrebonne Road." That same source continues, "A good guess for the spot of

Above, 1850 State Land Grant map, Bayou Terrebonne to Schriever

Right, Bayou Terrebonne origin in Thibodaux, La., 1846 Richardson Map

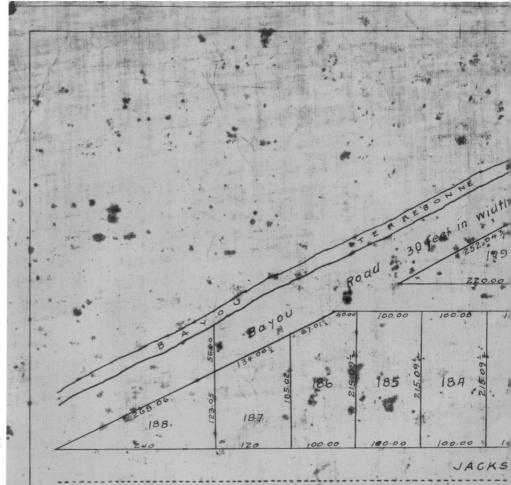

Joseph Malbrough's habitation would be centered on the former [Thibodaux] city hall, courthouse, and old jail squares, these being where the first lots were sold later by Henry Thibodaux, likely representing clearing by Malbrough between 1791 and 1812."

As were other area bayous, the head of Bayou Terrebonne was silted over long before Caucasians began to settle the area. To facilitate development of water transportation in the new Terrebonne Parish, Mrs. Henry Schuyler Thibodaux of St. Brigitte plantation in 1830 donated property 100 feet wide by 23 arpents length for the digging of a canal connecting the Lafourche and the deeper section of Bayou Terrebonne. The location of the canal is currently Thibodaux's Canal Boulevard from the bridge over Bayou Lafourche to the railroad tracks crossing Jackson Street and Canal.[3]

This water access no doubt benefitted residents of the old Malbrough settlement, which a Houma newspaper article in 1866 described as consisting of "little farms in a high state of cultivation, most of the labor being done by themselves." The article described the old folks as pleasant and the young ones sociable and intelligent.[4]

SOURCES

1. A Century of Lawmaking for a New Nation: U.S. Congressional Documents and Debates, 1774-1875, American State Papers, Senate
2. Louisiana State Land Office survey plat, 1856
3. Bayou Terrebonne History from *Historical Scenes of Thibodaux* book, Nicholls State University Archives
4. *The Civic Guard* newspaper of Houma, Terrebonne, Louisiana, June 16, 1866

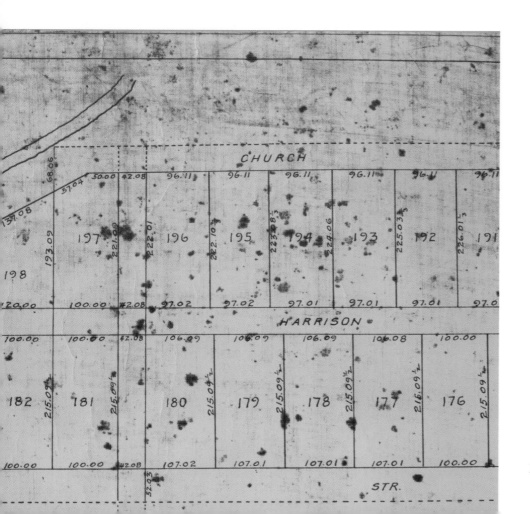

Armitage 2016

SARGEANT ARMITAGE

According to an 1810 map[1] indicating the properties of the first settlers in what was to become Terrebonne Parish, a more northerly plot near the headwaters of Bayou Terrebonne belonging to Joseph Malbrough and an adjacent plot owned by François Malbrough were the earliest roots of Sargeant plantation ownership. The Malbroughs' claim was registered with the U.S. government on March 23, 1816 and confirmed on May 22, 1820.[2]

Another chapter in Sargeant/Armitage's history began with the 1831 union of Judge George Seth Guion and Caroline Lucretia Winder at The Briars plantation in Natchez, Mississippi. Guion (born 1806) was the son of a Revolutionary War officer who had left New Rochelle, New York, to settle in Mississippi. Judge Guion was the second owner of record of the prior Malbrough claims.[3]

Caroline's stepfather, Dr. John Davidson Smith, gave her Ridgefield plantation's 1,200 acres of land in Lafourche Parish shortly after her marriage. The Guions donated land for St. John Episcopal Church and cemetery on what is now Jackson Street in Thibodaux. He gave land for a school for children ages six to 16, and began a process of town expansion.[3] George and Caroline's

Armitage stained glass window 2016

Armitage, front 2016

Detail of Armitage faux bois *above and* faux bois *door below 2016*

daughter Caroline Zilpha would become the wife of West Point graduate, attorney, Civil War brigadier general, and 28th Governor of Louisiana, Francis T. Nicholls.

Francis L. Mead of New Haven, Connecticut, prior to 1845 bought a strip of land in Terrebonne formerly owned by the Guions. Mead's first wife, Margaret Hicklin, had been married to Mead's employer Philippe M. Sargeant (died 1836).[4] Appraisers for Sargeant's succession described "A tract of land about 6 miles from Thibodaux on both sides of Bayou Terrebonne with 18 arpents front on the left side of the bayou and 16 arpents front on the right side, being separated on that side by 2 arpents belonging to Mrs. Gabriel Arsenaux and bounded above by R.G. Ellis and below by Widow Urbain Arcenaux, with the buildings: Sugar house, barn, gin, gin house, and Negro huts" as well as 36 men, women, and child slaves.[4]

Francis Mead had the house built on the right bank of Bayou Terrebonne at Sargeant plantation in 1852 after his marriage to the widow Sargeant in 1838. It is a story and a half, wood frame, in the Greek Revival style.[6] Margaret's daughter by Philippe Sargeant, Ellen Rebecca Sargeant, inherited two-fifths of Sargeant. She married Charles Armitage of Baltimore in 1843. Champomier's 1846-47 *Statement of The Sugar Crop Made in Louisiana* included an entry for "F.L. Mead & Armitage" as having produced 255 hogsheads of sugar that year.

Ellen Rebecca died in 1852, and Charles married Lucy Dean Foster of Hinds City, Mississippi. In 1859, Francis Mead sold the rest of his estate to the Armitage couple, who took up residence in the house that Mead had built in 1852. After Charles Armitage died in 1872, his wife Lucy lived on in the house for two more years until postwar conditions sent her back to her family in Mississippi. (The story from an Armitage descendant is that the Armitages' only son, who was then 13, settled the estate and made all arrangements for their move to Mississippi.)[4]

In 1883 Joseph Darden Hilache Roundtree bought the property. He was the father of 12 children, five by his wife Marie Evelina/Velina Boudreaux who later married Alfred Comeaux, and six by Eulalie Malbrough.[7] When he died in 1895, the land was subdivided into four sections.

One of the Roundtree children sold the narrow lot on which Armitage House stands in 1948 to Francis William "Frank" Wurzlow, Jr. and his wife Mary Elizabeth "Bettie" Chauvin Wurzlow of Terrebonne Parish. It is they who named the home Armitage for its former owner, and who renovated the house with painstaking care, completing the project in 1968. One of the first

Armitage lead waterhead 2016

Armitage, rear 2016

things they did was to center the house on the lot and move it far back from its location, which was too close to the highway.

The home is adorned with replica plaster rosettes and decorative lead waterheads fed by the house's gutters. Cornices are wood copies of the original and the stairway newel post is from Belle Grove plantation. Additional features are *faux bois* doors in the Greek Revival style and a chandelier in the library from the Old State Capitol in Baton Rouge. The gazebo incorporated into the house's rear grounds was salvaged from the Houma Academy/St. Francis School building on Point Street that was torn down in 1966. The gazebo was the crowning cupola on the old edifice. The original milk house with gingerbread trim is now a courtyard garden house.[4]

Renowned architect Henry Howard is credited with stylistic attributions in the design of Armitage.[8] Howard also designed the Houma Academy, the first St. Matthew's Episcopal Church and the first First Presbyterian Church in Houma. Ralph Gunn was the landscape architect. He is famed for designing the grounds of Rosedown plantation in St. Francisville.[4]

Armitage was named to the National Register of Historic Places in 1984.

Frank Wurzlow died in 1999 and Bettie in 2009. Current owners of the house, since 2001, are Keith Julian and Susan Poché Manning. The address at Armitage is 506 West Main Street, Thibodaux, Louisiana.

SOURCES

1. *First Land Owners and Annotated Census of Lafourche Interior Parish, LA (Lafourche & Terrebonne)*, 1995 by Audrey B. Westerman
2. A Century of Lawmaking for a New Nation: U.S. Congressional Documents and Debates, 1774-1875, American State Papers, Senate
3. Alcee Fortier, editor, *Louisiana: Comprising Sketches of Parishes, Towns, Events, Institutions, and Persons, Arranged in Cyclopedic Form*, Volume 3, published by Century Historical Association, 1914
4. Paul F. Stahls, Jr., *Plantation Homes of the Lafourche Country*, Pelican Publishing Company, 1976
5. *Terrebonne Life Lines, Vol. 24 No. 3*, Fall 2005
6. Louisiana Historic Standing Structures Survey, Terrebonne Parish
7. *Terrebonne Life Lines Vol. 16 Spring 1997*
8. Robert S. Brantley with Victor McGee, *Henry Howard, Louisiana's Architect*, Historic New Orleans Collection and Princeton Architectural Press, 2015

Top, Armitage staircase;
Above, Armitage plaster rosette medallion

Opposite page, Armitage gazebo interior and exterior, (originally cupola from Houma Academy, later St. Francis de Sales School) 2016

Armitage milk house with original gingerbread trim, now courtyard garden house 2016

BOURGEOIS-THIBODAUX-BARRAS HOUSE

A house in Schriever that is preserved only through a painting by Thibodaux artist Billy Ledet, and photos of its demolition, had ties to Armitage by virtue of its location on what was once Armitage land near the Carriage House.

The two-story house was built in 1853, the same year the Bourgeois Carriage House was constructed. It is thought to have been constructed by descendants of Antoine Barras (1757-1826), the father of Jean Valery Barras (1820-1845) and Judge Leufroy Barras (1798-1871). Leufroy was the owner of Isle of Cuba plantation besides serving as the second judge in Terrebonne Parish. The house's construction included mortise and tenon joinery, and its doors were given faux bois design treatment.

A 1944 conveyance record in Lafourche Parish describes part of the ownership history and particulars of the transaction.

"A certain tract of land situated in the Parish of Terrebonne, State of Louisiana, about 14 miles above the Town of Houma, Louisiana, on the right descending bank of Bayou Terrebonne, measuring one and one-half (1 ½ arpents front, more or less, by a depth of twenty (20) arpents, more or less. Bounded above by the lands of Willie Roundtree [Armitage], now or formerly, below by the lands of Mrs. Oscar Daigle and Valerie Bourgeois [Carriage House], in the rear by the lands of Ridgefield plantation, and in front by the paved highway known as the Thibodaux-Houma Highway....Being a portion of the same property acquired by the vendor herein by virtue of an act of sale from Mrs. Ulysse Barras Bourgeois on January 11, 1923...."

The February 11, 1944 sale of land and house was to Eula Mae and Percy J. Gros, Sr., at a cost of $5,000. According to artist Ledet's notes on the house in his painting "Schriever House and Land," "Percy's plan was to move his parents into the house. However, soon after the purchase, Percy's father died."

The old cypress house was demolished and materials salvaged by Dean Landreneau and his crew in 2007. Current address of the spot is 540 West Main Street, Thibodaux, Louisiana.

Bourgeois, Thibodaux, Barras house, mortise and tenon joinery 2015

Bourgeois, Thibodaux, Barras house painting c. 2000 by Billy Ledet (1934-2014)

Bourgeois house, former Armitage carriage house 2015

BOURGEOIS CARRIAGE HOUSE

The carriage house built on Sargeant/Armitage plantation grounds was constructed in 1853. The two-story structure was initially used to store plantation transport vehicles on the first floor, and to house slaves and other workmen on the second floor.

The old structure underwent alteration almost a hundred years later when a one-story addition was completed in 1932 by Valery Jean Baptiste Bourgeois who was born at Melodia plantation in Lafourche.

Bourgeois family members have had a longstanding relationship with the property, and still maintain a presence on the batture across from Sargeant/Armitage grounds in the form of Bourgeois Meat Market, a popular Terrebonne Parish business.

Valery was born on July 17, 1878 in Thibodaux, and married Julia Helene Thibodaux (1888-1973) on February 5, 1906. His parents were Ulysse Ursin Bourgeois (1829-1903), and Charlotte Barras (1850-1933) of Leufroy Barras lineage (Isle of Cuba). Julia's parents were Aubin P. Thibodaux (1840-1905) of Henry Schuyler Thibodaux lineage (St. Bridget and Balsamine) and Susanna L. Barras (b. 1847) of Leufroy Barras lineage (Isle of Cuba).

The family's Bourgeois lineage has been traced back 12 generations to Nicholas Grandejean Jacques Bourgeois, born 1575 in La Ferte, Haute-Marne, Champagne-Ardenne, France. He died on January 4, 1621 in Champagne, Dordogne, Aquitaine, France. His son Dr. Jacques Jacob Bourgeois, although born in France in 1621, emigrated to Port Royal, Acadia, Nova Scotia, Canada and died there in 1699. The next three generations of the Bourgeois family (Charles, Charles and Paul), remained in Canada, but Michel Bourgeois (b. 1741 in Nova Scotia) was the first of that Bourgeois line to emigrate to Louisiana, no doubt part of the forced exodus of Acadians by the British. He died in Lockport, Louisiana on June 27, 1811. Valery's father Ulysse was Michel's son.

Valery Jean Baptiste became a butcher in 1905 using a shop on Jackson Street in Thibodaux. In 1907 he used part of the establishment of Philip Maronge to process meat, and sold his product as an itinerant merchant, from Gibson to Bourg.

By 1929 the Bourgeois Meat Market was situated on the right descending side of Louisiana Highway 24, but in 1955 the market was established on the bayou side across from the carriage house and Armitage House. Valery Bourgeois, Jr. died in 1938, and his brother Lester took over the business when he returned from World War II service in 1945.

The current two-story carriage house structure is the home of Lester Bourgeois, Sr. and his wife, Rita Ethel Trahan. Current address is 540 West Main Street, Thibodaux, Louisiana.

Bourgeois Meat Market has been a fixture in that part of the parish for decades.

SOURCES
1. 2016 interviews with Lester Charles Bourgeois, Sr., Lester Charles Bourgeois, Jr., and Richard Joseph Bourgeois

Bourgeois house, former Armitage carriage house c. 1979

JOHNSON RIDGE

Above, Johnson Ridge resident Louis Robertson (1876-1987)

Johnson Ridge is a populated place in the northernmost reaches of Terrebonne Parish, composed to a great extent of African American descendants of former slaves at area plantations. It is located on the left descending bank of Bayou Terrebonne, across the bayou from Armitage House. The section of the small community known as the Old Ridge lies on the northern extreme of Terrebonne Parish on the southern outskirts of the city of Thibodaux. Old Ridge lies along Johnson Ridge Lane, which intersects West Park Avenue (Louisiana Highway 20) above Schriever.[1]

A more recent addition to the community is what has been known as New Ridge since before 1924 on the left bank of Bayou Terrebonne along Livas Lane below but adjacent to the older community. A further distinction locals make is what they call the Levee fronting the highway between the Old and New Ridge.

The village was settled on a tract of land formerly part of Sargeant plantation. A longtime anchor of the community was Magnolia Methodist Episcopal Church, which was built in 1870 on Magnolia plantation on Bayou Black.[2] The church was transported by a mule-drawn wagon in 1916 to its present site and reconstructed on Johnson Ridge Lane. The church's name later underwent a change to Magnolia United Methodist Church.[2]

Some pastors associated with the church were the Reverend J.D. Wilson, Rev. A.J. Hackett, who arranged for the first African Americans to perform on KTIB radio station, during which he preached a sermonette and the choir performed on Sundays. Rev. Matthew Byrd purchased 2,000 bricks for the church, and Rev. William Taft Bowie in 1972 had the wooden structure repaired using the bricks, added a multi-purpose building, and added air conditioning. Rev. Louis Streams, Jr. had a concrete floor added to the main sanctuary. Rev. Louis Augustine, Jr. had the Fellowship Hall air-conditioned and purchased an organ. He retired in 1991, and was followed by the church's first Caucasian pastor, Rev. Mel Zerger.[2]

Livas Lane nearby is the site of Morning Star Baptist Church. According to a written history supplied by the church, it was originally established in an old house on Waubun plantation on July 15, 1874. The Rev. Ben Fields was pastor. The church was relocated to its present address on Livas Lane on July 12, 1903. Rev. Ephon Smith was pastor from 1885 to 1940. Rev. A.J. Pharr began

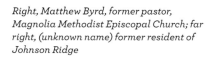

Right, Matthew Byrd, former pastor, Magnolia Methodist Episcopal Church; far right, (unknown name) former resident of Johnson Ridge

Morning Star Baptist Church, Johnson Ridge 2016

Magnolia United Methodist Church, Johnson Ridge and cornerstone Magnolia United Methodist Church, Johnson Ridge 2016

his pastorate in 1940, during which he had the church building renovated. He died on September 4, 1984, and was succeeded by Rev. Eugene R. Brown beginning April 21, 1985.

Prior to the Civil War, slaves walked from Acadia plantation to Johnson Ridge to attend church services via an elevated walkway through swampland called "Devil's Swamp Path," according to David D. Plater, a descendant of Acadia plantation owners.

Family names associated with the small community are Joseph, Livas, Round, Jackson, Harris, Johnson, Nolan, Streams, Thomas, Stripling, Toups, Redmond, Neville, Victor, Ballard, Randolph, Rives, Williams, Levron, Pharr, and Fields. Specific residents on record are Henry Joseph, James Thomas, A. Stripling, W. Victor, J.R. Joseph, M. Ballard, Lusignon Neville, James Randolph, Thomas Livas, T. Rives, James Williams, Jules Levron, and I. VanBuren.

Resident Louis Robertson of Johnson Ridge lived to the age of 112. He was born on September 18, 1876, and at an early age was baptized in Morgan City, but in 1916 joined the fellowship of Morning Star Baptist Church. When he died on November 15, 1987, he left 112 grandchildren, 95 great-grandchildren, and 68 great-great grandchildren. His motto was "clean life."[3]

SOURCES

1. Tobin map of 1938
2. "History of Magnolia United Methodist Church As It Happened," courtesy church records
3. Mr. Robertson's funeral program, 1987, supplied by Morning Star Baptist Church
   Lillian Joseph personal interview

Pew from original Magnolia United Methodist Church, Johnson Ridge 2016

DUCROS
JULIA (UPPER DUCROS)
HOME SITE (LOWER DUCROS)

The stately home on Ducros plantation has survived in its spacious grounds on the right bank of Bayou Terrebonne three and a half miles south of Thibodaux since the time its initial form took shape in the 1820s. It is a two-story frame Greek Revival structure with interiors featuring a combination of Greek Revival and Rococo Revival elements.[1]

The land surrounding the house has a known history dating to 1802, when the parish was, in effect, pioneer territory. In that year, Thomas Villanueva Barroso, born in Tenerife, Canary Islands, claimed a Spanish grant of 4,400 acres there on July 15, and the grant was confirmed by the U.S. government on April 12, 1814.[2] He was the first Caucasion to make use of Ducros property. Barroso was in 1801 the Commandante of Valenzuela (now Belle Rose) on upper Bayou Lafourche, in the district of Lafourche containing Lafourche, Terrebonne, and Assumption parishes.[3]

Major General Pierre Denis de la Ronde of Versailles plantation in St. Bernard Parish[4] obtained the Barroso tract on December 22, 1813. Author Paul F. Stahls, Jr. of *Plantation Homes of the Lafourche Country* wrote about "two slightly conflicting stories" of that transfer from Barroso to de la Ronde. "One has it that the home was raffled off in a lottery in New Orleans and that de la Ronde won it with a thirty-dollar ticket," but the other scenario was that it was "lost in a wager on a sulky race." De la Ronde was on the staff of General Andrew Jackson and served as commander of the Louisiana Militia during the Battle of New Orleans. The major died in 1824.

A Terrebonne Parish conveyance record dated March 1, 1828 (in New Orleans) documents "a donation between Miss Eliza White and P.D. de la Ronde, where Miss Eliza White, adolescent minor, assisted by her father and curator Maunsel White and her cousin and curator of the causes, George W. White and Pierre Denis de la Ronde, of St. Bernard Parish, in the inventory made in New Orleans, of the properties of the succession of Pierre Denis de la Ronde, their father and grandfather, deceased, whereas certain wastelands situated in parishes of Terrebonne, Feliciana and St. Tammany, belong to succession, and of little value, were by common consent.... adjudged to Pierre Denis de la Ronde for the sum of $3,150.00."

The namesake for the plantation was Pierre Adolphe Ducros de Lucinge, who was born in 1798 in Passy, Haute Savoie, France. He entered the Ducros ownership rolls after his marriage to Adelaide Adèle de la Ronde. After the death of Pierre Denis de la Ronde, *père,* his wife, Marie Elizabeth Eulalie Guerbois de la Ronde, sold all properties of her husband at public auction. Conveyance records[5] indicate that her son, also named Pierre

Pierre Adolphe Ducros de Lucinge (1798-1861) c. 1850

Opposite page, map of the Terrebonne Project, November 7, 1945 (Julia plantation); families of Rodney Shields and Richard Covington Woods on the front porch of Ducros plantation house c. 1895

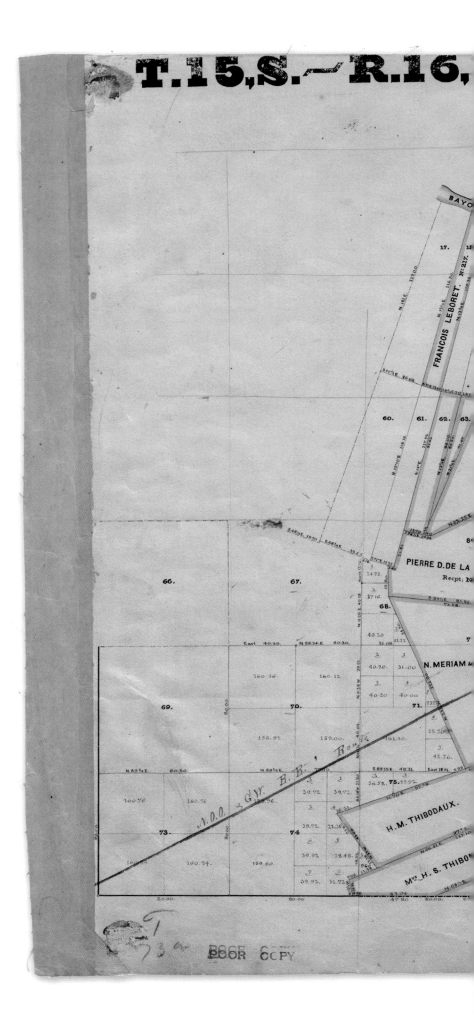

*Thomas Villanueva and Pierre Denis de la Ronde Claim
State Land Grant map December 24, 1831*

54 | HARD SCRABBLE TO HALLELUJAH

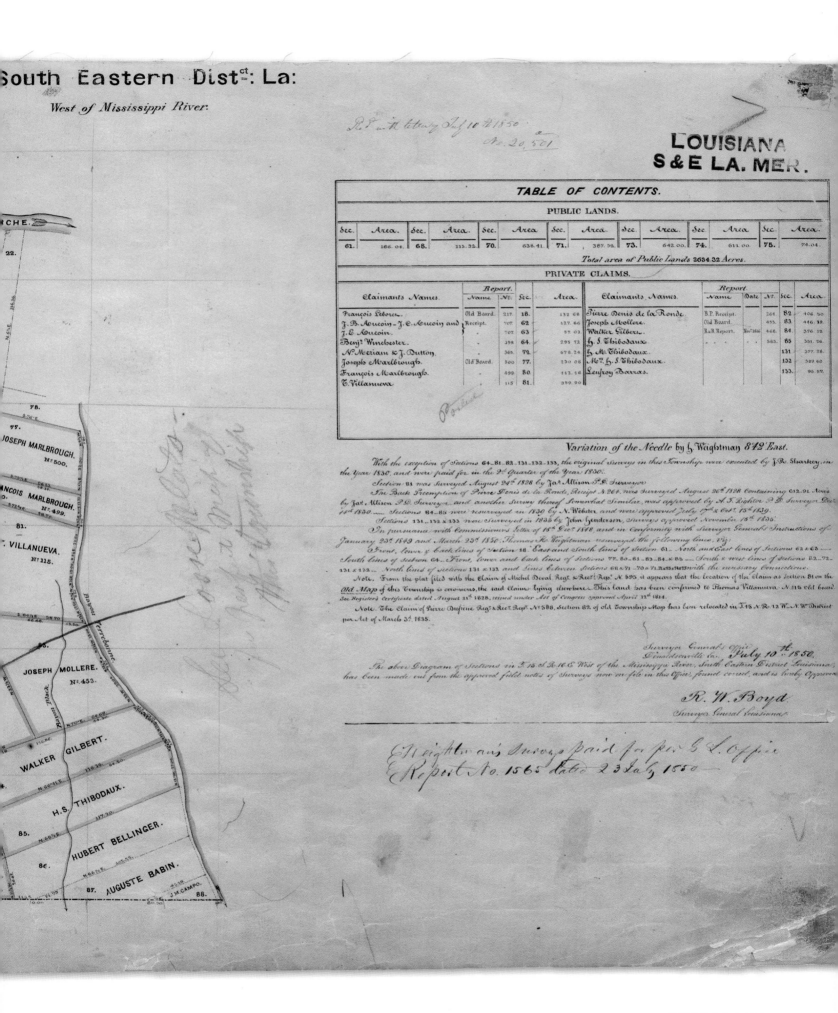

Denis, and Pierre Adolph Ducros, purchased the estate, "which they now own conjointly and undividedly owners [sic] of the land, each one half." The same record of June 23, 1830, continues, "Now they want to end their joint ownership, by dividing property in half, land in Terrebonne Parish 1224 acres superficies, more or less, with 20 arpents front on each side of Bayou Terrebonne, bounded North by land of Francois Malbrough, and vacant lands, and on the south side, land. . . . now belongs to Lemuel Tanner. The land was divided in half, with 10 arpents front on both side [sic]."

The joint owners chose a unique way to decide who would get which parcel. "By drawing lots, in ordinary manner and custom, the result, first choice was due Joseph Marcel Ducros, who chose the upper part of land, and Pierre Denis de la Ronde, *fils*, accepts the lower part of land."

Richard Gaillard (Gaillain) Ellis became the next owner of Ducros in 1833 by purchasing acreage from both Joseph Marcel Ducros and Pierre Denis de la Ronde, *fils*. Ellis was a successful planter and entrepreneur; by the time of his death 11 years later, he was the owner not only of Ducros but also of Ellerslie on Grand Caillou and Magnolia on Little Bayou Black.

Sugar house at Ducros c. 1902

The "Petit Terrebonne tract" (Magnolia, on Little Bayou Black) was "a tract of land on both sides of Bayou Little Terrebonne about 6 miles from Lafourche containing 35 arpents in front on each side of the bayou with such depth as to include 2,600 arpents, bounded above by the land of Evariste Porche and below by land of Thomas Butler, together with the overseer's house, sugar house & apparatus, barns, stables, Negro huts and other improvements valued at $80,000."[6] A later description of the tract for probate described it as "bounded above by Leufroy Barras and below by lands of the Succ. of Thomas Butler."

To connect his land on Bayou Terrebonne with his holdings on Little Bayou Black (obviously known at the time as the Petit Terrebonne), Ellis and his neighbor Evariste Porche petitioned the Terrebonne Parish Police Jury on October 7, 1844 to allow them to construct the so-called Ellis Causeway. That roadway is now the route followed by Louisiana Highway 311 as it veers right (northbound) at Magnolia plantation grounds, emerging aside what is now St. Bridget Church on Highway 24 and Bayou Terrebonne.

The Grand Caillou Tract was "a tract of land on Bayou Grand Caillou containing 1893 arpents; 36 arpents front with a depth of 40 arpents on the east side of the bayou bounded above by lands owned by Thomas Butler, below by vacant land. This tract formed a part of the original grant of 80 arpents front by 80 of depth made to Charles Jumonville de Villiers. Eleven arpents and one third on the west side of the bayou by 40 arpents in depth bounded above by Thomas Butler and below by Angus McNeill valued at $5,000."[6]

Ellis' cumulative property value in Terrebonne Parish at the time of his death in 1844 was $283,035; the accumulative total of his estate in Terrebonne Parish was $308,473.[6] He was one of the largest slaveholders in the parish.

At Ducros alone he left the dwelling house, sugar house and apparatus, barns, stables "Negro huts" valued at $38,000, 110 slaves, stock, wagons, carts, farming tools, blacksmith tools, 356 hogsheads of sugar, 356 barrels of molasses.[6]

His Magnolia estate included slaves, cattle, sheep, hogs, mules, horses, oxen, carts, farming utensils, 474 hogsheads of sugar, 474 barrels of molasses, library, household furniture, silverware, and carriage.[6]

The 1844 *Statement of The Sugar Crops Made in Louisiana* by P.A. Champomier lists two totals of sugar production for R.G. Ellis in northern Terrebonne (Ducros and Magnolia). One produced 354 hogsheads, and the other 528 hogsheads, the latter number among the top 10 producers in the parish that year. These were to be his last crops. He died on December 11, 1844.

Judge Thomas Butler, who was married to Ellis' sister Ann Madeline Ellis, acquired part of Ducros in 1835. When they occupied the house, the Butlers enlarged it by adding wings.

Mississippian Van Perkins Winder of Natchez bought 1,224 acres at Ducros in 1845, at a price of $60,000. Winder was the son of Dr. Thomas Jones Winder, a native Virginian who relocated

Judge Thomas Butler (1785-1847) c. 1835

Van Perkins Winder (1809-1854) c. 1845
Bottom left, Van P. Winder obituary, Thibodaux Minerva, *November 11, 1854*
Bottom right, Winder coat of arms

Ducros c.1935

Bell at Ducros 1904

Martha Ann Grundy Winder (1812-1891) c. 1845

to Natchez, Mississippi. At his death in 1812 the doctor left his children cotton plantations in Concordia Parish, Louisiana, immediately across the Mississippi from Natchez. His mother, Harriet Handy Winder, died in 1820. Van was only 11, his sister Elizabeth 13, and Caroline Lucretia, nine years old.[7]

As a young man attending the University of Nashville in Tennessee Van met Martha Ann Grundy. Martha was the daughter of the nationally acclaimed attorney, orator, and legislator Felix Grundy. Grundy would later serve as U.S. Attorney General under President Martin Van Buren.

Grundy withheld his permission for the couple to marry because of his daughter's age, so the couple eloped in December 1828 while he was away on a business trip to Washington. Winder family accounts recorded that Grundy was furious when he found out. Van was 19 and Martha 16.

Grundy's objections and initial dissatisfaction was the impetus for what Nina Winder much later wrote of her grandfather: "Grandpa Van Winder was a very genial, social, friendly person and had lots of friends and namesakes. Aunt Sallie said her mother (Martha Grundy Winder) told her he threw a lot of style and spent a lot of money when he went with his family to Tennessee because he said he wanted to show the Grundys that Martha had not done so badly in marrying a young college boy."

In 1832 Van's stepfather, Dr. John Davidson Smith, gave him and his sister Caroline each an interest in a sugar plantation in Terrebonne. Van and Caroline's sister Elizabeth had died years before, at the age of 14. Van's holding was "one-third interest in 1,000 acres fronting Bayou Black and running on both sides of Bayou Dularge," according to a Winder biography.

The couple moved to Louisiana after two of their children were born in Natchez while he worked on his deceased father's cotton plantations. Seven more children were born on Bayou Black. Van's next venture was to sell the Bayou Black plantation and purchase an interest in Southdown plantation from James Dinsmore in 1841 for $31,000. His partner was William John Minor of Natchez. While the Winder family lived at Southdown the Grundy grandparents moved in with them. Martha gave birth to one child, Louise, at Southdown.[7]

The Winders' house while they lived on Southdown plantation was located approximately where the USDA Agricultural Research Service Sugar Cane Research Unit building now stands at 5883 USDA Road across Little Bayou Black from Louisiana Highway 311 in Houma.

William John Minor purchased Van Winder's half-interest in Southdown on May 27, 1845. Winder then bought Ducros the same year, although the family had to wait for renovations to be made before they moved into the house. During that time they lived at Waverly plantation on Bayou Lafourche, north of Leighton plantation on Highway 1 in Thibodaux. Five more Winder children were born at Ducros.[7] The plantation remained in the hands of the Winder family until 1872. Over the course of his ownership, Van extended his holdings into a 3,300-acre tract.

The Winders lived in the house that was already situated at Ducros, but they had a vision of replicating The Hermitage, home of Andrew Jackson, in Nashville, Tennessee. Eldest child Caroline was married to John McGavock of Carnton plantation of Franklin, Williamson County, Tennessee, in the old house in 1848.

One conjecture is that an early structure may have been built as early as 1823 by de la Ronde or Ducros. Paul Stahls, Jr. wrote in *Plantation Homes of the Lafourche Country*, "It seems unlikely that the home would still bear the name Ducros had he not done some of the building."

That old house "could have been built as early as 1823, with wings being added in 1854," Stahls wrote, citing historian William Littlejohn Martin as that theory's proponent. Others theorize its date of construction to be between 1833 and 1835,

Dr. John D. Smith property on Bayou Black, State Land Grant map April 2, 1831

Below, portaits of Caroline Elizabeth "Carrie" Winder McGavock (1829-1905) and John McGavock (1815-1893); Thibodaux Minerva September 24, 1853 obituary of Martha Grundy Winder, 16-year-old daughter of Van P. and Martha Grundy Winder; Carnton plantation, Franklin, Tennessee, home of Carrie Winder McGavock; rear porch built c. 1850 in the image of Ducros porch

Milk shed/gas plant building 2016

Inside milk shed 2016

Ducros rose garden 1904

Above and right storage for root vegetables 2016

Stahls added. This was based on the knowledge that the same architect who designed Pierre Ducros' home in New Orleans also designed nearby Magnolia plantation. Because construction on the Ducros house had not progressed enough for installation of one of two identical winding French staircases ordered for the two homes when it arrived, the first went instead to Magnolia plantation, Nina Winder said. Stahls recorded that the second staircase did not arrive in New York until 1861 during the Civil War. It was never installed at Ducros. But Martha Winder received bills for the $1,000 staircase, and according to family memories she had to travel to Washington, D.C. to clear up the matter.

Another version of the Ducros home being built is that from the memories of Van and Martha Winder's daughters which Stahls documented. They said in a 1930s interview that their mother built the house after their father died of yellow fever in 1854.

Martha took over management of the plantation, and brought in an impressive 1858-59 crop of 701 hogsheads of sugar. The census of 1860 valued the Winder real estate at $205,000 and personal property at $207,000.

During the Civil War Ducros was used as a hospital, and before they were covered over after the war, Union soldiers' horses' hoofprints marked the central hallway where they rode through the house.[8]

The attractive white house the Winders finalized mid-nineteenth century still stands on Ducros grounds. Its foundations are of handmade bricks from Ducros kilns and its structure was constructed of heart cypress. The edifice's eight large square columns rise two stories high from the first-story gallery. The home has 18-foot-high ceilings.[8] On the first and second floors, passage from room to room is through panel doors. Through its long history, the majestic home sat amid oaks, magnolias, pecans, camellias, and a rose garden.

Another structure that remains on the grounds of Ducros is a building that served as a chicken coop on one side and a carbide plant on the other. Still existent in the small plant is a cylindrical metal container into which calcium carbide pellets were dropped into water to produce acetylene gas. The gas was then piped to the main house, where it provided fuel for acetylene lamps to light the home. The remnant of another structure on the grounds served as a storage place for root vegetables.

Carbide acetylene gas plant 2016

Inscriptions written during the Civil War on back of door frame:
"South Carolina seceded Dec 20 1860"
"Put up by P Bardons Jany 4th 1861"
"Springer & Evans Boss Carpenters"
"Hell of a time here sugar only 5 ¾ & 6 ½ & no buyers"

Inscriptions written during the Civil War on back of door frame:
"Fort Sumpter[sic] in S. Carolina occupied by the U.S. troops Jany 2nd 1860"
"Hurrah for S.A. Douglas"

Seceded Dec. 20. 1860
& Evans Boss Carpenters
5¾ & 6½ & no buyers

S. Carolina
Jan'y 2nd 1860
S. A. Douglas

Covington Barrett Woods c. 1903

Left to right, front fireplace main hall; ceiling medallion; French doors onto front gallery; panel doors; main hall ceiling medallion; left rear fireplace main hall, Ducros 2015

Rodney Shields Woods bought Ducros in 1872 from the Winder estate when the plantation "was sold for a division among the heirs." He paid $65,000 for the house and lands. In 1877 he sold half-interest in the Ducros estate for $37,500 to his brother Richard Covington Woods. The brothers were also owners of Rural Retreat plantation in Bourg. Rodney was married to Margaret Pugh, and Richard to Margaret's sister Fannie. The two women had been born into the planter society on Bellevue plantation in Assumption Parish. The two families lived at Ducros together 32 years until 1909, rearing their children there.[9]

As an elderly woman, Margaret Pugh Woods recorded the splendor of Ducros when she was a bride in 1877: "Few brides ever walked into such an exquisitely beautiful home, with all that one could wish for. It was in the Spring, the garden was aglow with flowers, the trees and grass were green—all seemed to wave a welcome to a young bride of eighteen. The stately mansion with its white columns never looked more inviting. As you approached Ducros in its banner days there was a long line of white plank fencing on the entire front. Next came the immense grove that you could see a distance off, and when the house was lighted with sugar house lamps for parties it could be seen as one crossed the railroad bridge at Thibodaux. By moonlight, where the glistening moonbeams wafted long shadows through the trees, there was nothing lovlier [sic]."

HARD SCRABBLE TO HALLELUJAH

Cane train Ducros c. 1901

In the years they owned Ducros, the Woods family produced many crops, but their post-war yields were considerably less than in the Winder years, for example. Early on, the total number of hogsheads totaled in the 30s, then climbed to 100, then to 255 in 1877-78, but never approached the abundance of pre-war years at Ducros. The sugar mill was dismantled after a flood in 1882, but the plantation's sawmill survived. (During the remaining sugar production years, harvested cane went to Waubun for grinding and processing.)

In January 1888, the Woods brothers sold a tract with ten acres frontage on both sides of Bayou Terrebonne (Upper Ducros) to McFarland, Baldwin & Company of Cincinnati. By 1897, that land was operated under lease by tenants Marculus Guillot, Pierre Prejean, Sylvere Guillot, Emile Lassage, Gustave Berthelot, Frank Hidalgo, Leoni Prejean, Maturin Adam, and others.

Captain John Thomas Moore of Waubun bought that tract in 1889 and renamed it Julia plantation after his wife and daughter. He then sold it to John T. Moore Planting Company in 1898. One unofficial description of Julia is that it ran along the north bank of a canal called the sugar house canal on the right bank of Bayou Terrebonne, a short distance above the garden of Ducros.

John T. Moore, Jr. died in 1907, his father in 1909, and his brother Charles in 1911. The John T. Moore Planting Co., Ltd., owned Julia uninterrupted until July 1912, when Col. John Baptiste Levert foreclosed on a $100,000 mortgage of Julia, Waubun and St. George. The company was able to pay off the debt, however; Moore heirs owned the plantations until 1921. David Steel bought the three plantations at a Sheriff's Sale that year. (Later, from 1936 through 1944, the U.S Farm Security Administration took over the property for a segment of its collective farm project named Terrebonne Farms.)

Captain John Thomas Moore, Sr. (1838-1909) c. 1900

Farm Security Administration map of the Terrebonne Project November 7, 1945 (Julia)

Hutchinson spring stopper bottle made by Polmer Brothers Louisiana Bottling Works c. late 1800s

Sam (1861-1940) (left) and Leon Polmer (1869-1951) (right) on porch swing Ducros c. 1930

Augusta F. Polmer death article
The Houma Courier *February 15, 1945*

Right, Interior of the Polmer Store, Ducros c. 1920

The Woods family owned Lower Ducros until 1909, when it was purchased by Austrian-born brothers Samuel and Leon Polmer, who relocated to Terrebonne from Donaldsonville. When the brothers sold the land, they excluded from the sale two tracts of land that were once parts of Ducros. These tracts had already been sold, one to C.P. Shaver and Whitmell P. Martin and the other to L.S. Toups.[10] Samuel Polmer became a longtime public servant, representing Ward 1 on the Terrebonne Police Jury from 1911 until 1941. His political clout extended to the state level through his cordial relationship with Governor Huey P. Long, supposedly one reason that the Houma-Thibodaux road was paved during that era.

The Polmers in Terrebonne first operated a soda pop factory in Thibodaux, and then ran four stores that catered to plantation laborers.[11] One store was at Ducros, one at Waubun, both in Schriever, a third at Ardoyne called Central (the "Old Green Store") on Highway 311 at the intersection of Bull Run Road, and one on Bull Run plantation.

Widower Samuel, and Leon with his wife Augusta Feitel Polmer, lived in the home at Ducros and became identified with it locally for decades because of their long domicile there. Samuel, whose wife had been Stella, sister of Augusta Feitel Polmer, died in 1940 and Leon in 1951. Polmer Brothers, Ltd. partnership legally passed into ownership by Irvin and Mervin, Leon's twin sons; their sister Estelle Polmer Slipakoff Rabin; and S. Cahlman Polmer. At that time, Polmer Bros. consisted of the home and 820 acres in Terrebonne Parish, Waubun Store at Waubun plantation, plus 448 acres in Lafourche Parish.

After the death of the twin sons of Leon, Mervin and Irvin, Dr. Jacob L. Fischman in 1973 inherited Ducros. He was a native of White Castle, a resident of New Orleans, and president of Polmer Brothers, Ltd. After his death, the Estate and Heirs of Jacob L. Fischman, M.D. owned the properties. His children were Debra J. Fischman, Kathy Rose Fischman, Diana Lynn Fischman, and Nathan Harvey Fischman, M.D. Ducros furniture was moved to New Orleans, where the majority of it was lost during Hurricane Katrina in 2005.

It was during the Polmer family descendants' ownership in 1985 that Ducros was named to the National Register of Historic Places.

Van Perkins Winder's cattle brand was the 144th registered in Terrebonne Parish, May 17, 1832.

An interesting fact about the lands of Ducros is that the headwaters of Little Bayou Black were just south of the main plantation house and slightly north of the railroad terminal, shown on an 1830s map delineating the course of that bayou.

The house remained vacant under the care of a local overseer for about 20 years before Richard Joseph Bourgeois purchased Ducros in 1994. Restoration work on the house by Bourgeois and Angela Cheramie is ongoing. Its current address is 147 Old Schriever Highway, Schriever, Louisiana.

SOURCES
1. Louisiana Historic Structures Survey, Terrebonne Parish
2. A Century of Lawmaking for a New Nation: U.S. Congressional Documents and Debates, 1774-1875, American State Papers, Senate
3. Catalina in Louisiana Slave Records 1719-1820; ancestry.com and *Instruments of Empire: Colonial Elites and U.S. Governance in Early National Louisiana, 1803-1820*, a dissertation by Michael Kelly Beauchamp submitted to the Office of Graduate Studies of Texas A&M University December 2009
4. *Creole Families of New Orleans* by Grace Elizabeth King via Louisiana Notables, La-cemeteries.com copyright 2007
5. *Terrebonne Life Lines, Vol. 19*
6. *Terrebonne Life Lines Vol. 25 No. 4,* Winter 2006
7. *Van Perkins Winder 1809-1854, Sugar Planter,* a factual account with reminiscences by granddaughter Nina Winder
8. Paul F. Stahls, Jr., *Plantation Homes of the Lafourche Country,* Pelican Publishing Company 1976
9. State Library of Louisiana, www.state.lib.la.us
10. Helen Wurzlow, *I Dug Up Houma-Terrebonne,* 1984
11. *Encyclopedia of Southern Jewish Communities, Houma, Louisiana,* Institute of Southern Jewish Life, copyright 2014

Jacob L. Fischman, M.D. (1917-1998) at left and Irvin Feitel Polmer (1899-1973)

Cattle brand of Van Perkins Winder May 17, 1832

Polmer Store, Ducros c. 1920

Schriever map February 14, 1923

Terrebonne Station c. late 1800s

SCHRIEVER
TERREBONNE STATION

The northwestern Terrebonne community of Schriever originated with the New Orleans, Opelousas, and Great Western Railroad's establishment of Terrebonne Station in 1855. The depot there served as an embarking point for more westward communities up to Opelousas in southwestern Louisiana, and as a disembarking point for travelers from New Orleans to Terrebonne, Lafourche and other communities reachable only by the rail line.

From the railroad depot, stagecoach or another horse-drawn express vehicles transported travelers north or south, since it was not until January 1872 that Houma was totally connected to New Orleans by rail with completion of the Houma Branch Railroad from Terrebonne Station. Such businesses were Houma's Price Hine & Co. and Berger lines, and Thibodaux's Holden line.[1]

Of course the railroad became a prime target for service interruption by Union forces during the Civil War, but also a means for transportation of their troops to south central Louisiana's bayou country.

"The major breakthrough in east-west rail service came after Charles Morgan reorganized the New Orleans, Opelousas, and Great Western Railroad into the Louisiana and Texas Railroad and Steamship Company in 1878. He bridged the Atchafalaya River in 1881 and completed the railroad line from Brashear (Morgan) City to Lafayette, which had been prevented by the Civil War. Simultaneous with this construction, the Louisiana Western Extension Railroad completed its rail line from Orange, Texas, across the Sabine River into Louisiana. One could now travel by rail, and ship freight, from New Orleans to Texas without interruption."[1]

These developments promoted railroad travel for greater numbers, and the village centered around Terrebonne Station thrived to the point that it was in the late 1800s the largest community in Terrebonne Parish, with a hotel and a number of stores and residences.

Terrebonne Station's name was changed in the decades after the Civil War. It was named for John George Schriever who was born in Jersey City, New Jersey on July 8, 1844, and became a New Orleans resident. According to his March 17, 1898 obituary in *The New York Times*, he "was Traffic Manager of the Atlantic system of the Southern Pacific Railroad, with headquarters in New Orleans. He was one of the most widely known transportation men in the South, having been connected with the Morgan Steamship Company for many years before it became a part of the Southern Pacific system."

Terrebonne Station c. 1910

Southern Pacific Depot, Thibodaux c. 1910

Waubun Refinery 1906

Waubun Refinery 1897

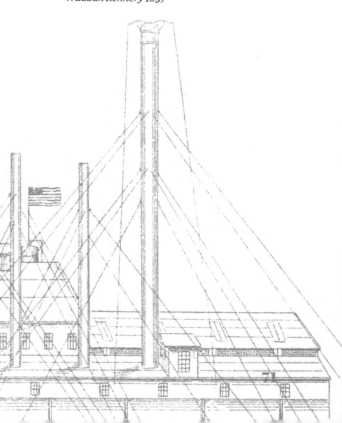

Schriever Hotel c. 1920

According to the *Times* obituary, "His connection with the Morgan's Louisiana and Texas Railroad began in 1869, and after working his way up through the various grades he became Vice President of that road....When C.P. Huntington bought the Morgan transportation lines and made them a part of the Southern Pacific system Mr. Schriever was appointed Traffic Manager in New Orleans." Schriever's prominence in New Orleans is indicated by his having been made Rex by the Carnival Krewe of that name during the 1889 season.[2] He also served on the General Finance Commttee of the World's Industrial and Cotton Centennial Exposition (1884-1885).[3]

Schriever's name began to be used for Terrebonne Station in the late 1800s, it is speculated, after he conducted an inspection at the depot during which he fell from a train and broke his leg.[4]

SOURCES
1. Christopher Everette Cenac, Sr., with Claire Domangue Joller, *Eyes of an Eagle: Jean-Pierre Cenac, Patriarch*, June 2011
2. William T. Nolan, current member of the Krewe of Rex
3. Mary Lou Eichhorn, Historic New Orleans Collection, 2016
4. Gayle Cope, *Schriever Memories: People, Places, & Stories*, 2015

Burlington Northern & Santa Fe railroad depot Terrebonne Station 2015

John George Schriever (1844-1898)
Death notice New York Times *March 7, 1898*

DEATH LIST OF A DAY.
John G. Schriever.

John George Schriever of New Orleans, a well-known railroad man, who had been sick at the Holland House, in this city, the past few days, died there at noon yesterday. Several members of his family, who reached here from New Orleans last Monday, were with him before he lost consciousness. The body will be forwarded at once to New Orleans, where the funeral services will be held.

John G. Schriever was Traffic Manager of the Atlantic system of the Southern Pacific Railroad, with headquarters in New Orleans. He was one of the most widely known transportation men in the South, having been connected with the Morgan Steamship Company for many years before it became a part of the Southern Pacific system. He was born in Jersey City, July 8, 1844, and entered railway service soon after he left school. His connection with the Morgan's Louisiana and Texas Railroad began in 1869, and after working his way up through the various grades he became Vice President of that road.

When C. P. Huntington bought the Morgan transportation lines and made them a part of the Southern Pacific system Mr. Schriever was appointed Traffic Manager in New Orleans. He was regarded as an exceptionally well-equipped man for that particular kind of business, and he enjoyed Mr. Huntington's confidence to a flattering extent. He was prominent socially in New Orleans, being a member of all the leading clubs and social organizations in that city.

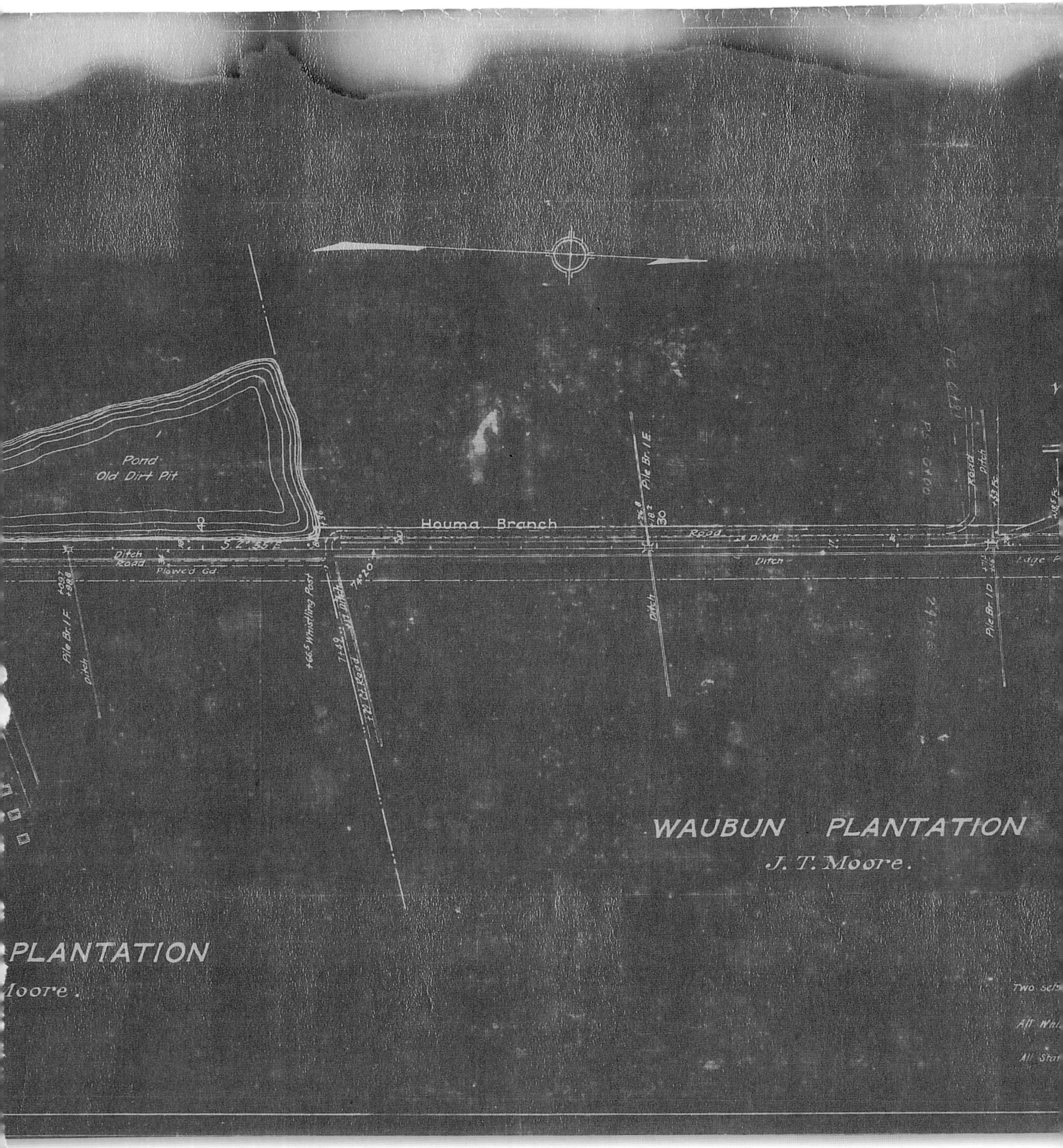

Pond
Old Dirt Pit

Houma Branch

Ditch   Road   Ditch
Road
Plowed Gd.

WAUBUN PLANTATION
J. T. Moore.

PLANTATION
Moore.

ST. GEORGE

J. T.

The John T. Moore Planting Co. Ltd. map 1903

Morgan's Louisiana & Texas Houma Branch Railroad map of Schriever, Waubun plantation, and St. George plantation 1909

Waubun 2015

Right, Celeste Belanger Tanner succession notice
La Sentinelle de Thibodaux *September 2, 1865*

WAUBUN
MAGNOLIA GROVE

Joseph Mollere and Walker Gilbert were the first claimants of a land grant registered with the U.S. Government on March 23, 1816 and confirmed May 11, 1820.[1] It was the core of a number of plantations in northern Terrebonne Parish, including Magnolia Grove/Waubun.

Lemuel Tanner of Screven, Georgia, purchased 20 arpents on both sides of Bayou Terrebonne from the widow and heirs of Henry Schuyler Thibodaux for $600, and named it Magnolia Grove.

Tanner had been a First Lieutenant in the Seventh Regiment of the Louisiana Militia during the War of 1812, and Thibodaux also served with the Louisiana Militia as Lieutenant of Volunteers attached to another regiment. Whether the two became acquainted in this way or not, they were connected by their marriages to Belanger family sisters. Tanner was the husband of Marie Agnes Celeste Belanger, sister of Henry Schuyler Thibodaux's wife Brigitte Emelie.

Magnolia Grove was six miles south of the town of Thibodaux and abutted Ducros plantation on Ducros' southern boundary. The Tanners' plantation was located on both sides of Bayou Terrebonne 13 miles north of Houma.

Tanner also established in 1840 what is now known as Ellendale plantation on Little Bayou Black, later owned by the McCollam family. The Tanners had ten children from 1818 until 1836. After her husband's death in 1843 and his burial in Halfway Cemetery, Celeste retained ownership of the Magnolia Grove tract until 1864. The 1843 succession described Magnolia Grove as 1,600 acres and buildings valued at $83,365. The Ellendale tract on the right descending bank of Bayou Black consisted of 988 acres valued at $14,820. Total value of the estate was recorded as $152,096.[2]

The 1844 Champomier sugar report records 708 hogsheads of sugar produced by Mrs. L. Tanner, with only James Cage of Woodlawn plantation and John Pelton of Dulac plantation exceeding her plantation's total sugar production in Terrebonne for the year. Seventeen years later, production remained close to that total, with the widow Tanner's mill producing 700 hogsheads in 1860-61.

Similar to most plantations, Magnolia Grove had its own hospital, blacksmith shop, cooperage, sawmill, sugar mill, overseer's home, field workers' cabins, stables,[2] and other buildings necessary to a working farm of its expanse.

In 1857, Celeste had donated a right-of-way to the New Orleans, Opelousas, & Great Western Railroad to benefit herself and other area planters who were paying excessive fees to have their sugar shipped via steamboat to New Orleans. She also

Waubun columns and eave front porch 2015

Waubun cistern foundation remains 2015

Above, Waubun chapel 2015
Left, Waubun rosewood and teak mantel on fireplace with Audubon-inspired tile on surround in the main dining room 2015
Bottom, Waubun fireplace clean-out access 2015

donated land for the railroad depot.

The first house predated the Civil War and faced the old U.S. Highway 90 and Bayou Terrebonne on its right bank. During the Civil War, Waubun grounds were used by Union soldiers as an army encampment.

William P. Dillsworth and William Otis bought Magnolia Grove in the year of Celeste's death in 1864. Four years later in 1868, John S. Dillsworth and George W. Woodworth became the owners. After two more years, George D. and Samuel Cragin of Rye, New York, bought Magnolia Grove for $15,000, reflecting a large post-Civil War drop in land value.

George D. Cragin had the second plantation home on the right descending bank built in 1874, designed by Colonel Thomas E. Beary, a New Yorker who had worked on the Cragin estate in Westchester, New York. *Plantation Homes of the Lafourche Country* by Paul F. Stahls, Jr. describes the house as "a weather-boarded product of Greek Revival architecture," with a "pedimented portico fronted by four fluted Ionic columns."

A full basement is enclosed by the brick foundation of the house, not a usual feature of south Louisiana houses. Cragin, a Catholic, had a chapel included on the top floor of the house, where many locals attended Masses, baptisms, and marriages. The house was constructed with 18 rooms in its 18,000 square feet of living area, each having its own fireplace. In 2015 it was revealed that the front bedroom fireplace features tiles on its surround painted in the style of John James Audubon. The house also has a wine

78 | HARD SCRABBLE TO HALLELUJAH

Waubun attic water system to combat fire 2015

cellar and an attic water system in case of fire.

George sold out to his brother Samuel in 1879, who worked the property until Captain John Thomas Moore bought it at a Sheriff's Sale in 1885. Moore and Samuel Longhorn Clemens (Mark Twain) had worked together on the Mississippi riverboat *John J. Roe* around 1856-1857. As did his famous friend, Moore became a river enthusiast, and eventually captained the *Ida Handy* steamboat. Captain Moore retired to become a merchant and planter in Terrebonne Parish.

Moore changed the name of Magnolia Grove to Waubun in 1885, from the name of Potawatomi chief Waubunsee, whose raids around Moore's Hannibal, Missouri hometown were usually at first dawn. (The Potawatomi word *Wabunseh* meant "dawn of day.") The mist of early dawn at Waubun reminded him of those facts, hence the derivative name.[3] Mark Twain visited his old

Schriever, La. 1906

Moore Cane Hook, Captain John T. Moore, Sr. on horse 1890s

friend at Waubun, according to one of Twain's letters.

Captain Moore married Julia M. Freret; they became parents of seven children. Besides the plantation, he owned the Waubun wholesale and retail store which he ran with his son John T. Jr., who was also the little town's first postmaster. In February 1888 all came under ownership of John T. Moore Planting Company, and in June 1889 under John T. Moore Planting Co., Ltd.

The *1897 Directory of the Parish of Terrebonne* gave Waubun's size as 3,500 acres, with 1,200 in cane and 450 in corn. The mill could then grind 500 tons of cane per day; Waubun employed 250 men during the season. Names of managers, overseers, their assistants, cooper, and refinery personnel are all listed in the same directory. In 1900, the mill ground 1,000 tons per day and five million pounds annually.[4]

Moore's invention, the Moore Cane Hook, was a labor-saving tool to pull seed cane from the windrow, and was widely used by sugar planters. The 1904-05 Bouchereau's *Louisiana Sugar Report* lists three plantations owned by John T. Moore Planting Co., Ltd. They were Waubun, Julia, the name of Moore's wife and daughter (Upper Ducros), and St. George (Windermere), all proximate to each other. The three plantations produced 5,737,567 pounds of sugar that year.

In 1907 Captain Moore joined with other regional planters and businessmen who convinced the U.S. Army Corps of Engineers to help with a project to deepen Bayou Terrebonne from Houma to the Gulf of Mexico. Moore wrote, "Every year we experience trouble and delays in putting our crop on the market, due to the inadequate railroad facilities at hand; we ship annually 15,000 barrels of sugar and 5,000 barrels of molasses, and every part of this crop could be marketed through Bayou Terrebonne and then through Barataria Canal, connecting Bayou Terrebonne with the Mississippi River." The Board of Engineers for Rivers and Harbors in 1908 affirmed the soundness of the proposed "channel 60 feet in width and 6 feet in depth at low tide, from the upper limits of the town of Houma to Bush's Canal, a distance of about 24 miles...."[5]

Moore, Sr., who died in 1909, was lauded by the *New Orleans Times Democrat* newspaper as "a merchant and planter and a

John T. Moore School c. 1915

Waubun Refinery 1906

John Baptiste Levert (1839-1930)

gentleman of high character and integrity." (April 23, 1909 edition)

John Thomas Moore, Jr. died of typhoid fever in 1907, two years before his father's demise. He had been in charge of the finances of their business, with his brother Charles Verhagen Moore responsible for agriculture; Charles was a graduate of the Audubon Sugar School with honors. He operated the three plantations alone until his death at his own hand on January 29, 1911 at age 36, his obituary recording that he was well-liked and praised in his capacity as Louisiana Sugar Planter's Association president.[6]

Colonel John Baptiste Levert, a sugar factor, foreclosed on a $100,000 mortgage in July 1912, but the John T. Moore Planting Co., Ltd., managed to pay off the debt. Moore's heirs continued operation from 1912 until 1921.

Subsequent owners included David Steel, who bought Waubun, Julia, and St. George for $85,000 at a Sherriff's Sale in 1921. A year later E. Gajan, Inc. and Charlton Beattie purchased Waubun lands for $110,000. It remained in their hands until it went the way of Julia when the Farm Security Administration of the United States developed Terrebonne Farms in 1936, an experiment in collective farming that lasted until 1944. The Waubun house was used as an administrative office during that time. The lands were liquidated and the refinery, which had closed in 1934, underwent conversion into a feed processing mill.

The Waubun Store was first leased in 1920 and purchased by Alfred P. Danziger in 1922. The Polmer brothers bought the Waubun store in 1924 and continued its operation until 1951. That store and the nearby mercantile of Leonard L. Toups had been the only two commercial buildings in Schriever to escape being consumed by a June 1904 fire that began in the Schriever Hotel and quickly spread throughout the little town. A few residences were also spared.[7]

Refinery foundation remains 2015

Below, left Waubun Store, later Polmer Bros. Store

Waubun trade tokens, 10 and 5 cents and Polmer Store ad, 1897

New owner Sam Bascle had the Polmers' Waubun store torn down in 1976.

Waubun House was bought by Dr. Thomas E. Powell, Jr., Ph.D., in 1945, again at a Sheriff's Sale. Biologist Horace L. Whitten and his wife lived in the home beginning in 1955, from which they shipped biological specimens worldwide for the Carolina Biological Supply Company. The same company continues to retain ownership of the homesite.

Biologist Whitten managed Waubun from 1955 until 1976. Dr. Thomas Powell's son, Joseph Powell, was manager from 1976 through 1978. Terry Philip Guidroz, an ichthyologist and herpetologist, managed Waubun from 1979 to 2015.

Waubun's current address is 102A Highway 24, Schriever, Louisiana.

SOURCES

1. A Century of Lawmaking for a New Nation: U.S. Congressional Documents and Debates, 1774-1875, American State Papers, Senate
2. *Terrebonne Life Lines Vo. 24 No. 4,* Winter 2005
3. Paul F. Stahls, Jr. *Plantation Homes of the Lafourche Country*, Pelican Publishing Company 1976
4. Henry Rightor, editor, *Standard History of New Orleans, Louisiana*, The Lewis Publishing Company, Chicago, 1900
5. *The Houma Courier*, July 29, 2015
6. *The Louisiana Planter and Sugar Manufacturer*, February 4, 1911
7. *The Lafourche Comet*, Thibodaux, Louisiana, June 4, 1904

Waubun interior panel door
Below, Waubun north side view
Opposite page, entry hall and staircase

WAUBUN | 83

Mural in St. George dining room, "Scenic America" by Jean Zuber, of Boston Harbor and West Point Military Academy; woodblock printed c. 1834, France

ST. GEORGE
WINDERMERE
LUTECIA
ERIN

St. George plantation home and grounds in Schriever descend from an extensive claim by Walker Gilbert which he registered on March 23, 1816 and which was confirmed by the U.S. government on May 11, 1820.[1] Those lands developed into several different planter estates. The Henry Schuyler Thibodaux estate of neighboring St. Brigitte plantation purchased part of the Gilbert claim on February 25, 1828; H.S. Thibodaux died in 1827. His widow and heirs officially owned the land as of August 17, 1832.

Felix G. Winder purchased the choice farming land from the Thibodaux family in the mid-1840s, and named the estate Windermere plantation. F.G. Winder produced 230 hogsheads of sugar during the 1860-61 season. Mrs. F.G. Winder retained the Windermere name as late as the 1869-70 growing season.

Hesse & Vergez owned the property from 1869 until 1872. In 1872-73, the plantation had a new name, Lutecia, which that year produced only 86 hogsheads as reported in the annual sugar report. Lutecia upped its production in 1876-77, to 200 hogsheads.

Molasses prices from Lutecia
The Times Picayune
October 4, 1878

St. George 2010

Top to bottom, left to right, mural in St. George dining room, "Scenic America" by Jean Zuber, of Boston Harbor and West Point Military Academy; woodblock printed c. 1834, France; a board under the bed signed Lucius J. Dupri of Opelousas, La., Confederate Congressman and philanthropist

Opposite page (page 87) top to bottom, left to right, rear exit center hall; bed in master bedroom; formal front parlor

Next spread (page 88), left hand page, top to bottom, left to right, kitchen stairway, dining room entrance pocket doors, dining room mural, chandelier; front entrance door and center hall; north side view; study

Next spread (page 89), top left, Dupri four-poster bed right rear bedroom; right top, view of the kitchen from the balcony landing; middle left, dining room; middle right, sleigh bed upstairs bedroom; bottom left four-poster bed left front downstairs bedroom; bottom middle, armoire master bedroom; bottom right, fireplace master bedroom

Sale notice of Oscar Crosier's Windermere plantation, The Weekly Thibodaux Sentinel and Journal of the 8th Senatorial District *July 21, 1877, and article on Thomas L. Winder's purchase of Windermere,* The Weekly Thibodaux Sentinel and Journal of the 8th Senatorial District *November 26, 1881*

Peter Berger (1837-1907)

Above, The Weekly Thibodaux Sentinel and Journal of the 8th Senatorial District, *January 12, 1884; below,* The Houma Courier, *September 29, 1900*

Bibolet & Oscar Crosier are listed as assuming ownership in 1877.

Thomas L. Winder, son of Van P. Winder, deceased former owner of Ducros plantation, bought Lutecia in 1881. Michel Shelly of Orleans Parish acquired the plantation by judgment from Thomas L. Winder, and sold the 1,437.86 acres to Peter Berger of Terrebonne Parish. The 1884 deed indicated that the plantation was known as Windermere, or Erin. The sale described it as "a certain plantation known as Windermere.... bounded above by Magnolia Grove Plantation of Samuel Cragin and below by lands of E.B. Hobert and Infonte Roussell."

Because Bavarian-born Berger was buying up so much land in the parish, some fellow Terrebonneans referred to the place as Peter Berger's Folly.[2] He seemed to embrace the name, since in 1886-87 reports of sugar production, his land produced 100 hogsheads on "Berger's Folly" plantation. (Each hogshead equals appoximately 1,000 pounds.)

N. Lorio is credited with producing 240,000 pounds of sugar on the Berger acreage in 1891.

John T. Moore Planting Company bought out Berger in 1895 and renamed the plantation St. George. The next year St. George produced 5,000 tons of cane. By 1898, John T. Moore Planting Company, Ltd. owned three plantations—Waubun beginning 1885, St. George beginning 1895, and Julia beginning 1898. The Company's total ownership in acreage was somewhere between 7,000 and 8,000 acres.[2] Captain John Thomas Moore, Sr.'s son Charles Verhagen Moore built the impressive house in 1905 on the right bank of Bayou Terrebonne. It is a fully raised, one-story, late Greek Revival plantation house.[3] It was supposedly modeled on the Gulf Coast's Longfellow House. In the last decades of the 20th century, owners described its colonnade with a full entablature as unique among homes in Terrebonne Parish, in their application for recognition on the National Register of Historic Places. Its 16-inch square pillars are a dominating feature, along with a 13-foot high gallery.

John T. Moore, Jr. died in 1907, and his father died in 1909. After Charles died at his own hand in 1911, sugar factor Col. John Baptiste Levert foreclosed on a $100,000 mortgage of Julia, Waubun and St. George in July 1912. Moore heirs subsequently paid off that debt and operated the John T. Moore Planting Co., Ltd., until 1921. That year David Steel bought the three plantations at a Sheriff's Sale for $85,000.

In 1936, the U.S. Farm Security Administration converted St. George lands into part of Terrebonne Farms, just as Ducros and Waubun had succumbed to this experiment in collective farming that ended in 1944. St. George house was during that time used as the Terrebonne Farms administrative office.

The home site came under the ownership of Warren A. Neubaurer from January until September 1945, when Charles Corbin purchased it. In 1962, Charles David Chauvin and his wife Merril Elaine Tucker purchased St. George and began to live in the home.

Current owners are Joseph John Weigand, Jr. and his wife

Tegwyn Murphy Weigand, who bought the home and its grounds in 1982. The Weigands completed the interior of the second floor, which had never been finished. The structure was named to the National Register of Historic Places in October 1982. The house is an outstanding example of restoration and renovation of an historic home. It is 12 miles north of Houma, at 1382 West Main Street, Schriever, Louisiana.

SOURCES
1. A Century of Lawmaking for a New Nation: U.S. Congressional Documents and Debates, 1774-1875, American State Papers, Senate
2. Helen Wurzlow, *I Dug Up Houma-Terrebonne*, 1984
3. Louisiana Comprehensive Statewide Survey, Terrebonne Parish

Charles V. Moore (1874-1911)

John T. Moore, Jr.'s marriage to Odette Ellis The Houma Courier, *June 16, 1900*

MOORE—ELLIS.
At the Church of the Immaculate Conception, New Orleans, Wednesday, evening June 6th, at 6:30 o'clock, Miss Odette Ellis and Mr. Jno. T. Moore, Jr., were wed. The wedding was attended by a large number of friends and relatives of the couple. Both parties are well-known here. Miss Ellis is the daughter of Mr. Wm. Ellis, who once resided in this parish, and since the family's removal to New Orleans the lovely young lady has made frequent visits to her many friends here. Mr. Jno. T. Moore, Jr., is a popular sugar planter of Terrebonne. He is an active member of the John T. Moore Planting Company, of Schriever, and possesses a host of warm personal friends who hope that the future has nothing in store, but happiness, for him and his charming bride.

John T. Moore, Jr. & Co. ad The Louisiana Planter and Sugar Manufacturer *1915*

(ix)

JOHN T. MOORE, Jr.   CHAS. H. WALK[ER]

JOHN T. MOORE, Jr. & CO.,

WHOLESALE GROCERS

AND

COMMISSION MERCHANTS,

Nos. 37, 39 and 41 TCHOUPITOULAS STREET,

P. O. Box 1806.   Warehouse, 14 NATCHEZ ST.,

NEW ORLEANS, LA.

DEALERS IN

SUGAR AND MOLASSES,

Planters' Supplies, Fertilizers, Cow Peas,

Sugar Barrel Material and Grain.

CASH ADVANCES MADE.

AGENTS FOR

EUREKA SUGAR PACKER.

Balsamine by Adrien Persac c. 1857

BALSAMINE

Balsamine was the residence of Henry Schuyler and Brigitte Emelie Thibodaux's sixth and youngest child Bannon Goforth, who was born at St. Brigitte plantation in 1812.

Bannon's home site was on the eastern portion of St. Brigitte, on the left descending bank of Bayou Terrebonne, across the bayou from his parents' home. The land was also known as St. Bridget East and the Left Bank. It was situated about six miles south of Thibodaux, in northern Terrebonne Parish, 12 miles from Houma.

As part of the Henry Schuyler Thibodaux properties, the Balsamine tract descended from an 1801 Spanish land grant to Hubert Ballenger (Belanger), which he registered with the U.S. government on November 20, 1816 and which was confirmed on May 11, 1820.[2]

After Henry's death in 1827, Bannon Goforth in 1828 assumed ownership of a portion of Thibodaux family land across Bayou Terrebonne from the St. Brigitte home, which was situated on the right descending bank.

It can be assumed, since his house was on his parents' land, that Bannon Goforth played some role in their plantation's management in the years after his parents' death. But his profession was the law, and his pursuits in public service. He served as a representative to the Louisiana constitutional convention in 1845 and 1852, and as U.S. Congressman from Louisiana's Second District in the U.S. House of Representatives from 1845 to 1849.[1]

After his Congressional terms, he resumed his law practice in Terrebonne and Lafourche parishes. His son-in-law Adam Eugene William Blake, husband of Bannon's daughter Cecile, joined his law practice. The Congressional biographical guide describes Bannon also as a sugar planter and manufacturer.

In 1840 Bannon married Justine Aubert, daughter of Pierre and Marguerite Barras Aubert of Plattenville, Assumption Parish. Justine gave birth to 12 children.[3]

The 1846 La Tourrette map delineates the large holding of "Thibodaux & Beaty." After Bannon's death in 1866, there was an 1877 Act of Partition with Taylor Beattie in reference to the large plantation south of the F. Mead property. The Thibodaux owners leased 70 arpents and six mules to Judge Adam Eugene W. Blake beginning in 1879. A notice by E.W. Blake in *The Weekly Thibodaux Sentinel* newspaper advertised for lease "The East St. Bridget Plantation with necessary implements and 6 mules—28 arpents first years stubble and seed in windrow sufficient to plant 70 arpents of cane. Some provender on the place and the Sugar House to be placed in order to take off next crop."

Andrew Price purchased 600 acres of the Bannon Goforth Thibodaux, Sr. property in 1903. The Thibodaux home and 45 acres were excluded from the sale.

Thibodaux Minerva *April 26, 1856 mention of Balsamine strawberries and* Houma Ceres *April 25, 1857 notice of Eugene W. Blake and Cecile A. Thibodaux marriage ceremony at Balsamine*

Bannon Goforth Thibodaux (1812-1866)

Bannon's cattle brand was the sixteenth registered in Terrebonne Parish, dated April 12, 1825.

Bannon died in 1866 and was interred, as was his father, at Halfway Cemetery. Justine preceded him in death, in 1853. The house where Bannon, Justine, and family had lived was torn down in the early 1900s. There is no proof when it was built, but an in-ground cistern marks the site of the structure. The land was subdivided among multiple heirs. The address where it used to stand is currently 1560 West Park Avenue, Schriever, Louisiana.

SOURCES
1. Biographical Directory of the United States Congress
2. A Century of Lawmaking for a New Nation: U.S. Congressional Documents and Debates 1774-1875, American State Papers, Senate
3. Louisiana Historical Association

Left, In-ground cistern at Balsamine 2016

Right, cattle brand of Bannon Goforth Thibodaux #16 dated April 12, 1825

Below, current site of former Balsamine house

HOBERTVILLE

A small community tucked between what had been St. George plantation and Isle of Cuba plantation in the late 1800s was named for Edgar B. Hobert. The area was on the right descending bank of Bayou Terrebonne, on land that had originally been part of the Henry Schuyler Thibodaux plantation, St. Brigitte.

Delineated on the 1938 Tobin map, the area is a substantial grouping of long, narrow properties fronting on what was then designated as Highway 69 paralleling Bayou Terrebonne. Hobertville was across the bayou from the Andrew Price School tract.

The map shows the plot designated "Mrs. A. Thibodaux" (a Henry Schuyler Thibodaux descendant's wife) immediately across Bayou Terrebonne from what is the present-day Andrew Price School building. That particular Hobertville spot was once the site of the Henry Schuyler Thibodaux home, upon which Frank and Mollie Brigitte Beauvais Olive built a house in 1914. Mollie was the great-great-granddaughter of Henry Schuyler and Brigitte Belanger Thibodaux. Current address of the home is 1922 West Main Street, Houma, Louisiana.

Hobertville's namesake, E.B. Hobert, was born in 1851[1] and died April 6, 1906.[2] In the 1880 U.S. Census he is listed as owner of a grocery and dry goods store, which probably was the centerpiece of the small community. The 1900 census describes his household as Edgar B. Hobert, owner of 44 acres, his wife Louise nee Goucheaux, mother-in-law Alexandra Goucheaux, boarder and clerk Alphonse Aucoin (age 21), nephew and clerk Charles Boudreaux (age 21), and niece Edna Goucheaux (age 6).

Mr. Hobert was listed as Police Juror for Ward 1 in 1903 *Houma Courier* newspaper minutes for the organization.

SOURCES
1. 1900 United States Federal Census
2. Gravesite inscription, St. Joseph Cemetery, Thibodaux, Louisiana

Map of Hobertville April 29, 1904

Tobin map 1938

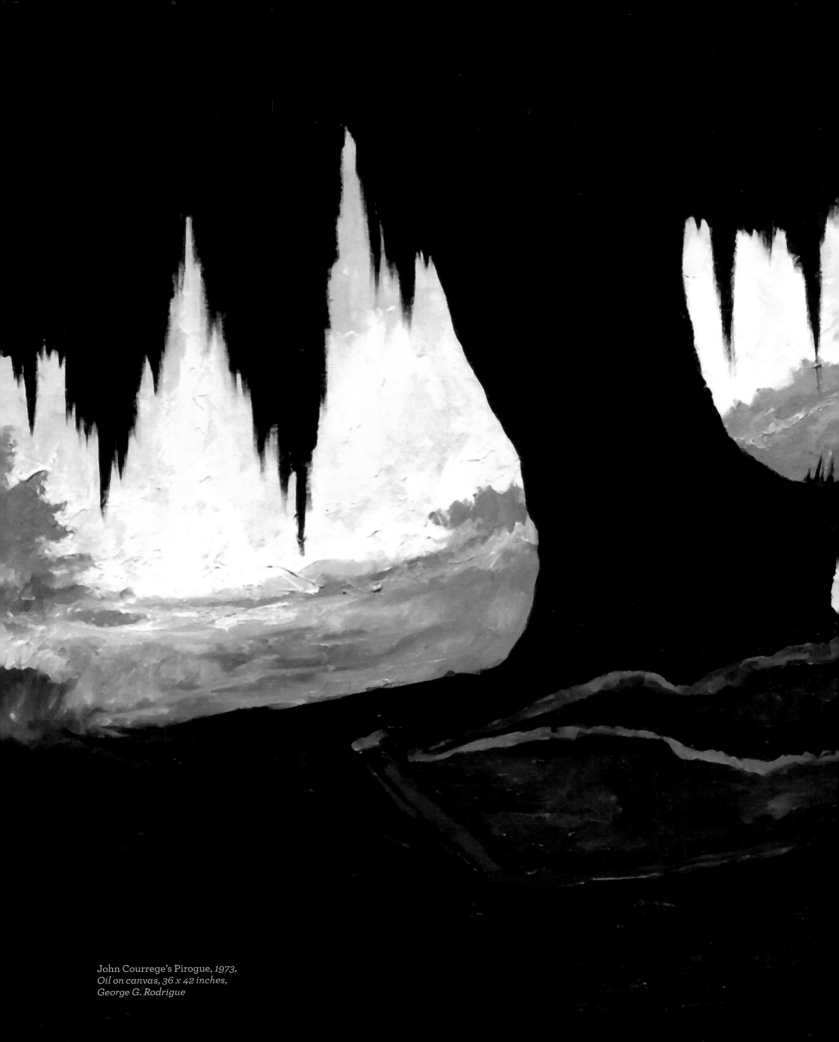

John Courrege's Pirogue, *1973,
Oil on canvas, 36 x 42 inches,
George G. Rodrigue*

ST. BRIDGET
ST. BRIGITTE

St. Brigitte by Adrien Persac c. 1861

ST. BRIDGET | 99

Henry Schuyler Thibodaux
(1769-1827)

Brigitte Emelie Belanger Thibodaux
(1775-1849)

Henry Claiborne Thibodaux
(1810-1885)

Mathilde M. Toups Thibodaux
(1817-1863)

One of the oldest sugar plantations in Terrebonne Parish, the holdings of Henry Schuyler Thibodaux and his residence became known as St. Brigitte, the French spelling of his second wife's namesake saint. Thibodaux began in 1810 to amass land about five miles south of the Lafourche Interior town that was later named for him, Thibodauxville.

The land he purchased was part of a Spanish grant by Governor Baron de Carondelet in 1801. These lands were claimed by Hubert Ballenger (Belanger) and Auguste Babin. They registered the claim with the U.S. government on November 20, 1816, and received confirmation of approval on May 11, 1820.[1]

The Ballenger-Babin claim gave rise to many planter estates, among them Thibodaux's land. A description in *U.S. Congressional Documents 1774-1875 (Claims in Eastern Louisiana)* gives details of his claim and confirmation of approval in 1822:

"Henry S. Thibodaux claims a tract of land situate on Bayou Derbonne, containing six hundred and two and thirty-seven hundredths superficial acres. It is proved that this land was settled, by permission of the proper Spanish officer, prior to the 20th December, 1803; and cultivation [sic]."

Those early purchases grew to the extent that when he died in 1827 his estate in Terrebonne was valued at $105,751. His plantation straddled Bayou Terrebonne, extending across both banks 11 miles north of Houma, but his house was on the right bank of the bayou. He was buried at Halfway Cemetery between Thibodaux and Houma.

Thibodaux was born in New York on January 1, 1769. He was orphaned at an early age and adopted by General Philip Schuyler of Albany, New York,[2] who became the father-in-law of Alexander Hamilton. Henry was educated in Scotland, and later settled in St. James Parish, Louisiana. There he married in 1793 Felicite Bonvillian, a French Acadian; they had three children together before her death in 1799.[2] His trade when he migrated to Louisiana was that of shoemaker.

His second wife was Pointe Coupee native Brigitte Emelie Belanger who was the granddaughter of the French explorer Jacques Cartier.[3] They moved to Lafourche Parish, where he began a career in public service. Their family grew to include five more children,[4] Henry served a term as a state senator 1812-1824, becoming president of the senate in the latter year and finishing out the term of Governor Thomas B. Robertson, who had resigned his office. He was thus the fourth Governor of Louisiana, although his term was short-lived. He was the only resident of Terrebonne Parish to serve as Louisiana governor. Henry Schuyler and Brigitte's fourth child, Henri Claiborne Thibodaux, was born in 1810 at "St. Bridgit."

Thibodaux was instrumental in establishment of Terrebonne Parish out of part of the Lafourche Interior, earning him the designation as Father of Terrebonne. When he died on October 24, 1827, he was campaigning for the Louisiana governor's term that would begin in 1828.

An 1861 painting of St. Brigitte plantation house and part of its grounds by Adrien Persac depicts an unpretentious large house set behind a white picket fence on cleared land dotted by large trees, the property fronting on Bayou Terrebonne spanned by a nearby bridge. No definitive record exists of the date of the house's construction. The St. Brigitte home Persac depicted was lost in a March 1892 fire. (Persac also painted a rendering of the Bannon Goforth Thibodaux house across Bayou Terrebonne from his parents' home, which the son had named Balsamine.)

St. Brigitte was one of many land holdings of "Henri" Schuyler Thibodaux, according to an 1809 appraisal made of his inventoried properties for the purpose of dividing their due among the three children of his first wife.

Other land he owned was identified as "the Plantation in Baton Rouge," "the Plantation on the Lafourche (Six acres front on the righthand side of the Bayou descending about sixteen leagues from the Mississippi)", "one lot in Donaldsonville," and "one Plantation 13 or 14 acres of land front on the Bayou Lafourche, more or less about ten leagues from the Mississippi on the left hand side descending." This latter could have been Rienzi plantation, for which he is listed as a onetime owner. The 1809 official inventory and appraisal recorded a value of $12,035 and debts due totaled $9,060, as recorded by Judge William Goforth of the Lafourche Interior Parish on March 29, 1809.

After his death in 1827, The Inventory & Appraisement of the Estate of Henry S. Thibodaux, a deed recorded as No. 484 in Conveyance Book D of Terrebonne Parish, describes his plantation as "one tract of land situated about 4 miles from Lafourche containing 12 arpents front on the right bank and 13 on the left bank of Bayou Terrebonne, with the depth of 40 arpents on the right side and 30 on the other, bounded above by the land of Leandre B. Thibodaux on the right bank & by land belonging to this estate on the left; and below by land of Jacques Verret: together with all the buildings thereunto erected excepting the canes in matelass" (windrowed canes) valued at $17,400.

The inventory also lists at least 27 more tracts of land under his ownership, plus the islands of Timbalier, Calumet, Casse-Tete, Caillou, and Labrosse, valued at $150 each, 56 slaves, 13 horses, 10 teams of oxen, conveyances, tools, interior furnishings, animals, bundles of official records and letters, and other necessities to plantation life, with a total value of $105,751. (recorded by Terrebonne Parish Judge Leufroy Barras January 23, 1828.) A separate inventory and appraisal was conducted for his properties in the "parish of Lafourche interior" including the town of Thibodaux named for him.

The pioneering couple had an interest in the betterment of the local area. Henry Schuyler donated two lots to Lafourche Interior Parish for a courthouse and market in Thibodaux in 1820. Brigitte was from the Belanger family that donated land for establishment of Houma, the parish seat of Terrebonne.

St. Bridget tenant house c. 1930s

Andrew Price School 1940

ANDREW PRICE SCHOOL

Andrew Price School is located on the left descending bank of Bayou Terrebonne on the onetime site of St. Brigitte plantation sugar mill.1 The school has its origin in a Terrebonne Parish School Board District 1 millage passed in 1908. The School Board authorized acquisition of the site on April 4, 1917 and proposed a purchase price of $440 to owner Mrs. Anna Gay Price, who rejected the offer. Subsequently the School Board authorized expropriation of the property in September 1917.²

Mrs. Price, the widow of Andrew Price, donated the property on July 22, 1919 with the stipulation that it be used for educational purposes only. Architects Torre & William T. Nolan designed the building, and Caldwell Brothers of Thibodaux, Louisiana, replaced Joseph A. Robichaux as contractors beginning April 3, 1919. Construction cost was $4,000.²

> We have placed the name of Hon. Andrew Price of Lafourche at our mast head as our choice for the 3d Congressional District, and there it will remain until the ides of November shall bring in the glad tidings of his re-election by an overwhelming majority.

Andrew Price nominated for Third Congressional District post, The Weekly Thibodaux Sentinel and Journal of the 8th Senatorial District September 22, 1894

The year after Henry's death, in 1828, Ambrose Gibson purchased 640 acres of the Thibodaux estate.

After Henry Schuyler's demise, his wife Brigitte proved herself to be a capable manager of the estate. Brigitte was described as "skillful, charitable, hospitable, and public-minded or spirited," managing the plantation in such a business-like manner that she left a small fortune to each of her five children."[5] In the Statement of Sugar Made in Louisiana report by P.A. Champomier for 1844, Brigitte's plantation is recorded as having made 608 hogsheads of sugar. Of the 42 Terrebonne sugar producers named, only four other parish planters exceeded her total production that year (her sister Mrs. L. Tanner at Magnolia Grove, Bisland & Watson at Aragon, James Cage at Woodlawn, and John Pelton at Live Oak).

In 1845, the plantation was divided into two segments for management. Brigitte took over the southern section of the plantation, and her oldest son Michel Henri Joseph Thibodaux had control of the northern section.

Brigitte Emelie Thibodaux died at age 74 in December 1849. St. Brigitte plantation was sold at auction for $133,325 to the three sons of Henry Schuyler and Brigitte Thibodaux.[5] St. Brigitte plantation thereafter came under division and ownership by many different individuals and entities, most of them of Thibodaux family lineage, until the plantation ceased to exist intact as it had been. Some of its lands were still being farmed as of 2016.

In 1860 the property was held by Thibodaux Brothers. The Freedmen's Bureau returned the property to the family in 1865 after it had been claimed by former slaves during the Civil War. The Bureau determined that the property had been maintained by its owners and therefore was ineligible for freedmen's claims. In 1872, Judge Adam Eugene Blake owned part of the original St. Brigitte lands. Blake was married to Cecile Adele Thibodaux, daughter of Henry Schuyler and Brigitte's youngest son, Bannon Goforth. In 1875 Blake donated the site for a Catholic church to be built. It became the Congregation of St. Joseph, which later formed St. Bridget Catholic Church in Schriever. (The church parish's official formation in 1911 occasioned the change of the spelling to the English form, St. Bridget.) Subsequently, the "Heirs of Henry Schuyler Thibodaux" were listed as holding the property.

St. Brigitte sugar mill on the left descending bank of Bayou Terrebonne closed c. 1905. Wallace Thibodaux bought family property in 1908-1909, and Frank Olive and his wife Mollie Brigitte Beauvais (great-great-granddaughter of Henry Schuyler and Brigitte) built a home on the site of the St. Brigitte plantation home in 1914; present address of the house is 1922 West Main Street, Schriever, Louisiana. Terrebonne Parish School Board built Andrew Price School on the former site of the sugar mill in 1919.

The 1846 La Tourrette map is indicative of the vastness of Thibodaux and associated properties in the northern part of Terrebonne. Brigitte Emelie's sister Marie Agnes Celeste was married to Lemuel Tanner, whose holdings are shown two plantations north of Mrs. H.S. Thibodaux's plantation lands that straddle both sides of Bayou Terrebonne. H.M. Thibodaux designates the portion of Henry Schuyler's land that went to his eldest son Michel Henri (or Henri Michel) Joseph Thibodaux, abutting his mother's estate on the north, on the right descending bank of Bayou Terrebonne.

The Hon. L. (Leufroy) Barras land adjacent to Brigitte's plantation on the south sat on both banks of Bayou Terrebonne as did his mother-in-law's; he was married to Henry Schuyler and Brigitte's third child Brigitte Emilie, named for her mother. The next descending tract, that of Evariste Porche, has a Thibodaux connection as well. Henry's second child with Brigitte, Elmire Marie Thibodaux, married Evariste Porche of Belle Farm on Big Bayou Black. He was the son of Joachim Porche, a relative of the Porche designated on the map.

Seven members of the Henry Schuyler Thibodaux family registered livestock brands in Terrebonne Parish. Henry's son Leandre Bannon's registered number 12 on April 4, 1825; son Henry Michael registered number 13 and Henry Schuyler registered number 14 on April 12, 1825; son Bannon Goforth registered number 16 and his sister Elmire Marie registered number 17 on April 12, 1825; daughter Eugenie registered number 22 on June 10, 1825; son Aubin Benoni registered number 27 on September 24, 1825.

The original St. Brigitte plantation house was built in 1796. It burned March 27, 1892.

The original St. Bridget Chapel was completed February 1, 1876 and continued as a mission of St. Joseph Church of Thibodaux until December 16, 1911 when it became an independent church parish. The present church was dedicated September 8, 1955 and exists on land that once saw sugar cane growing for miles around. Although some St. Brigitte lands are still being farmed, the church is the only landmark reminder of one of the first sugar plantations on Terrebonne Parish soil. The address is 100 Highway 311, Schriever, Louisiana.

SOURCES
1. *A Century of Lawmaking for a New Nation: U.S. Congressional Documents and Debates 1774-1875*, American State Papers, Senate
2. *Louisiana Governors, Henry Schuyler Thibodaux, Governor of Louisiana 1824*, Louisiana Cemeteries.com, copyright 2007
3. *Biographical and Historical Memoirs of Louisiana Volume 1*, The Goodspeed Publishing Company, Chicago 1892
4. Genealogy of Thibodaux children courtesy archivist Clifton Theriot, Nicholls State University Archives
5. *History of Terrebonne Parish to 1861*, Marguerite Watkins, a thesis submitted to the graduate faculty of Louisiana State University Department of History 1939

Andrew Price School 2016

The school opened for the 1920-1921 school session as Schriever Graded School, with Thomas J. Ellender as principal. The school replaced small plantation schools—John T. Moore, Jr., Minerva, Gray, and St. Bridget schools—which then closed. The name was changed in 1949 to Andrew Price School when the new Schriever School adjacent to St. Bridget Church was dedicated, then to Andrew Price Vocational School for the 1977 school session. In 1999, the school became Andrew Price Alternative Positive Placement for Students (TAPPS). The building, currently located at 1849 West Park Avenue in Gray, is at present being used by the Information Technology Department of the Terrebonne Parish School Board.[2]

SOURCES
1. David D. Plater of Acadia Plantation
2. Terrebonne Parish School Board Minutes Books 1-3

Andrew Price (1854-1909)

Above left, cattle brands of (top to bottom), Leandre Bannon Thibodaux April 4, 1825; Henry Schuyler Thibodaux April 12, 1825; Elmire Marie Thibodaux April 12, 1825; Aubin Benoni Thibodaux September 24, 1825; Eugenie Thibodaux June 10, 1825; above center St. Brigitte, home of Captain Joseph Tucker entirely destroyed by fire The Houma Courier April 2, 1892

Left below, St. Brigitte plantation home site; house built 1914 photo 2015

An Old Landmark Gone.
Last Sunday, at 12:30 o'clock p. m., near Shriever station, the residence of Capt. Joseph Tucker was entirely destroyed by fire. The furniture and clothing were also destroyed. This house was built by the late Henry Schuyler Thibodaux, at one time Governor of Louisiana, and was 96 years old. A large amount of valuable family relics, including a splendid library of priceless books, were reduced to ashes. The cause of the fire is unknown.

Isle of Cuba map,
Farm Security Administration
November 7, 1945

ISLE OF CUBA
LILAC
LYLAC AVENUE
ISLAND OF CUBA

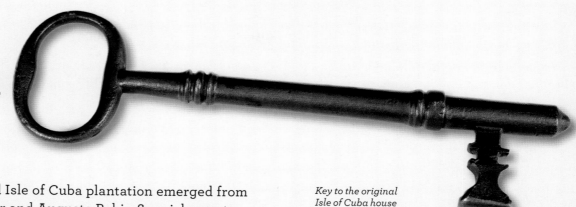

Key to the original Isle of Cuba house

The peculiarly named Isle of Cuba plantation emerged from the same Hubert Belanger and Auguste Babin Spanish grants which were registered with the U.S. government on November 20, 1816 and which were confirmed on May 11, 1820.[1] The Belanger-Babin claims spawned a number of other large planter estates in northern Terrebonne.

In 1822 Henry Schuyler Thibodaux of St. Brigitte plantation had ownership of the acreage, followed by his widow and heirs for a short time after his death in 1827. Thibodaux's son-in-law, Leufroy (Ludfrois) Emile Barras, in 1828 bought the land immediately south of what the 1846 La Tourrette map shows as Mrs. H.S. Thibodaux's plantation.

Barras' wife was Brigitte Emilie Thibodaux Barras, whom he married in 1819 when she was 15 and he was 21. Barras was one of nine children of Antoine and Julie Patin Barras of Pointe Coupee. Leufroy became the second judge in Terrebonne Parish, which accounts for the "Hon." title before his name on the map.

A 1939 newspaper account of former slave Augustus Coxen affectionately described the family at what was in the early years called Lilac plantation during Coxen's childhood there before the Civil War. Judge Barras "sure was a fine man. He was my first teacher, because while I was serving him he gave me a private education and much grammatical knowledge of the English language. When he died Mrs. Barrass [sic] sent me to the Straight University in New Orleans. . . . I did not finish, but I learned enough of the law course that I was taking up to become a justice of the peace. . . ."[2]

Court case involved governor as defense counsel The Weekly Thibodaux Sentinel and Journal of the 8th Senatorial District *February 29, 1896*

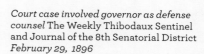

Notice of Isle of Cuba request to Terrebonne Parish Police Jury The Houma Courier *October 30, 1897*

Left, former Isle of Cuba plantation home site 2016

ISLE OF CUBA | 105

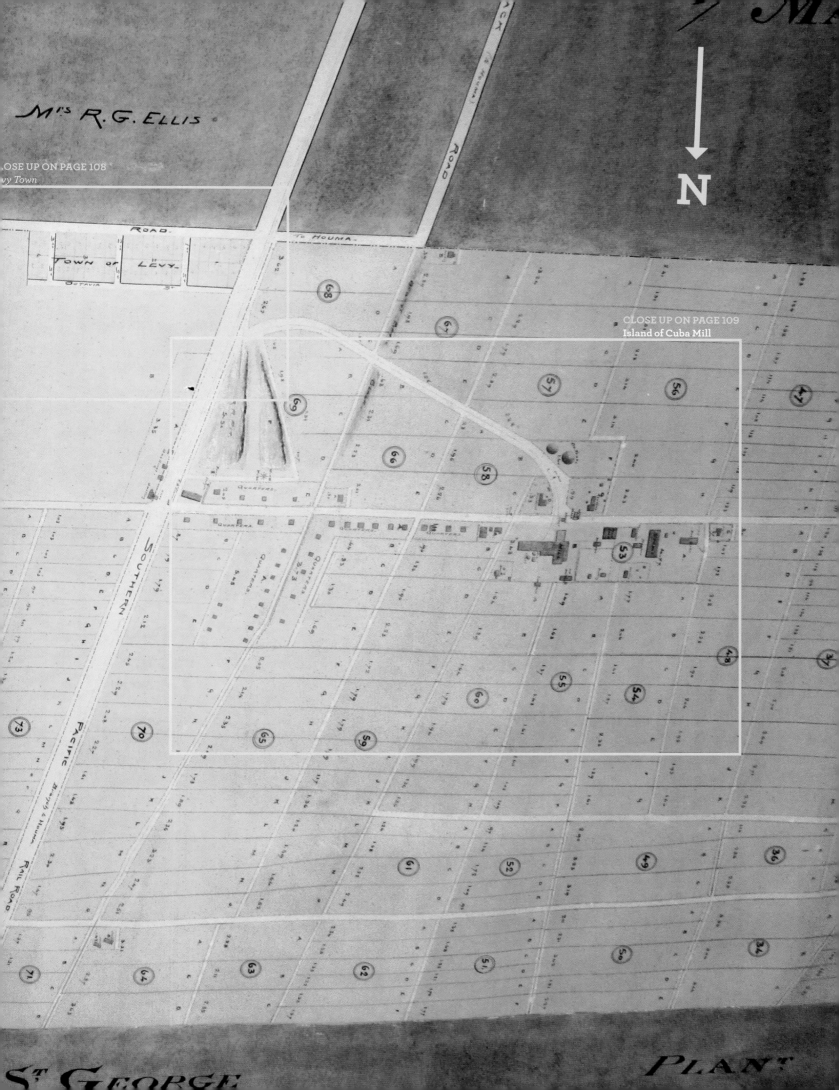

LEVY TOWN

A cluster of residences known as Levy Town is located on what was Isle of Cuba plantation land on the right bank of Bayou Terrebonne. From the direction of Houma along Louisiana Highway 311, parallel to the much-diminished Little Bayou Black, Levy Town is on the left side of Louisiana Highway 311 as it veers right toward Bayou Terrebonne. The highway then continues until it intersects with Louisiana Highway 20 at St. Bridget Church, but Levy Town is limited to a relatively small area facing the highway midway between Bayou Black and Highway 20.

The name of the area derives from David Levy and Max Mayer Levy who owned Isle of Cuba plantation from 1897 until 1908, and in the 1930s was part of Terrebonne Farms cooperative founded by the federal government's Farm Security Administration.

The community was founded by African Americans who were probably descended from ex-slaves from nearby plantations— Isle of Cuba, St. Bridget, Balsamine on Bayou Terrebonne and Magnolia on Little Bayou Black.

The group of houses is located across the highway from Caldwell Middle School and the New Magnolia Baptist Church. The school and church, although just across a highway, are on grounds that were once part of Magnolia plantation, the house of which still faces Highway 311 before it veers toward Bayou Terrebonne. Owners of Magnolia donated 0.22 acres to the church congregation for its edifice in 1942. It is located at 427 Highway 311-428 St. Bridget Road, Schriever, Louisiana.

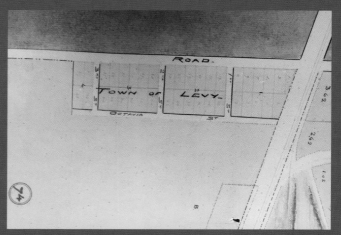

Levy Town between Bayou Black and Highway 20 January 1907

New Magnolia Baptist Church 2015

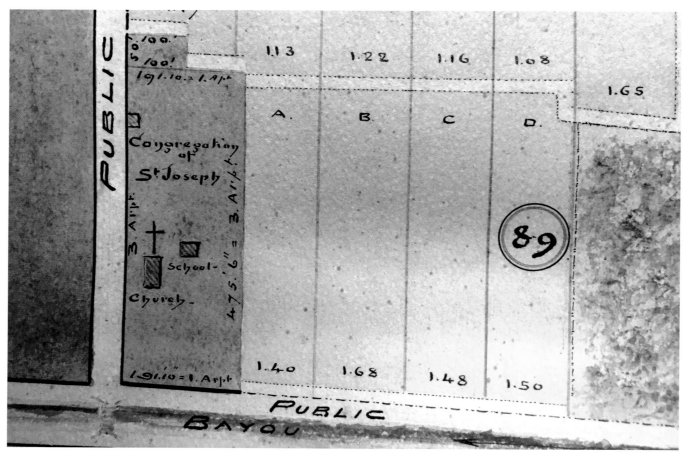

St. Joseph Congregation (later St. Bridget Catholic Church) January 1907

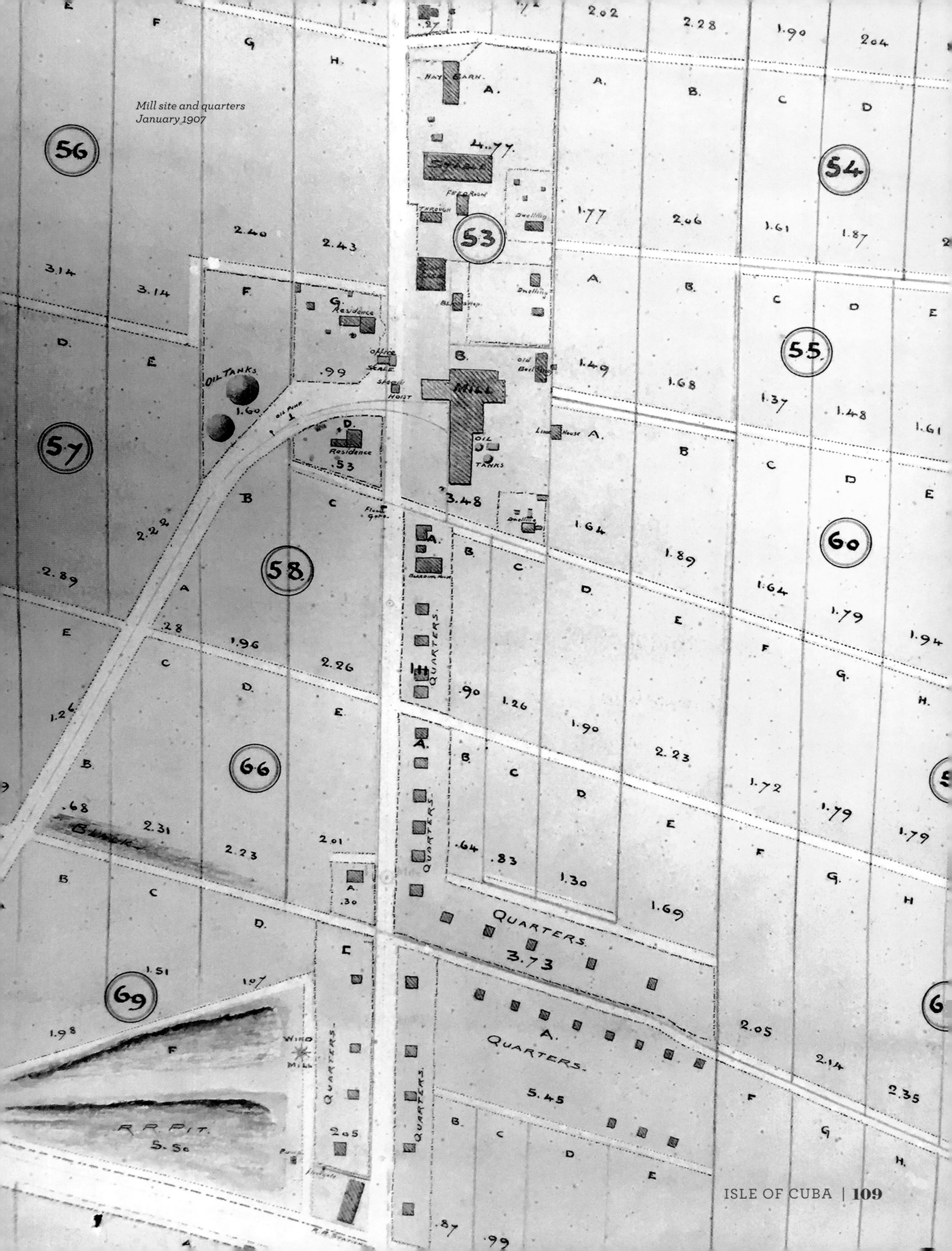

Mill site and quarters
January 1907

St. Bridget Church c. 1920s

St. Bridget, first chapel dedicated February 1, 1876

Right, marriage of Huey Cortez and Ruth Naquin Cortez, last wedding ceremony in the old church June 21, 1955

New St. Bridget Church during construction 1955

New St. Bridget Church dedicated September 8, 1955

ISLE OF CUBA | 113

Judge Leufroy E. Barras (1798-1871)

David W. Levy (1838-1908)

Arrest at Isle of Cuba The Houma Courier *May 8, 1897*

Below, shooting at Isle of Cuba The Houma Courier *June 16, 1900*

> Deputy Sheriff Henry Tucker made a very important capture this week on the Isle of Cuba plantation. John Allen a desperate negro who waylaid, shot and robbed Mr. William Roux at Wagaman, Jefferson parish, and for whose arrest a $100 reward was offered, was overpowered after a very desperate struggle. The negro had a spade and when called upon to surrender struck a fearful blow at the deputy which he caught on the barrel of his shotgun. Mr. Tucker was forced to shoot the negro in his legs with small shot and even then he would not surrender, making another attack with his spade. The deputy not wishing to kill him again avoided the blow, and then laid the negro out with his gun barrel.

> Joe Gordon, colored, on the night of May 26th, shot Cecil Collins, a colored girl about 12 or 13 years of age, in the leg. Both parties reside on the Isle of Cuba plantation, where the shooting took place. Gorden was arrested and brought to Houma, Tuesday, by constable Butcher, of the 1st ward.

The Barras home faced Bayou Terrebonne from its right bank, and the 750 acres of plantation lands extended west of what is now Main Project Road.

Although Lilac was not as expansive as some Terrebonne plantations by the time parish-by-parish sugar reports began, Barras' plantation produced respectable amounts of sugar for two decades. In 1844, Barras' total was 384 hogsheads, 260 the next year, 503 in 1858-59, 278 in 1860-61, and 555 in 1861-62. At some point before 1868, Barras was partnered with Evariste Porche in ownership of the plantation.

Before his death in 1871, Judge Barras sold Lilac to A. Menuet & Co., in 1868. The plantation name is recorded as Lylac Avenue in the 1869-70 sugar report, with the crop down to 114 hogsheads. Subsequently the plantation returned to ownership of Mrs. Emilie Thibodaux Barras, 1870-72. Under her ownership the plantation produced 152 hogsheads in 1872-73 and 26 the next year.

Emilie died in 1885, but not before selling her plantation to M.A. Piedra in 1876, when it was still named Lylac Avenue. Manuel A. Piedra, whose grandfather had been a wealthy plantation owner in Cuba, immigrated to the United States in 1875. The next year when he bought Lylac Avenue, he renamed it Island of Cuba in homage to his birthplace.[3] In 1878, J. Infante bought the holding and changed the name to Ile de Cuba, the French form of Island of Cuba. Jonathan J. Piedra held ownership from 1886 through 1892. The original house built before the Civil War burned either during or soon after the war.

David Levy purchased the plantation and changed the name to the more Americanized version, Isle of Cuba; he held it from 1897 until 1908. Levy was born in Schulzbach, Alsace, France; his trade was as an itinerant merchant and a commission merchant in New Orleans; he also served as vice president of two companies. Marcus M. Levy became proprietor of the Isle of Cuba store and named it Levy's General Merchandise. The Levys were among the first Jewish merchants in Terrebonne Parish, joined later in Houma by a number of families who established a variety of retail shops.[4] David Levy's estate retained ownership until selling to Crescent-Magnolia & Mfg. Co, Ltd. in 1908 "for a reported consideration of $90,000."[5] That company owned the property until 1916. Its manager was Senator John D. Shaffer, and vice president was Owen Walther.

In 1936 the federal government's Farm Security Administration used the former plantation lands, as well as Waubun lands and house, St. George lands and house, and Julia lands, for a collective farm operation named Terrebonne Farms. The houses were used as administration buildings. That venture ended in 1944.

The old plantation sugar mill closed in the early 1920s. It was west of what is named Main Project Road, a leftover term from Terrebonne Farms days. Present location of what was Lilac, Lylac Avenue, Isle of Cuba fronts at 2029 West Main Street, Schriever, Louisiana in northern Terrebonne Parish six miles below Thibodaux.

St. Bridget Catholic Church was built on part of what had been St. Brigitte plantation land after the property was donated in 1875 by Judge Adam Eugene Blake, son-in-law of Henry Schuyler Thibodaux and Brigitte Belanger Thibodaux. However, the church's location became identified with Isle of Cuba plantation, the lands of which eventually surrounded the church site.

Leufroy Barras is registered as having brand number 102 dated June 23, 1829 in Terrebonne Parish livestock brand records.

Founded by descendants of former slaves, the houses of the Levy Town community on Highway 311's northward leg toward Bayou Terrebonne are on former Isle of Cuba plantation lands. The .22 acres of New Magnolia Baptist Church (c. 1942) is located across the highway, at 427 Highway 311 - 428 St. Bridget Road, Schriever. The owner of Magnolia plantation donated the church site on December 24, 1942.

SOURCES
1. *A Century of Lawmaking for a New Nation: U.S. Congressional Documents and Debates, 1774-1875*, American State Papers, Senate
2. *The Houma Courier* newspaper, July 19, 1939
3. *Biographical and Historical Memoirs of Louisiana*, The Goodspeed Publishing Company, Chicago 1892
4. *Encyclopedia of Southern Jewish Communities*, Institute of Southern Jewish Life, www.isjl.org
5. *Louisiana Sugar News*, The American Sugar Industry and Beet Sugar Gazette, 1909

Island of Cuba plantation token good for one meal; Isle of Cuba trespass notice The Thibodaux Sentinel June 11, 1910; Levy's General Store c. 1920s

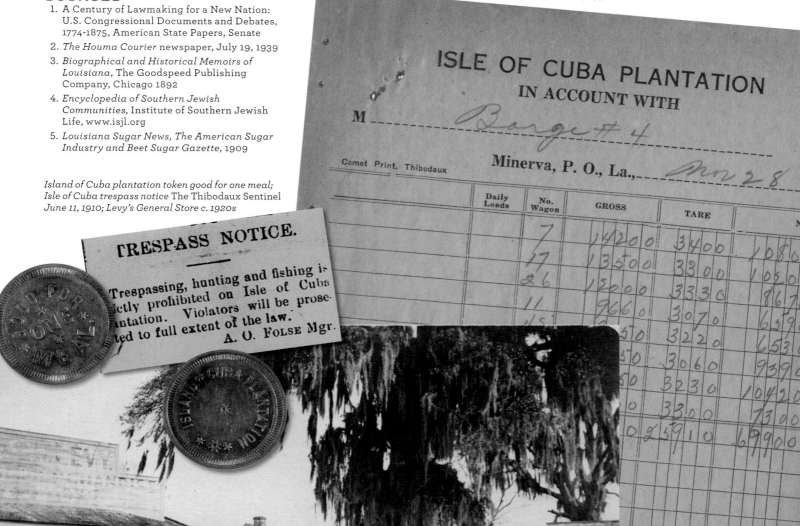

Cattle brand of Leufroy Barras June 23, 1829

Isle of Cuba plantation invoice November 28, 1930

Tombstone of Richard Gaillard Ellis (1800-1844) and Jane Ellis Towson (1805-1877)

CSA General Braxton Bragg (1817-1876)

EVERGREEN

The Auguste Babin claim confirmed by the U.S. government on May 11, 1820 and the John Mary (Jean Marie) Campo claim confirmed on February 28, 1823, spawned Evergreen plantation as well as other estates in the area.[1] On October 11, 1823 Henry Schuyler Thibodaux bought the part of the claim that became Evergreen property.

Thibodaux sold the land to Lemuel P. Tanner of Magnolia Grove plantation in 1825. (Thibodaux and Tanner were not only neighbors but also married to two Belanger sisters, Brigitte Emelie and Marie Agnes Celeste.)

Richard Gaillard (Gallain) Ellis, later of Ducros plantation, bought from Tanner in 1828 and built a home there on the right bank of Bayou Terrebonne c. 1830. After Ellis' death in 1844 his wife, Mary Jane Towson Ellis, inherited her husband's estate. Their daughter Eliza Ellis in 1849 married CSA General Braxton Bragg at the house her father had built 19 years before.[2] The Braggs lived at Bivouac plantation north of Thibodaux. A copy of Eliza and General Bragg's marriage certificate is held at Magnolia plantation on Bayou Black.

Mrs. Ellis died in 1877, after which Thomas J. Daunis took over Evergreen until the following year. Marioneaux and Henry Groebel were the titleholders beginning 1886-1887.

A "Sheriff's and Syndic's Sale of a Valuable Plantation in the Parish of Terrebonne, Known as the 'Evergreen' Plantation"

Below, Confederate States of America money 5 dollars, February 17 1864 front & back; Bottom right, June 7, 1849, wedding Certificate of Braxton Bragg and Eliza Ellis from Magnolia Plantation

Evergreen kitchen c. 1920s

Former Evergreen home site 2015

was conducted on May 5, 1888. The description of the property was advertised as "The whole of a certain piece or parcel of ground, together with the improvements thereon.... including sugarhouse, cabins, machinery, engines and fixtures, mules, cattle, wagons, carts, plows and all farming utensils thereto or used in the cultivation thereof, known and designated as the Evergreen Plantation, situated in the parish of Terrebonne in this State lying on both sides of bayou Terrebonne, about six miles from bayou Lafourche, measuring twenty-four and one-half arpents front on both sides of said bayou Terrebonne, with a depth of ten arpents on the west side of said bayou and thirty arpents in depth on the east side of said bayou Terrebonne, being bounded above by lands formerly belonging to the late Leufroy Barras [Isle of Cuba], and below by lands formerly belonging to John C. Potts [Hedgeford], excepting, however, therefrom that portion hereof acquired by Prosper Leblanc from Mrs. Mary Jane Ellis, and measuring one acre front by a depth of ten arpents; which said property was acquired by Henry Groebel and Olivier Marionneaux by purchase at the public sale thereof made by the United States marshal for the Eastern District of Louisiana on April 1, 1882... at the suit of Joseph J. Marion vs. Towson Ellis.... On March 29, 1888, Leo Leblanc reconveyed said one-third to Olivier Marionneaux, syndic of H. Groebel & Co. and of Henry Groebel and of Olivier Marionneaux."[3]

François Viguerie and Miss Marguerite Buron acquired Evergreen lands in 1892, followed by William H. Bland and then John Quimby. Ozemé E. Peltier owned Evergreen grounds from 1896 until 1908.

Ozemé Euzelien Peltier (1862-1933)

Left top, Evergreen trespass notice The Houma Courier *March 13, 1897; left center, notice of Celeste Belanger Tanner succession* La Sentinelle de Thibodaux *September 2, 1865; left lower, Evergreen sale or lease notice* The Weekly Thibodaux Sentinel and Journal of the 8th Senatorial District *October 9, 1880; middle, notice for Sheriff and Syndic's Sale of Evergreen* The Weekly Thibodaux Sentinel and Journal of the 8th Senatorial District *April 2, 1888; top right, newspaper notice for Peltier's Evergreen Plantation Fair* The Houma Courier *April 11, 1903; right center, newspaper ad for O.E. Peltier fresh meats* The Houma Courier *October 31, 1896; right lower, Levron & Peltier Butchers ad* The Houma Courier *January 23, 1897*

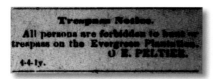

Tenant houses Evergreen 2015

Andrew Price bought 767 acres in February 1911 for $35,000. Price was implicated (through the journals and letters of Mary Pugh of Dixie plantation) in dragging a black man from a house during the Thibodaux Massacre of 1887.[4] He was not only a planter but also an attorney and politician who served as a U.S. Congressman from Louisiana from 1889 to 1897. He was a Democrat, and defeated Republican Judge Taylor Beattie in 1896 when he challenged Price for his Congressional seat.

Evergreen's sugar mill, which was on the left descending bank, opposite the homesite, closed c. 1905. The plantation was nine miles above Houma on both sides of Bayou Terrebonne; current address of the homesite is 2190 West Main Street, Schriever, Louisiana. The mill's crane foundation can still be seen on both banks of Bayou Terrebonne.

SOURCES
1. A Century of Lawmaking for a New Nation: U.S. Congressional Documents and Debates, 1774-1874, American State Papers, Senate
2. Nicholls State University Archives, courtesy archivist Clifton Theriot, 2015
3. *The Weekly Thibodaux Sentinel and Journal of the 8th Senatorial District,* May 5, 1888
4. Jeffrey Gould, *The Strike of 1887: Louisiana Sugar War,* Yale University dissertation

Evergreen cane hoist foundation remains on Bayou Terrebonne 2015

Above, Potts (Sonier) house 2016
Below, legend on Hedgeford plantation map April 20, 1872

Thibodaux, Parish of Lafourche, La
April 20th 1872

At the request of Mr Richard Lloyd, the proprietor of the Hedgeford Plantation I surveyed and I subdivided in the year 1871, a portion of said plantation as shown by this plan which I certify a correct report of my survey. All the measurements are given in links which are equal to 7 92/100 inches The distances and bearings written in red ink are taken from the United States

(Signed) R. A. Thibodeaux
Surveyor

HEDGEFORD
POTTS (SONIER) HOUSE

The first recorded ownership of what later became Hedgeford plantation were the John Mary (Jean Marie) Campo claim and the Laurent Pichoff claim, both registered with the U.S. government on January 6, 1861 and confirmed February 28, 1823.[1] The current site of Hedgeford land is in the vicinity of the Pitt Street Bridge between Thibodaux and Houma. Hedgeford properties went through several planters' hands until 1848.

It passed first from Campo-Pichoff to Philippe M. Sargeant in 1831 (of Sargeant /Armitage) but was sold two years later to Leandre Bannon Thibodaux. Richard G. Ellis, once of Ducros, became the subsequent titleholder in 1835. The next owner was Francis L. Mead, who bought the choice planting tract in the 1840s.

Major John Calvin Potts, who became synonymous with Hedgeford and the house he built, initially purchased northern Terrebonne Parish land in 1847, and adjacent property from Augustin Babin in 1848 to form his sugar estate of nearly 2,000 arpents on both sides of Bayou Terrebonne.[2]

Potts had been born in Philadelphia, the son of an immigrant Northern Ireland Presbyterian clergyman. His 1891 obituary in the *Weekly Thibodaux Sentinel* newspaper also detailed 21-year-old Potts' admission to the bar in Philadelphia and his migration to Natchez, Mississippi, two years later.

In 1836 he married Sarah Gustine, sister of Rebecca Gustine Minor, wife of William John Minor associated with Terrebonne Parish's Southdown plantation.[3]

He built what is now still known as the Potts (Sonier) house on the right bank of Bayou Terrebonne in the late 1840s and lived there 22 years.[2] While in Terrebonne, he became an advocate for planters' interests at the local, state, and national level; in 1876 he appeared with Lieutenant Governor Caesar C. Antoine before the U.S. Senate Committee on Foreign Affairs opposing duty-free sugar and rice proposed by the "Hawaiian Treaty."[4]

Potts' *Sentinel* obituary stated, "his home became famous for his elegant hospitality and refinement." Potts served in the Confederate army during the Civil War and moved to New Orleans at the war's end after declaring bankruptcy in 1868.[5] There he became a rice factor (commission merchant) and public servant. He died October 19, 1891.[3]

Henry S. Leverich bought Hedgeford in 1868, followed by Richard Lloyd in 1870. Lloyd had a survey conducted on a portion of Hedgeford plantation in 1871 by B.A. Thibodeaux, according to the legend on a map of that surveyed property dated April 20, 1872. The map shows the Hedgeford plantation subdivided into plots, thus making the plantation one of the earliest such estates to be subdivided.[6]

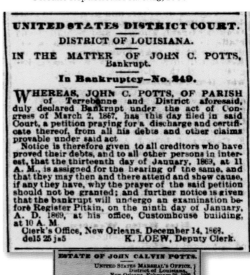

Notice of election of two directors for the Lafourche and Terrebonne Canal Co. at J. C. Potts (Sonier) house Thibodaux Minerva *December 23, 1854*

Article on Potts attendance at 1885 Opelousas convention of sugar planters Opelousas Patriot *December 8, 1855*

Notice of Potts bankruptcy New Orleans Republican *December 18, 1868 and article on John C. Potts bankruptcy warrant* New Orleans Republican *March 15, 1868*

124 | HARD SCRABBLE TO HALLELUJAH

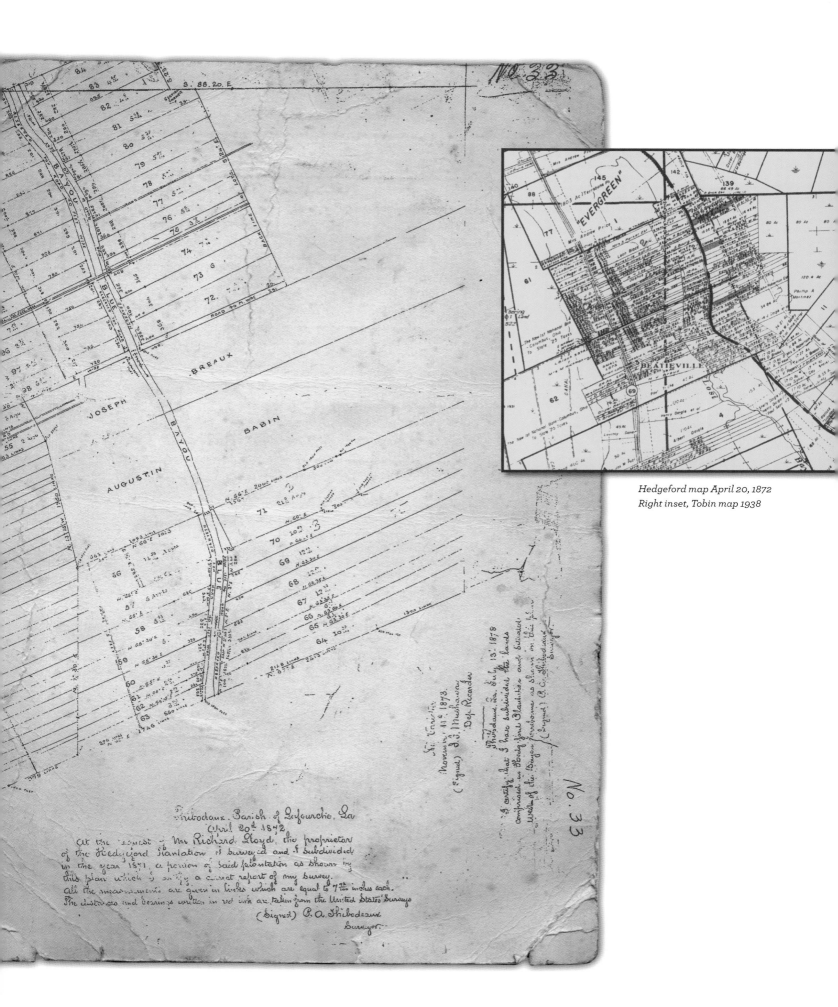

Hedgeford map April 20, 1872
Right inset, Tobin map 1938

HEDGEFORD | **125**

Lloyd sold a part of the property to P.A. Thibodaux in 1872, Frank Clemmon in 1877, Emile Gautreaux in 1878, Louis Turner in 1882, and Henry Gautreaux 1904-1912.

Joseph Ulger Boudreaux and his wife Rosalie Malbrough Boudreaux purchased the Potts house in 1878. The house became the home of the Sonier family in 1899 when Alcide Sonier and his wife Elise Crochet Sonier took title. Lily Sonier Rogers (d. 1979) lived in the house for a time, and Elmo P. Bergeron, Sr. (d. 1982) then inherited the property. Elmo P. Bergeron, Jr. and his sister Evelyn Bergeron Autin were owners of record in 2016. The estate has remained in the Sonier family line since 1899, but is now unoccupied.

The Potts house's walls are built of solid brick on the first floor and *briquette-entre-poteaux* (brick between posts) on the second floor. The inside walls are of plaster consisting of mud, lime, and animal hair. The original floors throughout the house are heart cypress.[7]

The Potts (Sonier) house is in disrepair now, but is still a welcome familiar sight at 2633 West Main Street, Schriever, Louisiana, on the right descending bank of Bayou Terrebonne between Thibodaux and Houma.

Philippe M. Sargeant registered his cattle brand as number 140 dated March 24, 1832 in Terrebonne Parish records. Major John Calvin Potts' brand was numbered 455, dated June 17, 1858.

SOURCES
1. A Century of Lawmaking for a New Nation: U.S. Congressional Documents and Debates, 1774-1875, American State Papers, Senate
2. Helen Wurzlow, *I Dug Up Houma Terrebonne*, 1984
3. *Weekly Thibodaux Sentinel and Journal*, October 14, 1891
4. *New Orleans Republican* newspaper June 4, 1876
5. *New Orleans Republican* newspaper March 15, 1868
6. *Plan of the Hedgeford Plantation* map by B.A. Thibodeaux, April 20, 1872
7. Owner Elmo P. Bergeron, Jr., 2016 interview

U.S. Senate Committee on Foreign Affairs, attended by J.C. Potts New Orleans Republican *June 4, 1876*

Left, cattle brand of Philippe M. Sargeant March 24, 1832; right, cattle brand of John Calvin Potts June 17, 1858

Rear yard of Potts (Sonier) house 2016

Smokehouse, Potts (Sonier) House 2016

Potts (Sonier) house brick front façade 2016

GRAY

The community that is strung along Bayou Terrebonne in upper Terrebonne Parish (identified to a great extent with the vicinity of current H.L. Bourgeois High School) was named for the first postmaster in what had before been called Bateyville. The previous designation came from nearby Batey plantation.

Walter Jasper Gray, Sr. assumed his post office duties January 4, 1898, according to the federal document *U.S. Appointments of U.S. Postmasters, 1832-1971*.

Gray was born on October 10, 1853 in Louisiana and married Mary Anne Hanagriff, born March 5, 1857 in Houma. They were married on February 28, 1876.

His son, Dr. Walter Jasper Gray, Jr., D.V.M., was the longtime veterinarian for Southdown Sugars, Inc., and lived in the Waterproof plantation home on Bayou Black Drive in the 1930s and 1940s.

Although Walter Gray served until his death on June 11, 1917, the postal service did not make the name change of the community until between 1925 and 1930. Gray's wife Mary Anne died on December 28, 1931, and probably lived to see her husband's name used for the community of Gray.

The current U.S. Post Office at Gray is located at 4442 West Main Street, Gray, Louisiana.

Gray School c. 1920

World War I registration card of Walter Jasper Gray, Jr., D.V.M. 1917

Leonard Bernard, Sr., Personality of the Week
The Houma Courier November 13, 1962

Bernard's Syrup Mill 1976
Opposite page, all photos Bernard's Open Kettle Syrup Mill from Good Earth Cookbook 1976

BERNARD'S OPEN KETTLE SYRUP & LACUITE

In the Sugar Bowl of Louisiana, it is logical that besides refined sugar, other commodities were produced from the abundance of sugar cane grown throughout south central Louisiana. Bernard's product was a staple of most area tables.

Born in Labadieville in Assumption Parish in 1920, Leonard Bernard was one of seven children of Valcour (Valcourin) John (Jean) Bernard (1879-1957) and Elvine Leonard (1881-1979). Elvine was born at Dixie plantation, and the couple married at St. Joseph's Church, Thibodaux. When Leonard was 12, his father began to teach him the process of making syrup, and eventually took Leonard on as partner. Leonard became the owner of the business at the young age of 22 after the death of his father in 1957.

Leonard moved to Gray because of availability of high pressure gas needed to fuel the plant's operations. In 1962 it was one of very few open kettle syrup plants in existence. Officially named Bernard's Open Kettle Syrup and LaCuite, the operation's first step was to secure hand-cut sugar cane stalks bought from local farmers. A worker then placed a few stalks of cane at a time onto a conveyor belt that led to a grinder. Sugary juices squeezed by the grinder were then pumped to an upstairs kettle for boiling. (Crushed stalks—bagasse—went on to a furnace for burning.) The juices passed through a tube in which sulphur was burning, the juices and sulphur dioxide fumes, as well as a small amount of lime, helping to clarify the syrup. The clarified juices were then boiled and impurities skimmed off the top, leaving the syrupy liquid at the bottom. The process consisted of heating juices until most of the water content dissipated.

From the boilers, juices cooled, then were delivered to open kettles, where juices were further heated by a fire directly beneath the kettles until the sweetness was perfect. A time in a cooling kettle followed, and a wait ensued for as long as three days for the cooling. Workers then packaged the sweet product.

The optimum time to produce syrup was 45 days, beginning the final days of October, because then sugar cane was most ripe. Production in those 45 days required Bernard to work from one o'clock in the morning until as late as eight-thirty at night.

Since the plant operated for only 45 days, Leonard Bernard was also a licensed plumber and a licensed electrician.

LaCuite, according to Ella Mae Baudoin Bernard (1922-2015), Leonard's wife, was a syrup cooked very hot, pumped into a smaller kettle with steam coils to cook the syrup more. A kind of glucose was added and heated to 190 degrees and cooked until very thick, when it was then jarred, sometimes with pecans added.[1]

The Bernard plant was opened in 1945 and closed in the mid-1970s. Leonard John Bernard, Sr. was named Louisiana Governor's Older Worker of the Year in 1997; he worked until age 90 and died at age 95.[2] The current address of where his plant was located on the right bank of Bayou Terrebonne is 3008 West Main Street, Gray, Louisiana.

SOURCES
1. *Houma Courier* "Personality of the Week" article, November 13, 1962 is the source for the above information.
2. Obituary, *Houma Courier,* November 2015

BERNARD'S OPEN-KETTLE SYRUP MILL

As I read this article about this book being published, I believe that this would be a very good article to put in it. The name of the syrup was "Bernards' Open Kettle Syrup and LaCuite". Now LaCuite was a syrup taken, once cooked, and as soon as cooked and while very hot, was pumped into a smaller kettle that had steam coils in to cook this syrup again more. A certain glucose was added and heated to 190 degrees and cooked until a certain stage where it was very, very thick but wouldn't turn to sugar. We'd jar this while very hot. At times we'd put our jars half full of pecans and fill the jar with "LaCuite". We could never make enough of this. There was a great demand for syrup and also LaCuite. When the LaCuite was jarred alone this was delicious on crackers or you'd catch some on a knife and then roll it yourself in pecans. I know about the "Lacuite" very well. I am Mrs. Leonard Bernard, Sr. and I was making it and also putting it in jars. My father-in-law Valcour Bernard showed me how to cook it.

Submitted by: Mrs. Leonard Bernard
Gray, La.

SYRUP MAKER HAS UNIQUE OPERATION

(Excerpts from a "Personality of the Week" article of Houma Courier, dated November 13, 1962.)

One of the few open kettle syrup plants in existence is still being operated by Leonard Bernard Sr. of Gray, Louisiana. The plant is small and functional, and is operated by only six people. But the syrup that is turned out can rank with the best — as many natives of the area and quite a few outsiders will readily tell you.

Bernard actually produces two types of syrup - the regular open kettle syrup and La Cuite, which has a heavier concentration of sugar and is almost candy-like in substance.

The production of the syrup is quite simple, although a lot of patience and hard work goes into the operation. The first step is to secure some good handcut sugar cane stalks, which Bernard buys from local farmers. From the heap of stalks, a man feeds a few at a time onto a conveyor belt system which leads to a grinder. At the grinder, the sugary juices are squeezed from the stalks and pumped to an upstairs kettle for boiling. The crushed stalks, or bagasse, continue on to a furnace where they are burned. Before the juices spill into the kettles, they are first passed through a tube in which sulphur is burning. The mixture of the juices with the sulphur dioxide fumes helps to clarify the syrup. A small amount of lime is also added for clarification.

After the juices have been boiled, a worker skims off impurities from the top, leaving the syrupy substance at the bottom. Actually, the whole process is one of heating the juices to a point where most of the water content is removed.

From the upstairs boilers, the juices are allowed to cool, then are delivered to the heart of Bernard's system — the open kettles. Here, the juices are further heated by a fire directly beneath the kettle to the point where the sugar content is just right.

White League sketch by Thomas Nast Harper's Weekly October 24, 1874

Cattle brand of Henry Claiborne Thibodaux April 12, 1825

Below, La Sentinelle de Thibodaux September 9, 1865

Above, La Sentinelle de Thibodaux September 2, 1865

Judge Taylor Beattie (1837-1920)

BEATTIE
BATEY

A name still familiar to Terrebonneans even after the turn of the 21st century was once associated with what later became Batey plantation in upper Terrebonne Parish. Robert Ruffin Barrow, Sr., who held vast acreages in all segments of Terrebonne and other parishes, purchased Batey lands from earlier claimant Constant Pitre. Pitre registered his claim with the U.S. government on November 20, 1816 and it was confirmed on May 11, 1820.[1]

However, the name overwhelmingly associated with the land is that of Beattie, Beatty, and Batey, different owners at different times, but none of whom were related. Further complicating the name situation is that Batey was alternately spelled Beaty by siblings of the same family.

The first of owners by the Beattie name was Dr. John C. Beattie, who bought the acreage from Barrow with Henry Claiborne Thibodaux in 1845. Their property is shown as Thibodaux and "Beaty" on the 1846 La Tourrette map, five plantations south of Thibodaux's mother's plantation, St. Brigitte. Thibodaux and Beattie's sizeable holding lay on both banks of Bayou Terrebonne, in what is now called Gray. (Thibodaux had cattle brand number 15 dated April 12, 1825 as recorded in the Terrebonne Parish brand registration book.)

Dr. Jesse Batey (1796-1851) bought the plantation nine miles above Houma in 1850, and held it until his death.

Jules Lepine & Victor Buron were the producers of record at Batey, as it was called at the time, in 1872. Four years later State District Judge Taylor Beattie took possession, in 1876. He was a Lafourche Parish native who married Fannie E. Pugh of Madewood plantation, Assumption Parish. He served as colonel in the First Regiment of the Louisiana Infantry in the War Between the States, as judge of the 15th District from 1872 until 1880, and judge of the 20th District from 1884 until 1888. A decade after his purchase, as a member of the White League and in his capacity as head of a "Peace and Order" committee in Thibodaux, he enabled the Thibodaux Massacre of 1887 in which vigilantes killed striking plantation laborers and others, most of them African Americans.[2]

Judge Taylor Beattie owned the Terrebonne plantation only two years; in 1878 Jules Lepine was the titleholder of record. Dr. Jesse Batey's long-term identity with the vicinity of the plantation's boundaries is that the area was known as Bateyville long after the plantation's demise. The name of the community was changed to that of its first postmaster, Walter J. Gray. He was appointed on January 4, 1898[3]; the name change of the community took place between 1925 and 1930.

A document dating to 1838 in the Jesuit Plantations Project from African Roots blogspot records the sale of 64 slaves to Jesse Batey of Terrebonne Parish, Louisiana, from Thomas Mulledy of Georgetown, District of Columbia. A March 1851 record from

Tobin map of 1938

BEATTIE | 133

REFORMATION.— Some people think that with Beattie at the head of the Radical party in Louisiana, he will bring about a reformation within the ranks of that party.

As well might a man go down into Hell and attempt to reform ten legions of devils, as to arouse any symptoms of honor, integrity and virtue in such scoundrels as Kellogg, Tom Anderson, Mad. Wells and their band of plunderers.

---

the week.
Judge Taylor Beattie and Judge W. E. Howell were in Houma for opening of Court, Monday.

—Says the *Baton Rouge Advocate* in regard to Judge Beattie's visit to that city.

Only one white man was on hand this morning to receive Col. Beattie, and that was Mr. ex P. M. O'Connor.

The colored man didn't rally much around Beattie this morning. The few who did put in an appearance did not meet with a very cordial greeting from him.

When we contrast the reception of Gov. Wiltz with that of Col. Beattie at Baton Rouge the outcome is easy to foretell. Wiltz was received by hundreds; Beattie retired to his boarding house accompanied by his valise boy.

Judge Beattie opened his remarks by explaining his emotions on this occasion, as, the last time it was his pleasure to appear beneath the walls of this grand old building was when the State seceded from the Union. He said they had met here to day to discuss the issues of the day, among which was the proposition to change the organic law of our State. Upon the Republicans fell a part of the duty of deciding whether the new constitution was better than the old one. He had heard that his political opponents had announced in their addresses at this place that the new constitution was the best the world ever saw. He begged to dissent from this opinion, for in his opinion it was the grandest conglomeration of nonsense he ever saw, a botch, a disgrace and a parody on the English language. Col. Beattie took high Radical ground in favor of the theory that this is a nation and not a Union of States.

With Taylor Beattie at their head,
The Radical hosts are forming—
But with LOUIS ALFRED in our lead,
We'll give that crowd a warning.
With all their little pranks,
And teach them there's no confusion
In the Democratic ranks.

Taylor Beattie or Betty, because proper names have no orthography; is the republican candidate to the office of governor.

But Taylor Beattie, shall never be an Excellency, although having excelled in jumping from one camp to another,

For Taylor Beattie, once a confederate, once a democrat, once a liberal, is now something like a republican.

But you say that he may not jump any more?

Don't believe it.

Taylor Beattie, who is a jumper by nature and feeling, could become again a liberal, a democrat and a confederate.

And if Taylor Beattie, to prove the marvellous elasticity of his muscles, was jumping into the imperial camp?

We cannot trust ourselves our destinies to a political grasshopper.

But it seems that Taylor Beattie, being back from the Globe convention and landing at Thibodaux, was not precisely received as a triumpher.

A colored brass-band a small cannon, an escort of two bar keepers and of about sixty colored men, woman and children, composed the whole ceremony.

The small cannon, borrowed from the Democrats, had a cold.

Poor Beattie!—*Louisiana.*

JUDGE TAYLOR BEATTIE 83, DIES AT THIBODAUX.

"Judge Taylor Beattie, one of the best known and popular members of the Louisiana bar, passed away at his residence in Thibodaux, Friday November 19th, 1920. He was 83 years of age at the time of his death. He served as district Judge for many terms and had been prominent in State affairs for a great number of years.

"Judge Beattie was a man of strong personality and was considered one of the best Attorneys in the State. He was a Republican during most of his career and was prominent in the affairs of his party.

"He had never been an active church goer but his daily life was an exponent of his religion. He accepted the last sacraments of the Catholic Church before he died. The burial took place at Madewood Plantation in Assumption parish. He is survived by his widowed wife and several children who reside in Lafourche Parish."—*Houma Times.*

TO A DEPARTED FRIEND.
By the Editor

Taylor Beattie has been nominated by the Republicans for Governor. What a difference between him and Wiltz. The one has for years been in close communion with the political vampires who well nigh destroyed the existence of our State, the other the grand and noble patriot whose voice loudly shouted for in the cause of liberty and pure government Taylor Beattie Louisianan but his friendly relation to Warmoth and Kellogg stamp him as the enemy of our people. Wiltz is also a Louisianan but all of his aims, his every ambition was to purge the country of the loathsome vermin that had invaded the State. Warmoth, Kellogg, Packard and Beattie! Well, let them go down history in infamy.—*Marksville Bulletin.*

☞ See T. Beattie's notice in another column of to-day's issue.

Newspaper items on the life and politics of Judge Taylor Beattie; Top to bottom, left to right: The Weekly Thibodaux Sentinel and Journal 8th District *November 8, 1879;* The Houma Courier *October 13, 1917;* The Weekly Thibodaux Sentinel and Journal 8th District *November 8, 1879;* Thibodaux Sentinel and Journal 8th District *November 8, 1879;* Thibodaux Commercial Journal *December 4, 1920;* The Weekly Thibodaux Sentinel and Journal 8th District *November 8, 1879;* The Weekly Thibodaux Sentinel and Journal 8th District *November 8, 1879;* The Houma Courier *April 2, 1892;* The Houma Courier *May 7, 1892;* Thibodaux Commercial Journal *December 4, 1920*

Hon. Taylor Beattie, of Lafourche is an independent candidate for District Judge, and Mr. L. C. Moise an independent candidate for District Attorney.

Judge Taylor Beattie, of the district composed of Lafourche and Terrebonne, who, although a Republican, has had a sort of "lead-pipe cinch" on the judgeship ever since "reconstruction times," was active and aggressive in his support of Foster for the governorship, and thereby came in violent contact with the Democratic buzz saw in the hands of the supporters of Justice McEnery, who were in large majority in the district. The result was disastrous; and District Attorney Caillouet, who represented the Thibodaux Sentinel at the annual meeting of the Louisiana Press Association held at Opelousas in May, 1888, will for the next four years fill the office which Beattie had almost come to consider as his private property. We congratulate Judge Caillouet.—*St. Landry Democrat.*

Memorial Exercises For Judge Beattie

Exercises were held last Wednesday in the 20th Judicial District Court suant to assignment, in memory of Judge Taylor Beattie. Judge opened court in the regular order the committee appointed to draft resolutions for the bar, presented the through its chairman, Judge W. E. Howell, in behalf of the bars of Assumption, Lafourche and Terrebonne following the introduction of the resolutions, with a few remarks and a review of the deceased. He was followed by Judge-elect Sam LeBlanc of Assumption, on behalf of the Assumption bar, then by Mr. Harris Gagnier of the Terrebonne bar, who in turn was followed by the different members of the bar, as also by Mr. Adrien Caillouet of the Terrebonne bar, all eulogizing and paying tribute to the worth, merit and character of the deceased, as a soldier and jurist. Some excellent eulogies were offered, and that of Mr. Harris Gagnier, deserves special mention being an oratorical effort most classic.

The members of the bar in attendance were: Judge W. E. Howell, Judge L. P. Caillouet, Messrs L. H. Pugh, J. A. O. Coignet, L. Edwin Anet, N. F. Montet, Camille Moise and F. L. Knobloch, Judge Sam LeBlanc, representing Assumption, Messrs Harris Gagnier, Adrien J. Caillouet and Judge H. M. Wallis Jr. representing Terrebonne. Messrs Coignet, R. Beattie and Taylor Beattie Jr. besides several lady members of the family, and a number of friends, acquaintances and the officers of the court parish were also in attendance.

The resolutions were as follows:

WHEREAS Divine Providence has called JUDGE TAYLOR BEATTIE from his field of usefulness—from the sphere in which he was peculiarly well adapted—the law, and

WHEREAS we who know him well, realize that in his death. there has been removed one who was an ornament to bench and bar; one who was a zealous and untiring champion of right, who at all times measured up to the highest standards, and ever maintained that the scales of justice be held at even balance; whose incumbency as Judge of this district was a record of duty well done; whose conduct as a practitioner was an example to the younger members of the profession.

ty, and his absence shall be noted in our ranks, therefore

BE IT RESOLVED that while voicing our loss, and the sorrow and grief of those left to mourn him, we find satisfaction, consolation, and solace in the knowledge of the devotion he held for those higher ideals of character through his years of toil, which enabled him to give his best effort to the upbuilding of country, home and profession, and bid those who feel and deplore his loss remember that a life spent as was his, must find the reward eternal, therefore

BE IT FURTHER RESOLVED as a mark of our esteem and respect for the deceased, that a copy of these resolutions be spread on a special page of the minute book of this court, a copy furnished the bereaved family, in testimonial of our sympathy for them and that the local press be furnished copies for publication.

Respectfully submitted
(Signed)
W. E. Howell,
L. P. Caillouet,
L. H. Pugh,
J. A. O. Coignet,
F. L. Knobloch

Attest P. J. Auc——
Clerk of ——

——Get W. S. S.

Reports have it that the Lafourche Transfer Co. Inc., propose placing operation a passenger service be the same source is an inventory of nine slaves he held (Nace Butler, his wife Biby, a child of a year and a half, Martha Anne, Gabe, child Biby, Henry, Tom, and Mary, all seemingly of the same family). The Plantations Project does not specify where in Terrebonne Parish Dr. Jesse Batey resided in 1838.

However, a more recent effort, the Georgetown Memory Project, gives a more complete picture of the Batey purchase. An April 16, 2016 article in the *New York Times* revealed that Mulledy was the Rev. Thomas Mulledy, then president of the Jesuit-owned Georgetown University. In tracking the sale in 1838 of 272 slaves from Georgetown, the Georgetown Memory Project discovered that "in June 1838, he [Father Mulledy] negotiated a deal with Henry Johnson, a member of the House of Representatives, and Jesse Batey, a landowner in Louisiana..." who had land in Terrebonne Parish, Maringouin, and in Ascension Parish near Donaldsonville. The long-ago Jesuit university's sale of the slaves has caused descendants in Louisiana to consider asking Georgetown to consider reparation or recognition of some type for the slaves sold by the university.

A glimpse of the magnitude of Dr. Jesse Batey's properties was given in the will of his brother Robert Beaty, [Batey] Sr. when he bequeathed to his heirs "my interest in three tracts of land situated in the State of Louisiana, being the real estate of which my brother Dr. Jesse Beaty [Batey] died seized and possessed, containing in all about sixty four hundred acres."[4]

Jesse Batey was the undoubted namesake for a steamboat built for the Lafourche and Terrebonne trade. The book *Sugar Country* by J. Carlyle Sitterson quoted an 1861 traveler who wrote that the *Dr. Batey* was "a very neat and substantial craft of medium size, and the most magnificently furnished of any steamer we have seen for a long time."[5] A steamer *Dr. Batey* was built in 1850 in Louisville, Kentucky and was part of a unit of the Confederate expeditionary force in 1860.

In 1880 the ever-entrepreneurial R.R. Barrow, Jr. purchased Batey plantation property in partnership with his sister Volumnia Roberta Barrow Slatter Woods after Jules Lepine's ownership, thus placing a second-generation Barrow ownership stamp on the plantation their father had sold to Dr. John C. Beattie in 1845. They commissioned surveyor C.H. Buford to subdivide the property, and sold parcels on the right bank of Bayou Terrebonne in 1883. The location of the "big house" was then across the bayou from Halfway Cemetery, which is located at 3842 West Main Street, Gray, Louisiana, nine miles above Houma.

Thomas Mulledy slave sale to Jessie Batey 1838, Washington, D.C. (Conveyances and Records Book O, entry M-5, 13/2327-28; and 4893)

SOURCES
1. A Century of Lawmaking for a New Nation: U.S. Congressional Documents and Debates, 1774-1875, American State Papers, Senate
2. John DeSantis, "A page torn from history: Bayou area's hidden past emerging from the shadows" published in *Tri-Parish Times* newspaper July 14, 2015
3. Federal document *U.S. Appointments of U.S. Postmasters, 1832-1971*
4. Will of Robert Beaty, Sr. (1800-1890), Union County, South Carolina, recorded June 29, 1878
5. J. Carlyle Sitterson, *Sugar Country: The Cane Sugar Industry in the South, 1753-1950*, University of Kentucky Press, 1953

*Main Street entrance gate Halfway Cemetery
Background Tobin map 1938*

136 | HARD SCRABBLE TO HALLELUJAH

HALFWAY

The earliest claimant on record of what became Halfway was named John Mason. He registered his claim with the U.S. government on November 25, 1816, which was confirmed on May 11, 1820.[1]

As early as November 1828, D.D. Downing, M.D. of Port Gibson, Claiborne County, Mississippi, purchased ten and a half arpents by depth of that land from Brigitte Belanger Thibodaux on both banks of Bayou Terrebonne for $3,800. He expanded his holdings a few days later by purchase of two arpents by depth from Norbert Bonvillain on both banks for $600.

When Dr. D.D. Downing died intestate in 1830, he left behind that property in Terrebonne Parish described in the parish's civil records as "One tract of land nine miles from Thibodauxville twenty-five arpents front on the left bank of Bayou Terrebonne and twenty-three and one half arpents front on the right bank, bounded above by Jacques Bonvillain and below by E. Lyman with buildings."[2]

His only heir was William V. D. Downing, a minor ten years old who lived in Mississippi.[2] The sisters-in-law of Dr. Downing asked that Dr. Jesse Batey be appointed curator of the estate, which included the land, 53 slaves, 80 pounds of cotton, a crop of sugar, three carts and farming equipment, 19 head of cattle, three and one half pair of oxen, three horses, one mule, and 25 sheep. The total estate was valued at $32,767, and notes due were $5,048.25.[3]

In 1845, W.V.D. Downing was of age and gained ownership of his relative's property. He became a medical doctor, and owned Halfway for more than 20 years, after which Jules Lepine purchased it in 1869. In relatively quick succession, Halfway was then owned by Lepine & Buron (beginning 1872), Lepine & Ferry, and J.A. LeBlanc from 1878 to 1886. That year the titleholder was Mrs. Albert Caillier, who owned it until M.W. Whitehead purchased the estate in 1897.

The *1897 Directory of the Parish of Terrebonne* recorded for the plantation 1,700 acres, 125 in corn and 75 in rice.

J.E. Himel was the first 20th century owner of Halfway, beginning in 1902. Two years later some of the acreage became property of J. Paul, Foret T., and Eno Landry. The Landrys retained those grounds until 1916. Ferdinand Daigle held another parcel from 1908 until 1912.

J. Paul Landry, Sr. had a livestock brand numbered 666 dated August 18, 1903 in Terrebonne Parish registration records.

Halfway plantation was seven miles above Houma, the house situated on the right bank of Bayou Terrebonne. Halfway Cemetery is on land that was the plantation. The cemetery was

Death at Halfway Plantation The Weekly Thibodaux Sentinel and Journal of the 8th Senatorial District *March 25, 1876*

Middle, Henry Schuyler Thibodaux tomb at Halfway Cemetery 2015

Cattle brand of J. Paul Landry August 18, 1903

Henry Schuyler Thibodaux historical marker 2015

Halfway early 1900s

established in the late 1700s when Louisiana was under Spanish control, before the Louisiana Purchase and before Louisiana became a state.

Henry Schuyler Thibodaux, original owner of Halfway lands, his wife Brigitte of St. Brigitte plantation, and Curtis Rockwood, the first postmaster of Houma, are interred there. Also interred there are the families of H.S. Thibodaux's wife's brother-in-law Lemuel Tanner and Brigitte's sister Marie Agnes Celeste Belanger Tanner, and his son-in-law Leufroy Barras (of Isle of Cuba plantation) who was married to Brigitte Emilie Thibodaux, Henry Schuyler and Brigitte's daughter.

The current H.L. Bourgeois High School is located on what was once part of that estate's grounds. The school address is 3489 West Park Avenue, Gray, Louisiana.

SOURCES

1. A Century of Lawmaking for a New Nation: U.S. Congressional Documents and Debates, 1774-1875, American State Papers, Senate
2. *Terrebonne Life Lines Volume 22*, Civil Suits and Probate Matters, Terrebonne Parish, Louisiana, Fall 2003
3. Terrebonne Parish Civil Records, Succession of Dr. D.D. Downing

Joseph Paul Landry, Sr. (1853-1943)

The Houma Courier *November 13, 1897* and The Houma Courier *January 30, 1897*

Halfway Cemetery 2015

Charles A. Favrot, architect c. 1920

Left, Edward L. Theriot School Board member and teacher, right, H. L. Bourgeois c. 1920

AYO
BON AMI
EVERGREEN

Evergreen Junior High School 2016

Ayo plantation developed in the area between Halfway and Pilié plantations, on both sides of Bayou Terrebonne seven miles above Houma. The current Evergreen School is situated on part of what was Ayo plantation grounds on the right bank of the bayou. Earlier names were Bon Ami and Evergreen.

The Ayo estate descended from a Peter Watkins (Williams)-Charles Bergeron claim which Soloman C. Lawless purchased in 1830. *The Statement of Sugar Crop Made in Louisiana* for the year 1844 shows that S.C. Lawless produced 286 hogsheads of sugar that year, and 300 hogsheads the next year. Richard C. Lawless paid $89,000 for the land in 1846.

Ten years later, in March of 1856, Col. Robert Carter Nicholas and John Becemis bought the sugar plantation. Nicholas was a native Virginian who served in the War of 1812, during which he achieved his military rank. His father Wilson Cary Nicholas served as Virginia Governor from 1814 to 1816. Robert attended the College of William and Mary in Virginia, moved to Terrebonne Parish, Louisiana, c. 1820, and became a sugar planter.[1]

He was married to Susan Adelaide Vinson of Tennessee in 1840 at White Hall plantation in St. James Parish. Mrs. Nicholas is an ancestor of the Plater family of Acadia plantation in Lafourche Parish.[2]

Nicholas became a political force in his new state. He was a U.S. Senator from 1836 until 1841, and became chargé d'affaires to Naples on appointment by President Andrew Jackson. Nicholas went on to become Louisiana Secretary of State from 1843 to 1846, and later served as Louisiana Superintendent of Public Education.[3]

Col. Nicholas and his partner purchased the plantation in March 1856 and Nicholas died in December that same year. His wife, Susan Adelaide Vinson Nicholas, and R.C. Lawless became titleholders in 1857. However, Nicholas and Becemis are on record in the 1858-59 sugar report as having produced 265 hogsheads. Mrs. Nicholas produced 64 hogsheads in 1860-61, but that number climbed to 340 in 1861-62.

Citizens Bank of Louisiana seized the plantation in 1868, and the holding was offered at Sheriff's Sales in 1868 and 1870. The land in question was Evergreen plantation, 30.29 arpents front on both banks of Bayou Terrebonne, consisting of about 989.43 acres. A second tract of land was Lot 4 Section 82 Township 16 South, Range 17 East, consisting of 110 acres. Still another was a tract of land, the south half in Section 69 Township 16 South, Range 17 East, containing 69.42 acres.[4]

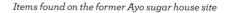

Robert Carter Nicholas (1787-1856) and an article on his death The Feliciana Democrat *January 3, 1857*

Items found on the former Ayo sugar house site

Opposite page, home of Ivy and Rosalie Ayo Trahan stands on the site of the former sugar house of Ayo plantation 2015

Opposite page inset, Joseph Odressi Ayo family; Odressi is the gentleman seated, wearing a derby hat and cane. Joseph is standing at the back fourth from the left. His other children as listed in his succession, were Abel B., Ivy G., Ernest, Eddie, Charles E., Sidney J., Mary, and Adele Ayo Darcey.

143

The owner of record in 1871 was Kentuckian John I. Adams. The sugar records of 1870-71 give the production of "Adams & Viala" as 158 hogsheads. Joseph Odressi Ayo in 1882 bought the plantation which became his namesake. CSA Corporal Ayo was a Lafourche Parish native who had married Donatille Zimorie Olivier of Assumption Parish.

Ayo's estate went to his wife and ten children when he died in 1927. Dennis A. Falterman and Holton J. Hornsby purchased the original homesite in October 1940 from Abel, Ivy G., and Mary Ayo. One tract was 651 feet on the left bank by depth, the second

Southland Raceway WCS, Houma, La. National Dragster Magazine *June 9, 1972*

Bottom, Candies & Hughes Southland Raceway WCS, Houma, La. National Dragster Magazine *June 9, 1972*

Dennis Adam Falterman, Sr. draft registration card 1940

was 651 on the left bank by depth, and the third 531 feet on the left bank by depth.[4] The two-story plantation house in which the Ayos resided was dismantled. Falterman's residence which he had constructed on the same site has since also been dismantled. (Much later, part of the left bank Ayo land was the location of the now-defunct Southland Dragway.) The current address of that site is 4609 West Park Avenue, Gray, Louisiana. Hornsby sold his half-interest to Dennis Falterman in 1942.

Joseph John Ayo inherited his share of the estate in 1927. The remainder of the estate was divided and sold by his brothers and sisters. Ella Williams Ayo, Joseph John's widow, sold two parcels to Eulice A. LeBoeuf in 1955, and a tract in Sections 7 and 82 to Robert R. Wright in 1967.[4] Her estate in 1969 had 12 residuary legatees. All sold their shares except her adopted daughter Roselia, known as Rose.

Rose was bequeathed, in Ella's will, a tract in Section 7 which was the plantation's sugar house site on the bayou's left bank. Rose inherited the seven-room house built in 1926 by her adoptive parents and the two and a half acres surrounding it.[5] Her husband

War stamp rations c. 1940s

Ivy and Rosalie Ayo Trahan c. 1934

Ivy Trahan bought six acres from the other heirs, and at Rose's death her children inherited the eight and a half acres and the homesite at 4607 West Park Avenue, Gray, Louisiana. Garland Anthony Trahan and his wife Gail Hebert Trahan now reside at 4519 West Park Avenue, where the plantation blacksmith shop once stood.

Two owners of Ayo had livestock brands registered in Terrebonne Parish records; Soloman C. Lawless, number 320 dated February 23, 1835, and Corp. Joseph Odressi Ayo, number 509 dated July 21, 1865.

The most recent transaction having to do with Ayo property was that Michael Gene Burke bought the original 1926 residence of J.J. Ayo and two and a half acres of once-plantation land in 2015.[4]

SOURCES
1. *Biographical Directory of the United States Congress*
2. David D. Plater, descendant of the owners of Acadia Plantation
3. Obituary in *The Evening Picayune*, New Orleans, December 29, 1856
4. Terrebonne Parish conveyance records
5. Helen Wurzlow, *I Dug Up Houma Terrebonne*, 1984

*Soloman C. Lawless cattle brand
February 23, 1845*

*Joseph Odressi Ayo cattle brand
July 21, 1865*

Bayou Blue Presbyterian Church 1972

ORANGE GROVE
PILIÉ
GLENWOOD

Pierre Bergeron held title to what would become Pilié in 1816, before Terrebonne Parish became a civil entity. He registered his claim with the U.S. government on March 28, 1816, and it was confirmed on May 11, 1820.[1]

Dating to the antebellum year 1858, the plantation was under ownership of Taylor Beattie & S. Beattie. The large farm first known as Orange Grove was located on the left descending bank of Bayou Terrebonne in the area that is now known as Bayou Cane around the present-day Buquet Street Bridge site. The earliest Terrebonne Parish police jury meetings were held in the small community of Williamsburg, located in the Bayou Cane area.

In 1861 Taylor Beattie & Co. was listed as titleholder, but from 1869 through 1874, Armand Pilié took over the property. Pilié gave his name to the plantation and made 115 hogsheads of sugar each year of his possession of the plantation, a fair amount in post-war years.[2]

Taylor Beattie & Co. had another short stint as Pilié's possessor, before Mutual National Bank took possession of the property in 1876. S.P. Michel & Co. became owners of record in 1877. Mastero, Miche & Co. owned the tract in 1887. A map from 1903 indicates the name of the plantation as Glenwood.[3]

In 1908-1909 Elphege Daigle (1862-1923) and his wife Marie Molly Josephine Pitre Daigle became owners of a part of the original plot, which extended east to Bayou Blue. During Daigle's years and some time before, a small narrow gauge railroad along the bayou hauled sugar cane to a mill for grinding, probably to Evergreen's sugar house.[4]

Judge Taylor Beattie (1837-1920)

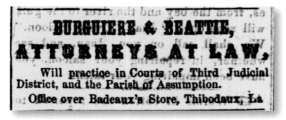
La Sentinelle de Thibodaux September 9, 1865

Daigle also owned half-interest in a dance hall, and offered use of the hall for a fledgling Presbyterian congregation for their worship. According to a history by Mrs. Farrel Robertson, "Soon Elphege dissolved his partnership and had the hall literally cut in half and his half moved to a piece of land owned by him on Bayou Blue," where services were conducted at first in both French and English. It was the forerunner of the Bayou Blue Presbyterian Church.[5]

The church's address is 3200 Highway 316, Gray, Louisiana.

SOURCES

1. A Century of Lawmaking for a New Nation: U.S. Congressional Documents and Debates, 1774-1875, American State Papers, Senate
2. P.A. Champomier, *Statement of the Sugar Crop Made in Louisiana* annual reports
3. A map by surveyor A.C. Bell dated July 28, 1903
4. Billy Earp Robertson in *Lafourche Country, The People and the Land* by Philip D. Uzee published by the University of Southwestern Louisiana Center for Louisiana Studies 1985
5. *Houma Daily Courier and the Terrebonne Press* newspaper article, October 8, 1972

Some of the founders of Bayou Blue Presbyterian Church, from left, are the Rev. and Mrs. John N. Blackburn, the Rev. M.R. Paradis, and Mr. and Mrs. V. G. Ballard
April 23, 1924

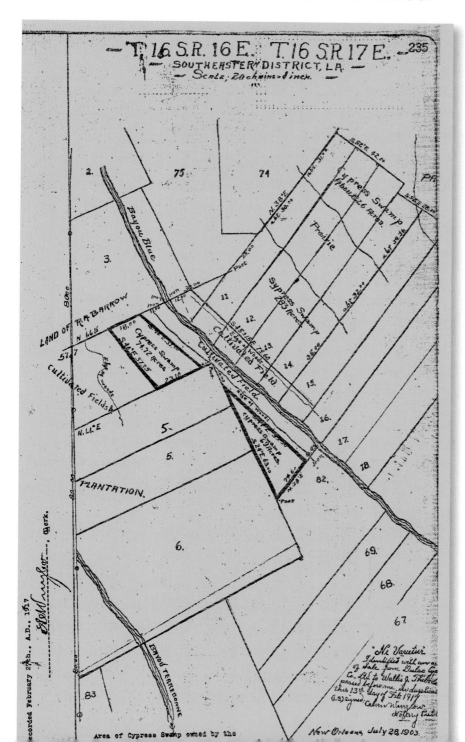

Map showing Pilié and Glenwood
February 27, 1917

Elphege (1862-1923) and Marie Molly Josephine Pitre Daigle (b.1862) photo c. 1920

ORANGE GROVE | 147

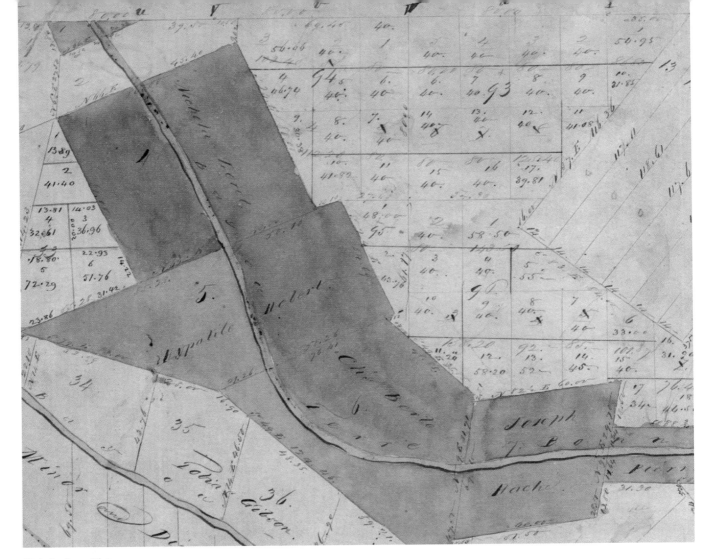

Nicholas Leret Claim
State Land Grant map December 24, 1831

BAYOU CANE
NICHOLAS LERET CLAIM

Along both banks of Bayou Terrebonne below Pilié and Williamsburg lay fertile lands first owned by Nicholas Leret, his ownership registered with the U.S. government on March 28, 1816 and confirmed on May 11, 1820. Later, that property became identified with the Lirette family.[1]

Alexander J. Lirette was the first of his family to own the land. He was Terrebonne Parish Sheriff from 1838 until 1844[2], besides being a sugar cane planter. In 1886 the place was known as Alexander Lirette & Sons, and in 1890 the property passed to Volcar Lirette. Alexander J. Lirette's cattle brand was number 161 dated June 29, 1833 in the parish registration book.

Alexander J. Lirette (1886-1921)
and his cattle brand June 29, 1833

SOURCES
1. A Century of Lawmaking for a New Nation: U.S. Congressional Documents and Debates, 1774-1875, American State Papers, Senate
2. Terrebonne Parish records

The Houma Courier April 2, 1892

> Miss Lena Lirette, the pretty daughter of Mr. Franklin Lirette, of Bayou Cane, gave a birthday party at her home last Wednesday evening.

SHERIFFS OF TERREBONNE PARISH

CALEB B. WATKINS	**1822**
P.H. DARCE	1832
REUBEN BUSH	1836
ALEXANDRE LIRETTE	**1838**
J.M. VORIS	1844
ALEXANDRE LIRETTE	**1846**
MARTIAL VERRET	1850
JOHN H. FIELDS	1854
JOSEPH AUGUSTE GAGNE	1854
AUBIN BOURG	1855
LEO LIRETTE	1862
ROBERT W. BENNIE	1864
ALBERT G. CAGE	1866
FREDERICK MARIE	1868
WILLIAM H. KEYS	1870
	(dismissed July, 1872)
AMOS SIMMS	1872
	(appointed July 27)
GENERAL LYONS	1872
	(December 27)
GENERAL LYONS	1875
	(1874 election contested; installed April '75)
JORDAN STEWART	1876
ALFRED KENNEDY	1878
THOMAS A. CAGE	1880
JOHN B. BUDD	1884
OSCAR DASPIT	1888
A.W. CONNELY	1892
OLIVIA HEBERT	1915
ERNEST A. DUPONT	1916
F.X. BOURG	1924-1940
PETER G. BOURGEOIS	1940-1951
ABEL P. PREJEAN	1951-1968
CHARLTON PETER ROZANDS	1968-1980
RONNIE DUPLANTIS	1980-1984
CHARLTON PETER ROZANDS	1984-1987
JERRY J. LARPENTER	1987-2008
L. VERNON BOURGEOIS, JR.	2008-2012
JERRY J. LARPENTER	2012-

BAYOU CANE
FILICAN ALEXIS DUPLANTIS

Filican Alexis "Tecon" Duplantis
(1867-1955)

Filican Alexis Duplantis, a planter who moved from Houma to Bayou Cane during Reconstruction years, became an important figure in the parish by virtue of instituting highly successful diversification of products on his acreage, as well as establishing a Mardi Gras parade tradition locally.

Tecon (also Tican), (b. November 21, 1867) recounted the history of the Bayou Cane settlement in a 1939 *Houma Courier* newspaper interview. He credited two Frenchmen, "one by the name of Leaneaux," and a man named Beattie, Joseph Buquet, and Alexander Lirette with founding the new settlement, which was already about 140 years old in 1939.

Originally intended to be the parish seat, Bayou Cane (Williamsburg) was the site of the first courthouse which Tecon described as "nothing but a wooden affair with no floor to it."[1] A jail across the road on the bayouside was "just a flimsy building which anybody could have broken out of at any time," according to Duplantis.[1] The little community at the intersection of Bayou Cane and Bayou Terrebonne was at the time three miles north of the town of Houma.

Duplantis married Ada Daigle Duplantis (1870-1965). They were the parents of Jean Felix Duplantis (1893-1967) and Adley J. Duplantis (1895-1983).

The family's sugar cane plantation on its 1,400 acres, later 600 acres, at Bayou Cane owed its success to the principle of diversification, according to a *Houma Courier* newspaper article of January 30, 1926. As early as 1910, the article stated, he laid the foundation for producing not only sugar cane but also what

Filican Alexis "Tecon" Duplantis (1867-1955)
Mardi Gras c. 1930s The Houma Courier *October 8, 1972*

Aerial photo of Bayou Cane and Duplantis land on Bayou Terrebonne, and Hollywood and Crescent Farm on Bayou Black 1953

Mardi Gras Queen Lela Leboeuf 1927 inset Mrs. Ned (Lela Leboeuf) Picou 1972

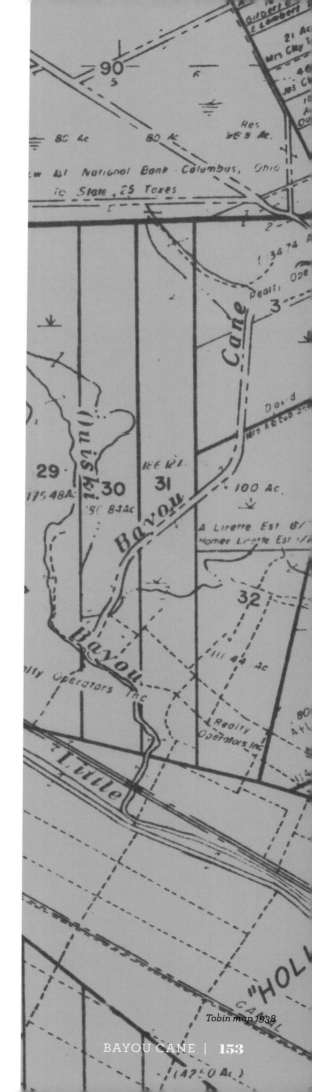

Tobin map 1938

amounted in 1926 to 73 different kinds of products, from varieties of vegetables and fruit to self-butchered beef and pork, to handmade preserves and filé, self-manufactured corn meal and grits, geese, turkeys, ducks and other wildfowl, sheep and goats.

"One relaxation from the long, hard work year that offered little or no entertainment," wrote one reporter,[2] was the Mardi Gras celebration Filican and his wife Ada founded in Terrebonne Parish as early as the 1880s.

His parades began at his plantation and early on consisted of maskers on foot making their way to and through Houma. According to newspaper accounts, his and his wife's parade grew through the years to "some 200 floats drawn by oxen and horses. . . . Decorations were all native products of his own realm—graceful draperies of Spanish moss, swaying stalks of sugar cane, palmettos from the swampland, and wildflowers from the bayou banks." Acrobats, bands, trained animals and other entertainment became part of the parades. Tecon (Tican) rode at the head of the parade astride horses and eventually convertibles.[3]

"Mr. Carnival" died on November 8, 1955. The present-day address of his homesite is 6008 West Main Street, Houma, Louisiana.

SOURCES
1. *Houma Courier* article, 1939
2. *Houma Daily Courier* article by Betty McMillan, 1969
3. *Houma Daily Courier* Houma Sesquicentennial edition, October 8, 1872

BAYOU CANE
WILLIAMSBURG

William Sternin Watkins, an Englishman who arrived in Louisiana via Chesterfield County, Virginia, settled in Louisiana in 1812. He made his mark on Terrebonne Parish, at least partially, in the name of Williamsbug at Bayou Cane, where the first governmental meetings of parish officials were held at the home of Alexandre Dupre in 1822.[1]

Watkins fought in the Battle of New Orleans as a major with the 6th Regiment Landry's Louisiana Militia, and was appointed by then-Governor William C.C. Claiborne as the first sheriff of the 2nd Superior Court District in Donaldsonville, Ascension Parish.[2]

According to Bayou Cane longtime resident Filican Alexis Duplantis (1867-1955) in a 1939 *Houma Courier* newspaper article, the community of Williamsburg was founded by two Frenchmen, "one by the name of Leneaux," and a man named Beattie, Joseph Buquet, and Alexander Lirette. The settlement was originally intended to be the parish seat.

In 1822 William Sternin Watkins became president of the Terrebonne Parish Police Jury and director of the school at Bayou Cane. He was subsequently the first U.S. Deputy Surveyor in Terrebonne, and the first senior justice of the peace.[1] Watkins also owned Wright's Bayou Place plantation two miles below Houma.

In the spring of 1829, Williamsburg lands along the right bank of Bayou Terrebonne were divided into lots 74½ feet by 60 feet. The name Williamsburg no long exists in the geography of Terrebonne Parish.

William Sternin Watkins' son, Caleb Baker Watkins, was born in Donaldsonville, and became the first sheriff of Terrebonne Parish in 1822 and postmaster of Williamsburg from 1831 until 1833.[3]

A reporter exploring Terrebonne Parish Courthouse records dating back to 1822 wrote, "A relic of the War of 1812 is the affidavit of Mrs. Caleb B. Watkins requesting bounty land for the service rendered by her husband in the war, according to the act of September, 1850. After serving in the company commanded by Captain Hicks, Watkins had received his honorable discharge in New Orleans, March, 1815."[4] Terrebonne Parish brand registration records show Caleb applied for number 133 dated July 16, 1831.

There is no definitive record of when the name Williamsburg passed into history. The Williamsburg post office was discontinued in 1833.[5]

Above, reporter's account of oldest (1822) Courthouse records and a Watkins relic of the War of 1812, The Houma Courier April 11, 1946

Right, cattle brand of Caleb B. Watkins July 16, 1837

SOURCES
1. Marguerite E. Watkins, "The Watkins Family History" date unknown
2. Valerie Hartnett, written Watkins family details, August 8, 2011
3. U.S. Government records of post office appointments
4. Dorothy Laux, *Houma Courier*, April 11, 1946
5. U.S. Postal Service records

Gilbert Luke Watkins (1863-1924), right, and Joseph W. Watkins (1871-1948)

Home of G. L. Watkins, Main Street c. 1890

Cecile Salome (Lilly) King Watkins (1870-1963)

Judge J. Louis Watkins, Sr. (1889-1978)

John Washington Watkins, overseer Southdown and Ridgefield plantations (1871-1948)

Bayou Cane Elementary School 2015

BAYOU CANE SCHOOL

The Bayou Cane School consisted of two classrooms, two coatrooms, and a small lobby on its completion on April 10, 1928, at a cost of $5,000. Architect for the structure was William T. Nolan, and contractors were C.C. Duplantis, carpenter; Teles Babin, metal work; A.A. Bonvillain, masonry; and Frank Kurtz.

The school in District 3 of the Terrebonne Parish School Board opened for the 1928-29 school session, and is currently used as the Bayou Cane Adult Education center. The current address is 6484 West Main Street, Houma, Louisiana.

SOURCES
1. Terrebonne Parish School Board Minutes Books 1-3

William Thomas Nolan, architect of the school (1877-1969)

BAYOU CANE | 155

GABRIEL MONTEGUT II RESIDENCE

Gabriel Montegut's 40-acre home site, three quarters of a mile above Houma in his day, was not very remarkable compared to the large sugar plantations of the parish. But it is noteworthy for the man who established it after moving to Terrebonne Parish in 1868.

Montegut was born in 1839 into one of the oldest Creole families in New Orleans.[1] His French immigrant paternal great-grandfather Dr. Gabriel Montegut was the first resident surgeon of Charity Hospital in New Orleans, and his paternal great-grandmother Francoise Dupart's family came to Louisiana with Bienville.[2] Gabriel married Lizzie Willis[3] (1842-1896) of New Orleans, their marriage bond dated October 12, 1878.

After serving with the Confederate army, Montegut first worked as a clerk in the mercantile business in New Orleans. In Terrebonne, he established a collection bureau with August Wurzlow and the Montegut Insurance Agency with Jasper K. Wright in 1875.[4]

Governor John McEnery appointed Montegut parish tax collector in 1872,[4] and his career hit its peak during the years 1885-1891 when he served as superintendent of the U.S. Mint in New Orleans. His appointment was by President Grover Cleveland.[5]

From 1891 to 1894 during the second administration of Grover Cleveland, he served as the appointed Chief Deputy Naval Officer in New Orleans.[4] It is then that he received the honorary title of Colonel. At both the parish and state levels, Montegut was active in public affairs most of his adult life.

When The People's Bank in Terrebonne was chartered in 1896, Montegut returned to Terrebonne Parish to become its cashier,[4] tantamount to being chief executive officer. Later in life Montegut indicated his pride in the fact that he had a role in dividing and selling into small tracts eight different plantation estates through his notarial business with Aubin Bourg.[4] The *1897 Directory of the Parish of Terrebonne* stated that "Col. Montegut as agent for land owned by the Citizens Bank of New Orleans, divided Rural Retreat plantation and other places into small tracts, and sold them to industrious white and colored men who are to-day [sic] prosperous farmers."

Some of the plantations he helped to divide and sell into small tracts were Semple plantation in Bourg, Wade Estate above Houma, Smith & Barrow plantation in Little Caillou, Rural Retreat on Bayou Terrebonne, Mechanicsville below Houma, Radical Ridge Farm for African Americans, part of Argyle plantation named Louise, and Halfway plantation.[4]

Opposite page, left to right, Candidate for State Treasurer at nominating convention 1892; Gabriel Montegut (1839-1924); Special Deputy, New Orleans Naval Customs Office July 21, 1893; Gabriel Montegut home; Montegut's inventory of estate 1923

Gabriel Montegut tomb St. Francis Cemetery #1, 2016

Above, People's Bank & Trust Co. check December 16, 1922

Below, Gabriel Montegut Notary Public and Land Agent letterhead, c. 1890

Terrebonne High School on former Gabriel Montegut home site

The same 1897 directory further described Montegut: "He has at all times encouraged and assisted worthy enterprises among the negroes, and has been ever willing to give them counsel and advice by which a better feeling would spring up between the races."

His residence, according to the directory, was on the right descending bank of Bayou Terrebonne "comprising about 35 arpents of land under a high state of cultivation, a portion used as a truck farm in which is raised a large variety of vegetables." He also cultivated flowers, shade, ornamental and fruit trees.

When he died with no forced heirs in 1924, his library was donated to the Terrebonne Parish School system[4] and his estate sold at a Sheriff's Sale in 1924 to Paul Lawrence Adam. Paul and his wife Mabel Nellie Adoue there raised vegetables, chickens, and fruits for use at Houma's Durand Hotel which had been built in 1898, later the City Hotel (1913) owned by Paul and Mabel Adam.[4]

The Terrebonne Parish School Board bought the Montegut/Adam site (275 feet frontage along Main Street just north of Houma) on June 14, 1938. That year construction began on the current Terrebonne High School building and an athletic field which were completed in the summer of 1940 at a cost of $775,637.[6] The school's address is 7138 West Main Street, Houma, Louisiana.

Montegut was so influential in Terrebonne Parish and in the state that the community of Montegut in lower Terrebonne Parish was named for him via the postal service's designation of the earliest post office in the area in honor of him.

Paul Adam and Mabel Adoue Adam c. 1920s

An August 1885 newspaper article attributed to the *New Orleans Times Picayune* about the Montegut Post Office reads: "The people of Lower Terrebonne have long felt the need of better postal facilities, and have made repeated attempts to secure a post office in their locality, but always in vain. Business men have been compelled to send twenty miles to Houma for mail, while other neighborhoods in the parish have had a post office for the asking. After the election of a Democratic Congressman, these staunch 'defenders of the faith' thought to make another effort to secure their due in this respect. Enlisting the good offices of Mr. Gabriel Montegut, whose zeal and energy in the service of Terrebonne's people is never known to flag, Mr. Gay. [Congressman Edward James Gay's (1816-1889) daughter was married to Andrew Price of Schriever.] was appealed to and the post office as [sic] promptly secured. Desiring to testify his appreciation of Mr. Montegut's worth as a man and as a worker in the good cause, Mr. Gay insisted that the new post office should receive the name of Montegut, a fitting exponent of the faith that prevails among the good people of Lower Terrebonne, and a deserved compliment to the gentleman who lends the name. The new post office will be located at Mr. Klingman's with Eugene Fields as postmaster, and a mail service twice a week established."

SOURCES

1. *1897 Directory of The Parish of Terrebonne*
2. Laura Browning, draft, *Gabriel Montegut and How Montegut Got Its Name*, 2016
3. *Terrebonne Life Lines, Volume 16 No. 1,* Spring 1997
4. Helen Wurzlow, *I Dug Up Houma-Terrebonne*, 1984
5. *Biographical and Historical Memoirs of Louisiana,* Goodspeed Publishing Company, Chicago, 1892
6. Terrebonne High School History, http:/ths-tpsd-la.schoolloop.com, Terrebonne High School copyright
7. "Montegut Post Office" article, *New Orleans Times Picayune,* August 3, 1885

Note: Additional information provided by L. Philip Caillouet, Ph.D., FHIMSS

Article on the naming of Montegut Post Office New Orleans Times Picayune, August 3, 1885

Letterhead of The Montegut Insurance Agency Est. October 1875

Paul Adam in front of the City Hotel 200 block of Main Street c. 1930s

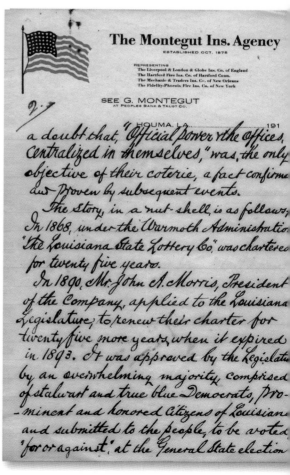

GABRIEL MONTEGUT II RESIDENCE | **159**

WADE CLAIM

Charles Bertaud (Berto) was the first claimant of land that became known as the Wade Claim. His claim was registered with the U.S. government on November 20, 1816 and confirmed on May 11, 1820. Hypolite Hebert was the certified owner of land adjacent to the Bertaud claim on both banks of Bayou Terrebonne in the environs of present-day Wright, Central, and Maple Avenues off Park Avenue and across the bayou on Main Street in Houma. His ownership was certified in 1816.[1]

The more northerly Hebert land abutted the Charles Bertaud claim, certified in 1820. Bertaud land was immediately north of the Haché grant upon which the town of Houma originated.

Planter and landowner in various parts of the parish, Lemuel Tanner, subsequently owned property above Houma which N.C. Wade purchased in 1828. Details of the sale are recorded, "On November 5, 1828 Lemuel Tanner made a conditional sale to N.C. Wade, Natchez, Mississippi, to wit, Lemuel Tanner sells, abandons, and delivers unto N.C. Wade first, a tract of land 12 ½ arpens [sic] front on Bayou Caillou [by which name Bayou Terrebonne was variously called at the time] and joining the land sold by Tanner to Mr. Walker and above land with 40 arpens [sic] depth." Dr. Nathaniel C. Wade bought that tract and another disputed, unguaranteed tract, for $5,800.[2]

Dr. Wade later expanded into the Hebert and Bertaud claims. In 1837 he donated a half arpent front on Bayou Terrebonne for the Barataria Canal to be built.

The Park Avenue home of Dr. Thaddeus Ignace—known as T.I.—St. Martin (1886-1968), who is famous as the author of *Madame Toussaint's Wedding Day*, is located on the Wade Claim. His book was published in 1936 by Little, Brown, and Company, and became a bestseller. The current address of the house he and his wife, nee Gladys Davidson (1889-1979), lived in is 7731 Park Avenue, Houma, Louisiana.

The home of Albert Deutsche O'Neal Sr. (1906-1989) and his wife, nee Mary Sanders (1911-1965) is also a landmark of the Wade Claim area. The house's current address is 7605 West Park Avenue, Houma, Louisiana.

SOURCES
1. A Century of Lawmaking for a New Nation: U.S. Congressional Documents and Debates, 1774-1875, American State Papers, Senate
2. Terrebonne Life Lines, Volume 18 Number 1

Opposite page top, Dr. T.I. St. Martin home, 2016; bottom, Charles Berto Claim c. 1823

A. Deutsche O'Neal, Sr. home 2016

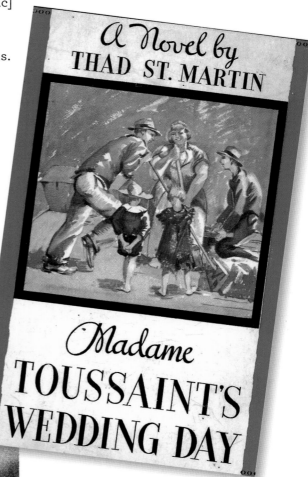

Madam Toussaint's Wedding Day published July 1936

Author Dr. T. I. St. Martin

Cajun Bride of Oak Alley,
1974, Oil on canvas, 24 x 32 inches,
George G. Rodrigue

The paddlewheeler Phyllis *with barges in the Barataria Canal c. 1910*

BARATARIA CANAL

Barataria Canal was dug in the 1830s to connect Bayou Terrebonne to Bayou Black as part of an inland waterway system intended to connect New Orleans to the Atchafalaya River and the Attakapas District of Bayou Teche, in an effort to facilitate westward expansion in Louisiana.[1] In the absence of good roadways in antebellum southern Louisiana, water travel was the more convenient way to transport people and goods at the time.

Shippers of goods needed water access connecting Jefferson, St. Charles, Lafourche, and Terrebonne parishes. The founders of the Barataria and Lafourche Canal Company (Robert R. Barrow, Sr., Dr. Walter Brashear, and Judge Charles Derbigny) planned the B & L Canal Company to "run from the Mississippi River across Lafourche and Terrebonne parishes to Bayou Teche."[2] The Company Canal in Bourg was also a link in the system of which Barataria Canal in Houma was a part.

The Barataria Canal formed the western boundary of Houma. Near that point was the locus of express lines that provided overland coach transportation between Houma and Thibodaux to the Schriever railway depot, the closest railroad access to the heart of Terrebonne Parish. The canal was 17 miles below the origin of Bayou Terrebonne.

"RAILROAD AVENUE," HOUMA, LA.

Houma Branch Depot, Railroad Avenue c. 1920; below, The Terrebonne docked in New Orleans, Louisiana photo by Charles L. Franck c. 1930

BARATARIA CANAL | 165

The Ohio on Barataria Canal c. 1910

166 | HARD SCRABBLE TO HALLELUJAH

The Houma *on Barataria Canal 1910*

BARATARIA CANAL | 167

Top row, left to right, view of houses on Railroad Avenue from the canal c. 1920; train depot c. 1920; Main Street Bridge c.1914; The Phyllis in the Barataria Canal passing through Main Street Bridge going to Bayou Terrebonne c. 1914

One of the Barataria Canal's original functions was to provide a place to turn around at its juncture with Bayou Terrebonne for steamboats and large boats ready to make their return trips "down the bayou" toward the south.

All parts of the B&L canal system were completed by 1847.[2] Robert Ruffin Barrow, Jr. assumed ownership of the B&L Canal Company No. 2 in 1859.[2]

A federally-controlled intracoastal waterway across the Louisiana coast became an interest of the U.S. Government by the late 1880s.[2] It was a point of discussion at a national convention in New Orleans in 1910. Robert Jr. began negotiations with the U.S. Corps of Engineers for the sale of the B&L No. 2 in 1921,[1] which resulted in the federal government's purchase of the company on July 3, 1925 for $84,000.[2]

The demise of the Barataria Canal began with the digging of the Houma Canal, which connected the southeastern-most point of Bayou Black to Bayou Terrebonne. A part of the Houma Canal formed the eastern edge of the City of Houma, looping around the "Back of Town" area to meet Bayou Terrebonne. When the section of the Gulf Intracoastal Waterway from Bayou Black to

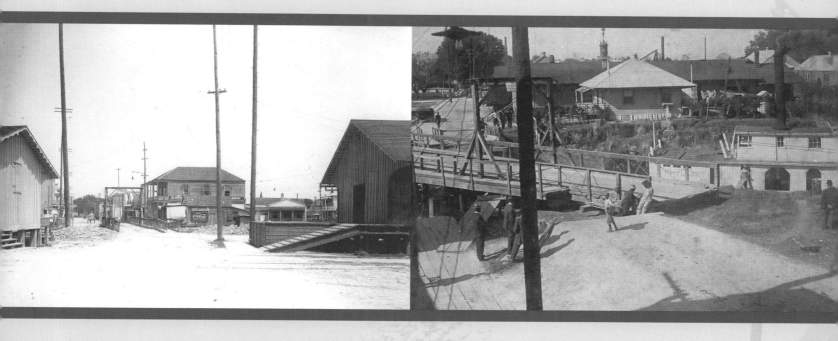

Bayou Terrebonne was dug, it "captured" and widened the Houma Canal where it was configured east-west. The GIWW stretch was completed on August 30, 1924, making the Barataria Canal obsolete. It was filled in during the 1930s and is now the narrow strip of land between Barataria and Canal Streets at the western edge of the business district of Houma.[3]

In March 1932 the U.S. Dredge *Delatour* filled the Barataria Canal with spoil from Bayou Terrebonne, as well as dirt and debris from the city garbage dump, creating Canal and Barataria streets and the land between them. Robert Ruffin Barrow, Jr. sold the right of way in 1925 to C.A. Ledet, Lee P. Lottinger, and Allen J. Ellender, Sr. (then the city attorney), who developed the land along the two streets and sold lots after the canal was filled in.[4]

Banana boat on Barataria Canal c. 1910

SOURCES

1. Thomas A. Becnel, *The Barrow Family and the Barataria and Lafourche Canal: The Transportation Revolution in Louisiana, 1829-1925*, Louisiana State University Press, Baton Rouge and London, 1989
2. "Historic Habitat Changes" by the Lafourche Parish Game and Fish Commission
3. *Eyes of an Eagle: Jean-Pierre Cenac, Patriarch (An Illustrated History of Early Houma-Terrebonne)* by Christopher Everette Cenac, Sr., M.D., F.A.C.S., with Claire Domangue Joller copyright 2011
4. Helen Wurzlow, *I Dug Up Houma Terrebonne*, 1984

Opposite, top to bottom, St. Louis Cypress Company Ltd. dredge; St. Louis Cypress Company Ltd. sawmill; Oscar C. Sundbery with a wide cypress board c. 1920; Houma Cypress Co., Ltd. was first St. Louis Cypress Co., Ltd. c. 1920

ST. LOUIS CYPRESS COMPANY, LTD.
HOUMA CYPRESS CO., LTD.

Frederick Wilbert of Plaquemine in Iberville Parish began to deal in Terrebonne Parish by buying cypress timber in Terrebonne just after the turn of the 20th century. He was president of the St. Louis Cypress Company, Ltd. on April 1, 1903 when it bought all the cypress trees and timber from the Ellenders' Hope Farm near Bourg for $35,000. In July 1903 the company bought C.P. Smith & Company's mill and timber holdings. St. Louis Cypress Company, Ltd. then built a sawmill in 1905 on Main Street in what was north of downtown Houma at the time.

The company ran their large sawmill on West Main Street until the year 1913, when the mill changed owners and the name became Houma Cypress Co., Ltd. Oscar C. Sundbery and Herman Albert Cook were the chief stockholders in the new company. One source of their cypress was the swamp at Klondyke.

Their large mill which could produce 65,000 feet a day closed in 1926. In that year the company built a smaller mill on the bayouside to saw the cut-over timber, with a capacity of 15,000 feet a day. The smaller mill closed in 1931, a planing mill closed in 1946, and the lumber yard closed in 1972. The smaller mill and lumber yard had operated as the Houma-Terrebonne Lumber Company, and the lumber yard subsequently did business as Home Building Supply Company.

The Houma Cypress Co., Ltd. mill's location was at the current location of Sundbery Shopping Center at 7482 Main Street in Houma.

SOURCES
1. Helen Wurzlow, *I Dug Up Houma Terrebonne*, Vol. 5
2. Personal communication Sidney C. Sundbery 2016

St. Louis Cypress Company Ltd. c. 1920

Sundbery House, home of Houma Cypress Company owner, burned fall of 1990, photo 1980

170 | HARD SCRABBLE TO HALLELUJAH

William Jean-Pierre Cenac, Sr.'s cypress shutters made by Houma Cypress Co. Ltd.

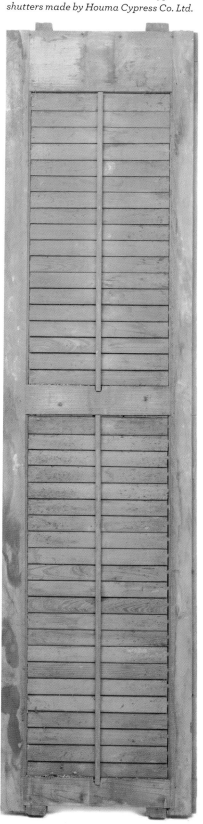

HOUMA CYPRESS COMPANY | **171**

IN ACCOUNT WITH

HOUMA CYPRESS CO., Ltd.

MANUFACTURERS OF
Louisiana Red Cypress Lumber
Shingles and Lath
HOUMA, LA.

★ MR G MONTEGUT
HOUMA LA

My dear friends,

I have one share (unpledged) of Peoples Bank Stock. Take it at $130.00; deduct above am't & send me check for the balance, $67.00. Don't allow the value of the Stock to be demoralized, that is a crime. The Stock is cheap at $130.00.

Mortgage creditors may not be able to settle their indebtedness in full, but all accrued interest must be paid, and the indebtedness put in shape. That is business, and that is banking. The second freeze killed the stubble cane, and it has been raining nearly three months, — Result, poor prospect for crop in this section, for all of which, planters are not responsible. They have done their best, and must be assisted. That is what Banks are created for.

Success to Houma Cypress Co, with kind regards for Mr Cook & Mr Sunberry.

Aug. 24/23.

G.P. Montegut

172 | HARD SCRABBLE TO HALLELUJAH

HOUMA BRICK AND BOX COMPANY, INC.

Chester A. Morrison and Fernand "Mr. Pete" Robichaux founded Houma Brick and Box Company, Inc., in 1923. Prior to that time, the two men owned a sawmill in Charenton, St. Mary Parish, where they resided. The mill was destroyed by a flood in 1914.

The year before opening Houma Brick and Box, on February 22, 1922, they bought all the cypress timber, "both standing and cut down, on a 70.8 acre tract of land 17 miles below Houma," from Red Cypress Lumber and Shingle Company, Ltd. For a time Morrison and Robichaux operated a small sawmill at what is now Morrison Avenue, on St. Louis Canal. In 1923 the company bought two tracts of land from Senator Allen J. Ellender on November 12, and Morrison became president of the company.[1] Houma Brick Company, Ltd. was located on lot Number 1.

In 1927 the company bought from Lee P. Lottinger, Allen J. Ellender, and Charles A. Ledet "a certain tract of ground in the City of Houma, being a portion of the Barataria Canal, including both banks thereof, formerly belonging to R.R. Barrow."[1]

Houma Brick and Box Company, Inc. moved to the site of present-day Morrison Terrebonne Lumber Center in 1927.[1]

The building they constructed straddled the canal, appearing much like a covered bridge with the canal beneath it.[1] Despite its name, the company did not make bricks, and made relatively few boxes. Senator Allen J. Ellender, one of the stockholders in the original company, did use some boxes to ship his Irish potatoes and other produce to market.[1]

In 1973, the son of the original owner, and president of the company, Chester F. Morrison, changed the name of the company to Morrison Home Center. In 1998, the business became Morrison Terrebonne Lumber Center after a merger with Terrebonne Lumber & Supply Company.[2]

Ownership changed again in 2002, when owners Chester F. Morrison, Robert J. "Bobby" Vice, Burnelle P. Landry, and Ray C. Voisin sold the company to the son-in-law of one of the four owners, and to one son of each of the other three owners: Douglas A. Gregory, David G. Vice, Gregory J. Landry, and Keith M. Voisin.[2]

The business site is located on the side of and over the filled-in Barataria Canal at 605 Barataria Street in Houma.

SOURCES
1. Helen Wurzlow, *I Dug Up Houma Terrebonne, Vol. 5*
2. Chester F. Morrison in a 2016 interview

Opposite page, Houma Cypress Co. Ltd. correspondence; Houma Cypress Company token good for one dollar; Houma Cypress Co. Ltd. aerial; note that Main Street is elevated to allow logs to pass to the mill

Above, top from right, clockwise, Chester A. Morrison (1880-1943), Fernand (Mr. Pete) Robichaux (1881-1986), J. Farquhard Chauvin (1878-1959) Peter H. Hebert, Sr. (1926-1976)

Houma Brick & Box Co. Ltd. spanned the Barataria Canal like a covered bridge c. 1930s

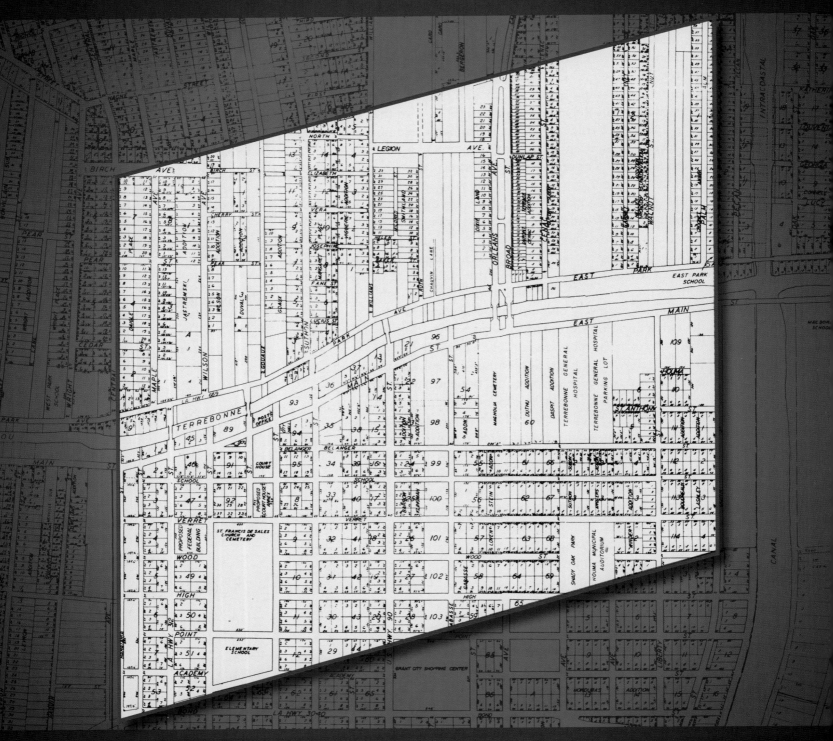

Map of Joseph Haché Grant, City of Houma 1960

JOSEPH HACHÉ GRANT: CITY OF HOUMA

The parish seat and largest town of Terrebonne Parish, Houma, was founded in 1834 on land known as the Haché Grant. When Louisiana became a state, Joseph Haché applied for a claim of land that he owned prior to that time. His claim was confirmed in 1820[1]; it consisted of land on both sides of Bayou Terrebonne centered at present-day Houma. The Haché grant consisted of 338.72 acres.

Haché sold the property on March 2, 1828 to Henry Schuyler Thibodaux's widow, Brigitte Belanger. Five years later, she traded a part of it to François Belanger. The tract straddled Bayou Terrebonne and stretched eight arpents along the bayou. François sold part of that tract (four arpents front by ten arpents deep on the right bank of the bayou) to Hubert M. Belanger and Richard H. Grinage in 1834.[2]

Property owners Belanger and Grinage donated, on May 10, 1834, one arpent front by ten arpents deep to the parish to relocate the parish seat from Williamsburg in the Bayou Cane area north of the Belanger/Grinage tract. The rest of their property was divided into lots for sale.[2]

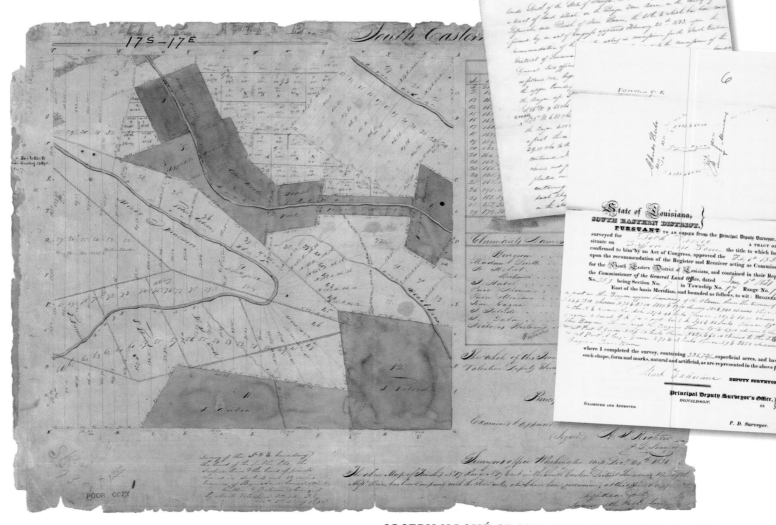

Below right, Joseph Haché Land Grant recorded February 6, 1823

Below left, State Land Grant Map December 24, 1831

Map City of Houma 1940

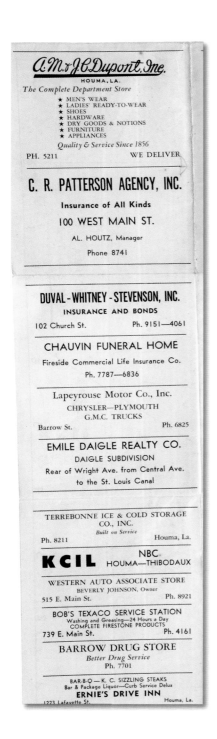

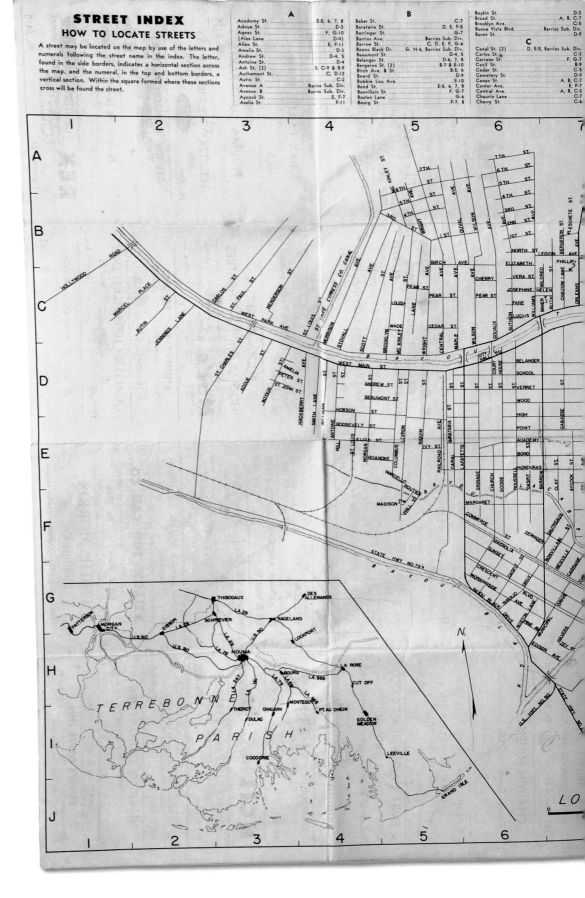

176 | HARD SCRABBLE TO HALLELUJAH

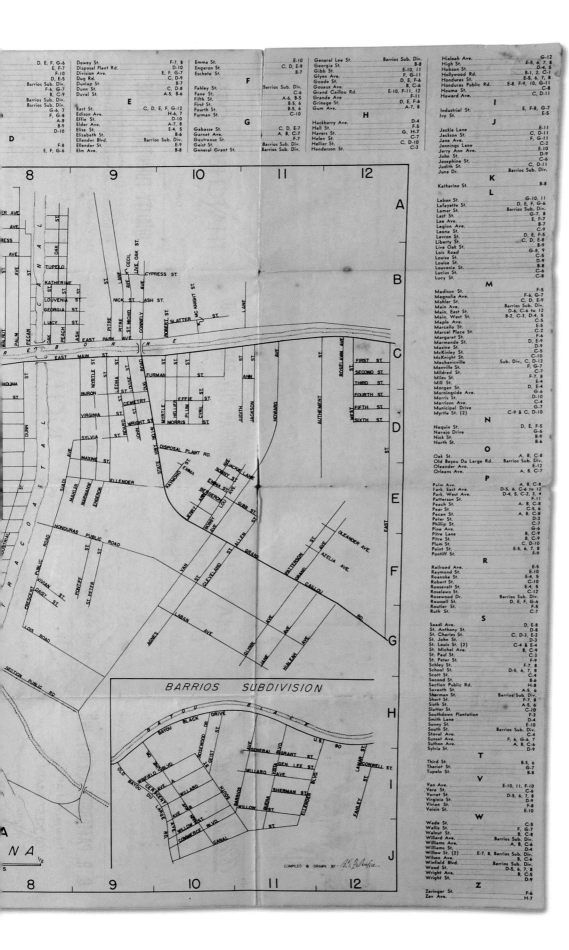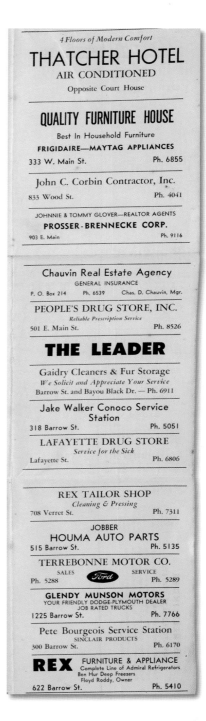

Bernard Bazet Davis aerial photo of Houma 1945

Article in New Orleans Times Picayune July 23, 1929

Houma Academy, later St. Francis de Sales School late 1800s

Article on Houma in New Orleans Times Picayune *July 23, 1929*

Plaques in Terrebonne Parish Courthouse square about Cenac oaks

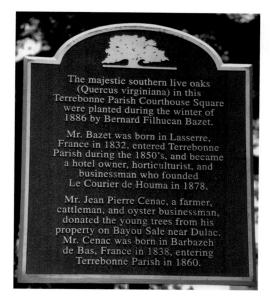

The donated property is the site of the present courthouse and other public buildings. The land was valued at $150 at the time of the donation.[3] Grinage bought out Belanger and sold the first lot to Francis W. Jones, who established the first inn of the parish at the southeast corner of Main and Goode streets. J.B. Grinage drew the first map of the town in 1847. Houma was incorporated on March 16, 1848.[2]

The *1897 Directory of the Parish of Terrebonne* gives a different version of the first structures in what is now Houma. E.C. Wurzlow wrote, "The first building erected in Houma is said to have been built at the corner of Court and Church streets; two or three others were constructed nearly at the same time."[3]

Highlights of Houma's history include the completion of the Houma Branch Railroad in 1872, connecting Houma to New Orleans by rail. An "elegant" Opera House was erected in 1896 by Houma Fire Co. No. 1 (organized in 1872), at a cost of $8,000.[3]

By 1897, hotels, an ice factory, a saw and shingle mill, and a steamboat and transportation line between Houma and Lower Terrebonne followed. Oyster shipping became an important industry, shipping to all parts of the United States west of the Mississippi River; 25 million oysters were shipped from Houma in 1896. The 1897 directory counted among the town's assets five churches; four public schools, two schools for people of color, and two private schools; two banks; two neswpapers, the *Houma Courier* and the *Terrebonne Times*; music and other opportunities. E.C. Wurzlow in 1897 gave the population of Houma and its suburbs as about 3,000.[3]

The Courthouse Square, part of the long-ago Belanger-Grinage donation, still dominates the city center. The square is still shaded by 16 of the original 24 oak trees from Jean-Pierre Cenac's Bayou Salé property, brought to Houma by Cenac and his sons, and donated to the parish. The trees were transplanted during the winter of 1886-87 by "Lafayette" Bernard Filhucan de Bazet. But the City of Houma has grown far beyond its early boundaries. The city's population from the 2010 U.S. Census was 33,727.

SOURCES
1. A Century of Lawmaking for a New Nation: U.S. Congressional Documents and Debates, 1774-1875, American State Papers, Senate
2. *The History of Terrebonne Parish, Louisiana, Part 4* by Tim Hebert, copyrighted 2001 (information here either quoted or paraphrased from Hebert's work)
3. *1897 Directory of the Parish of Terrebonne* by E.C. Wurzlow

Back of Town, Houma City map 1960

HOUMA ADDITIONS

- Newtown 01/25/1870
- Celestin Addition to Newtown 01/05/1899
- Deweyville-05/19/1899
- Oscar Daspit Addition to Newtown 08/01/1899
- Honduras Addition-04/1923
- Crescent Park Addition-04/03/1924
- Breaux-Morrison Addition-08/19/1932
- Daspit-Breaux Addition-05/27/1937
- Clark Place Addition-02/04/1938
- Houma Colored School
- The Alley
- Silver City
- Good Templars Hall
- Odessa's Place

HOUMA ADDITIONS

The *1897 Directory of the Parish of Terrebonne* by E.C. Wurzlow, among other information about Terrebonne Parish, gives a short history of the town of Houma. Wurzlow included information from the U.S. Census of 1890, when Houma's population was 1,220, and "Newtown in the rear of Houma had a population of 456." Census figures gave the population of Houma and its suburbs as about 3,000.

"Back of Town" describes the area immediately south of the Haché Grant, consisting of Newtown and Deweyville. Newtown was described as "in rear of Houma" on the January 25, 1870 map from a survey by A. Jolet. Today's parameters of Newtown extend from its western boundary at Canal Street and its eastern edge at Aycock Street. The northern border of Newtown was the slanting "Back Line of Corporation of Houma, the Joseph Haché Grant." The southern boundary of Newtown was roughly the "covered bayou" which was Bayou LaCarpe.

Newtown was the subject of a Sheriff's Sale by A.W. Connely to be held on July 24, 1897. The public notice of the impending sale followed the court suit brought by Jack Bisland against Noah J. Daspit.[1]

The property to be sold was, first, "A certain lot of ground situated in 'Newtown' in the rear of Houma, La., measuring 60 feet front on Clay Street, with a depth of 125 feet; with the buildings and improvements thereon; being Lot 4, in Block 82."[1]

Second was "A certain lot or parcel of ground in 'Newtown' in the rear of Houma, Parish of Terrebonne, La., and measuring 7.26 chains front on Clay Street with a depth of 125 feet designated on plan 'B' of 'Newtown,' executed by C. L. Powell, Surveyor, as the East half of Block 82; Lot 5, East half included; bounded North by property first above described, East by Clay Street, South by Bayou LaCarpe and West by property of Louis Arcement; together with all the buildings and improvements thereon."[1]

Another expansion of Houma on the right descending bank of Bayou Terrebonne was Deweyville, its expanse surveyed in a map by C.E. Smedes on May 19, 1899. It was the property of Isaac Daspit. Deweyville is shown on the accompanying map as having its southern boundary on Dupont Street, forming an irregulary-shaped tract with Harmon Park on its west, Barringer Street on its east, and the "covered bayou" as its northern extreme.

The same year, 1899, Oscar Daspit made an "addition to Newtown" with a map of the area filed on August 1, prepared by surveyor C.E. Smedes. This rectangular tract's northern boundary is Willow Street and its southern boundary is Miles Street. On the west it is conscripted by Aycock Street, and on the east, Lee Avenue.

A notice in the March 4, 1903 edition of the *Houma Courier* advertises 100 town lots which "must be sold at once" in Newtown and Deweyville, belonging to Oscar Daspit. Applications were to be made to Isaac Daspit or Harry L. Wilson.

The area known as Back of Town has Aycock Street as its limits on the west and on the east at Dunn Street. The Haché

Jack Bisland vs. Noah J. Daspit Newtown court case article The Houma Courier *June 26, 1897 and ad for Deweyville town lots sale* The Houma Courier *March 4, 190*

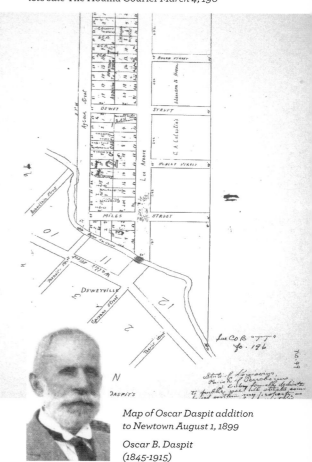

Map of Oscar Daspit addition to Newtown August 1, 1899

Oscar B. Daspit (1845-1915)

Map of Newtown January 25, 1870

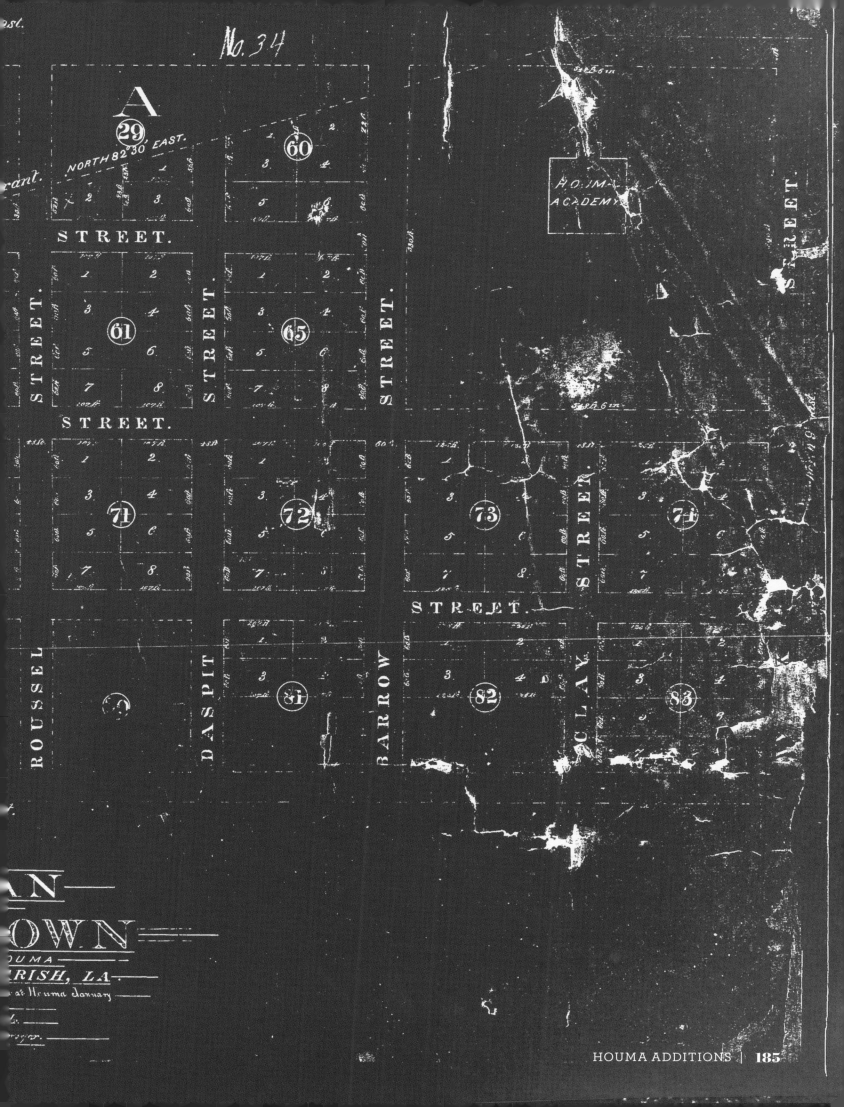

HOUMA ADDITIONS

A celebration at The Alley, located on Church Street opposite the Houma Colored School; Mrs. Flossie Myles front row sixth from left c. 1950s

Article on war stamps sale by the Houma Colored High School The Houma Courier *October 21, 1943*

Grant's slanting southern line is the area's northern boundary, and the "covered bayou" is its southern edge.

The Celestin Addition map dated January 5, 1899 has Lee Avenue as the western boundary, Division Avenue as its eastern edge, and Academy Street as its northernmost border. Bayou LaCarpe is its boundary on the south.

The Honduras Addition, City of Houma, was described in a map dated April, 1923. The addition was named for Honduras plantation lands from which the addition was carved; the plantation had extended to the "Back Line of Corporation of Houma, the Joseph Haché Grant" line. The subdivision's western demarcation is Division Avenue, and its eastern boundary is the Intracoastal Canal. On the north, the Haché Grant line conscripts Honduras Addition, as does Bayou LaCarpe on the south. Bayou LaCarpe was not yet filled in, or "covered."

The Alley is the local African American community's name for a relatively small section of town bounded on the north by the Haché line and on the east by Church Street. The Alley consists of ten lots facing east. The houses in that block facing Grinage Street are part of Newtown.

The Crescent Park Addition to the City of Houma dates back to an April 3, 1924 map laying out the subdivision, by J.C. Waties. It extended from "Barrow's Canal" (Barataria Canal) on the west to the Intracoastal Canal on the east, and from Magnolia Avenue on the north to Bayou Black on the south.

A map dated August 19, 1932 by T. Baker Smith C.E. delineates the Breaux-Morrison Addition with Bayou LaCarpe (not as yet filled in) as its northern boundary and forming a point along the Ashland Extension railroad tracks below Commerce Street. Lafayette Street is its western boundary, and a slanting line abutting the Daspit-Breaux Addition is its eastern extreme.

The Daspit-Breaux Addition is adjacent to the Breaux-Morrison Addition to the east along a slanting line from the Ashland Extension railroad on the south to Bayou LaCarpe on the north, with Barrow Street and Harmon Park's conventional fence line as the eastern boundaries. The subdivision was formed according to a May 27, 1937 map by T. Baker Smith C.E.

The Clark Place subdivision dates back to a February 4, 1938 map by T. Baker Smith C.E. The nine lots are bounded by Lafayette Street on the east and the old Barataria Canal on the west and Bayou LaCarpe on the north and the Ashland Branch Railroad on the south.

Although there seems to be no official status for the name, Silver City is a more recent designation for a neighborhood on the eastern side of Barrow Street, occupying several blocks along Aycock and Clay streets. It was so named for the shiny silver tin roofs the owners of the rental properties built onto the houses. Owners of those properties were the grandchildren of Bannon Bonvillain of Boykin plantation and children of his daughter Louise Bonvillain Breaux.

SOURCES
1. Public notice in *The Houma Courier*, June 26, 1897

Clifford Percival Smith residence 1930

BURGUIÈRES, SMITH, ST. MARTIN HOUSES

One Houma home not associated with present-day sugar planters deserves mention for its landmark status on the left descending bank of Bayou Terrebonne where Barrow Street's northern end intersects with Park Avenue. The land is also notable for its history with an early French family with multiple familial connections in Terrebonne Parish.

The Smith house at 7947 West Park was completed in 1905 by Clifford Percival Smith, a pioneer lumberman in Terrebonne Parish. It is a large Queen Anne residence which also "displays significant elements of the Colonial Revival style."[1]

The commanding presence of the white house can be seen, unimpeded, for blocks away on Barrow Street, which dead-ends on the northern side of the bridge situated perpendicular to Park Avenue. The house is now owned by the Kenneth W. Smith family; Kenneth is the great-great grandson of the builder Clifford P. Smith and his wife Claire Marie Pauline Dupont Smith, familiarly known as "Aunt Clara."

Clifford Percival Smith (1858-1943)

BURGUIÈRES, SMITH, ST. MARTIN HOUSES | 187

C.P. Smith Insurance Agency letterhead, invoice of and newspaper ad for C.P. Smith & Co. lumber company

Clifford P. Smith bought a piece of land and home in Houma on April 4, 1894 from Jules Martial Burguières, who by that time had moved to St. Mary Parish and was building the Burguières sugar plantation empire. Smith's purchase price was a total of $3,500, $1,000 paid in cash and two notes of $1,250. The land was situated on Bayou Terrebonne, "facing the town of Houma" according to the volume *A Southern Neo-Colonial Home, Houma, Louisiana and The partial story of its owners: Mr. and Mrs. C.P. Smith and C. Mildred Smith*.[2]

The book goes on to say that "With other tracts added he had a convenient center both for home and his business," a lumber business and sawmill situated on the left descending bank of Bayou Terrebonne. The Smith book locates the C.P. Smith Company operations "on land in the back of Clifford Smith's home at the end of what is now Ruth Street," and by mentioning that "Children crawfishing where the American Legion Park is today could see the logging and bustle."[2]

C.P. Smith registered cattle brands at two different times in Terrebonne Parish records, Number 656 on May 7, 1906 and Number 846 on April 9, 1919.

The Smiths developed a subdivision called Smithland behind their Park Avenue home, from 1905 to 1907, encompassing present-day Ruth Street, Baker Street, Helen Street, and Mildred streets, named for the couple's four living children.[2]

Clifford Percival Smith, 1930

Cattle brands of C.P. Smith, May 7, 1906 and April 9, 1919

Left, Lois Ruth Smith (1903-1983), right, Clara Mildred Smith (1899-1986)

Below, Smithland map December 9, 1905

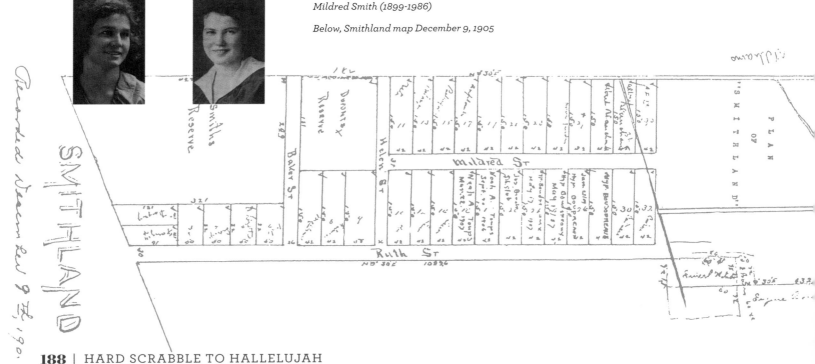

Clifford P. Smith house 2016 (home of the Kenneth W. Smith family)

An important part of the Smith property history is that of the Burguières family, who were among the early settlers of Terrebonne Parish. Eugène Denis Burguières, from Montauban, Midi-Pyrénées, France, migrated to Terrebonne in 1831. He married widow Marie Marianne Verret Delaporte, who was from a prestigious French Louisiana family, in 1837, and they made their home where the Smith home now stands.[3] Marianne's great-grandfather, Jacques Cantrelle (d. 1778), was a "member of the Louisiana Legislative Council that governed the territory."[4] Her grandfather, Nicolas Verret (d. 1775), "served as lieutenant governor of the Louisiana Territory as the first Spanish commandant and as a judge on the Acadian Coast in St. James Parish. Her father, Jacques Verret (1752-1834)... owned vast real estate holdings, including an eight-thousand-acre indigo plantation" as well as being a pacesetting engineer.[4] Among Marianne and Eugène's holdings in Terrebonne was Flora plantation on Big Bayou Black.

Marie Marianne Verret Delaporte Burguières (1812-1883), wife of Eugène Denis Burguières

BURGUIÈRES, SMITH, ST. MARTIN HOUSES | 189

Helen Snow Smith
(1885-1983)

T. Baker Smith, Sr.
(1869-1962)

Clara Marie Pauline Dupont Smith
(1867-1944)

Above, Burguières crest; below, letter of J.M. Burguières to Miss Alice Broussard February 24, 1896

The Burguières house was divided into three parts when the Smiths planned to build their house there. According to C.P.'s daughter, Helen Snow Smith, "Part of the structure was moved across the street to the bayou side to serve as an office for her father. Another part became what is today a seafood mart on the bayou side of East Main."[3] Clifford P. moved the front part of the house east, adjacent to the new home site on West Park Avenue, and his family lived there until the current home was built.

Official possession of the former Burguières property was transferred to the Smiths on January 1, 1895. The house was under construction for 10 years, completed in 1905. C.P.'s grandson William Clifford Smith recalled that his mother (Mrs. T. Baker Smith) told him the house was designed after C.P. and Clara Smith visited the Chicago World's Fair in 1893, when Clara saw architectural details she wanted incorporated in the Smiths' new house. The house was elevated, because the last Mississippi River flood in 1890 swelled Bayou Lafourche and Bayou Terrebonne above their banks. From the land's 11 feet above sea level, the Smith house was built high so that it sat at 16 feet above sea level, five feet above the flood stage. It was the 1890 flood that prompted the government to dam Bayou Lafourche at Donaldsonville, to prevent further such floods on the bayous. The dam took from 1903 to 1906 to complete.

St. Martin house Park Avenue, Houma 2016

*Dr. Hugh P. St. Martin, Sr.
Charity Hospital Staff 1907-1908
(1884-1971)*

Smith first rented the older house aside his new home to Ernest and Yvette St. Martin Dupont, where their son Jules (later Dr. Jules S. Dupont, Sr.) was born. When the Dupont family moved to a larger house across the bayou in Houma, Clifford P. sold the older home at 7949 West Park to Dr. Hugh P. St. Martin, Sr. (1884-1971), who lived there until his death in 1971. Dr. Eugene C. St. Martin (b. 1920) was born in the house his family owned for more than 100 years. Dr. H.P.'s daughter, longtime high school mathematics teacher Rhea Marie St. Martin (1912- 2002), resided in the house until her death in 2002. The St. Martin estate sold the home to Kenneth W. Smith, C.P.'s great-grandson, in 2014.

Rhea Marie St. Martin (1912-2002)

Left to right, Dr. Eugene C. St. Martin (b. 1920), Dr. Hugh P. St. Martin, Sr. (1884-1971), Hugh P. St. Martin, Jr. (1916- 1986), Dr. Roy J. St. Martin (1913-1975) c. 1960

BURGUIÈRES, SMITH, ST. MARTIN HOUSES | **191**

C. P. Smith house and C.P. Smith Cypress Co. office across Park Avenue c. 1930

Ernest Denis Burguières (1838-1878) c. 1878 and his wife Aglaé Bonvillain Burguières (1836-1904) c. 1891

Cattle brand of Jules Martial Burguières July 2, 1868

Jules Martial Burguières (1850-1899) c.1897 and his wife Marie Corinne Patout Burguières (1852-1890) c.1872

Eugène Burguières' son, Ernest Denis Burguières, moved to St. Mary Parish in 1868 where he developed four plantations—Cypremort, Ivanhoe, Richland, and Crawford, covering 10,000 acres, 7,000 of which were in sugar cane. Ernest Denis and Onezime Theophile Aycock also owned Forest Grove in Chacahoula, Terrebonne Parish; he died at age 40 when he was thrown from a horse at Alice B. plantation in St. Mary Parish. His wife was the former Aglaé Bonvillain of Terrebonne Parish, whose family were large landowners.[3] Another son of Eugène and Marie Marianne, Jules Martial, married Marie Corinne Patout from "another prominent dynasty, the Patouts."[4] of Iberia Parish. In 1872 Jules was owner of Pointe Farm plantation on Bayou Terrebonne in lower Terrebonne Parish. His cattle brand was registered as number 526 in Terrebonne Parish records on July 2, 1868.

Some of the Terrebonne families connected to the land-rich Burguières individuals, in "the tradition of marrying well (and marrying French)" were the Vigueries, the Bonvillains, and the Duponts.[4] Eugène's stepdaughter Marie Elvire Delaporte married Jean Pierre Viguerie, who with his brother François "became business tycoons, acquiring thousands of acres in St. Mary, Terrebonne, Vermilion, and Lafourche"[4] parishes. A further connection of the two families came later when François (Frank) Camille Viguerie, (son of Jean Pierre Viguerie, and a merchant at the Waterproof Store on Big Bayou Black), married Ernestine Louisiana Burguières of Orange Grove plantation on Big Bayou Black on January 13, 1885.

"Both a son and a daughter of Eugène and Marianne married into the Bonvillain family, another French pioneering clan whose ancestors had been sugar planters in Terrebonne Parish long before Eugene arrived."[4] They owned Argyle, Boykin, and Crescent plantations on Big Bayou Black, as well as Mulberry and Ridgeland on Bayou Dularge. Besides Aglaé Bonvillain marrying Ernest Denis, his sister Pauline Camilla Burguières married Alphonse Bonvillain, associated with Laurel Farm (Laura) plantation on Big Bayou Black in Terrebonne Parish.

Aglaé and Ernest Denis' daughter Alice married J. Cyrille Dupont, "who became one of Terrebonne's great political leaders" who "served in the Louisiana House of Representatives, as Houma's mayor, and as an assistant in the writing of the state constitution."[4]

The Smiths became connected to the Burguières family via the Duponts when Clifford Percival Smith married Claire Pauline Dupont, daughter of French natives Jean Marie Dupont and Lydie Marie Tasset, of Magenta plantation on lower Bayou Terrebonne, in 1878.

Donna McGee Onebane, the author of a book on the Burguières family, wrote, "Not until the end of the Civil War did one of Eugène and Marianne's children, Marguerite, marry an Acadian, Augustave Theriot (Terraiu) (1839-1923)." He served with the Lafourche Creoles as a Confederate soldier and fought in the Battle of Shiloh with the 18th Louisiana Infantry Regiment.[4]

Jules Martial Burguières, after the death of his first wife, Marie Corinne Patout, married Ida Laperle Broussard (1868-1948) of St. Martin Parish, who was descended from the legendary Joseph Broussard *dit* Beausoleil who had fought the British to remain on their Nova Scotia homeland before expulsion of the Acadians. That family was among the first Acadians to settle in Louisiana.[4] Besides his holdings in St. Mary Parish, Jules Martial once owned Oak Wood plantation in Gibson.

Even today, the J.M. Burguières Co., Ltd. is a significant private company in the state of Louisiana. The year 2017 will mark the 140th anniversary of the enterprise. The slogan for a celebration on March 17, 2017 will be "Taking Care of Business and Family for 140 Years."

SOURCES
1. Historic Standing Structures Survey, Terrebonne Parish, by Division of Historic Preservation
2. National Register staff, 1980
3. *A Southern Neo-Colonial Home, Houma, Louisiana, and The partial story of its owners: Mr. and Mrs. C.P. Smith and C. Mildred Smith* by C. Mildred Smith, M.A. and G. Portre-Bobinski, Ph.D., copyright 1970 by G. Portre-Bobinski
4. Helen Wurzlow, *I Dug Up Houma Terrebonne*, 1984
5. Donna McGee Onebane, *The House that Sugarcane Built: The Louisiana Burguières*, University Press of Mississippi, Jackson, 2014

Joseph Cyrille (J.C.)Dupont (1863-1929)

Senator Alcide J. Bonvillain (1873-1937)

Hon. Arthur H. Bonvillain (1862-1944)

Felix A. Bonvillain 1865-1946)

Martial J. Bonvillain (1868-1918)

Bottom,1908 Houma officials, seated, from left, Chief of Police F.X. Zeringer, Mayor Calvin Wurzlow, Councilman C.P. Smith, Town Clerk Drew Angers; standing, from left, Police Official Winslow Hatch, Policeman Felix Webber, Councilman Celestin Cunningham, Town Treasurer Alphonse Dupont; Policeman Lewis Chauvin, Councilman Arthur J. Bethancourt, and Councilman J. Harry Hellier

Cora's Restaurant, 1975,
Oil on canvas, 36 x 48 inches,
George G. Rodrigue

Right, Main Street arched entrance to Magnolia Cemetery 2016

Notice for annual meeting of the Auxiliary, Magnolia Cemetery Association The Houma Courier *October 13, 1917*

A. W. Connely (1847-1915) in his 5th term as sheriff

Peter Berger (1837-1907)

Dr. Philip L. Cenac, Sr. (1918-1990)

MAGNOLIA CEMETERY
TERREBONNE MEMORIAL PARK

Magnolia Cemetery on Main Street in Houma was formed in the late 1800s primarily for the purpose of burial of non-Catholics. "Under the Magnolias, a History of Magnolia Cemetery" by Agnes Daspit Kennedy explains the historical circumstances that preceded its formation.

She wrote, "Settled by mostly French Catholics, the town of Houma boasted a Catholic Church with a cemetery at its rear. None but Catholics could be buried here. Most other burial places were on the individual sugar plantations with a small plot of ground set aside for burials."

According to Mrs. Kennedy, as transportation improved, people of many denominations moved into the area, and "There arose a need for a cemetery for the deceased non-Catholics." The group of citizens who had the corporation recorded and chartered by A.W. Connely, Clerk of Court, were Peter Berger, John Hubbard, Lucius Suthon, John N. Winder, John Berger, William McCollam, and William P. Tucker. All but Tucker became members of the Houma Magnolia Cemetery Board of Directors.

Land for the Magnolia Cemetery Association was purchased from Herman Loevenstein on February 7, 1883 and from Josephine Loevenstein Hubbard on February 16, 1895.

Magnolia Cemetery was sold to Houma Magnolia Cemetery after a charter was drawn up on February 9, 1884.

On June 22, 1893, Magnolia Cemetery sold a batture lot between Bayou Terrebonne and Main Street to the Houma Fish and Oyster Company, Limited.

The cemetery today occupies a space at 8064 West Main Street as its main entrance and Belanger Street as its rear boundary. In its early years, cemetery board and auxiliary members dealt with drainage issues and a lack of good record keeping. However, those problems were resolved.

The large magnolia trees that gave the place its name no longer exist. The cemetery sits between Terrebonne Memorial Park mausoleum built beginning 1967 by Dr. Philip L. Cenac, Sr. on its south, and the City Court of Houma building on its north. Terrebonne General Medical Center and an adjacent Medical Arts building are to the immediate south of the cemetery and mausoleum.

Terrebonne Memorial Park 2016

HOMESTEAD

Homestead pre-1890

Arthur Warren Connely's Civil War pass to return home June 4, 1865

Clara Himel Connely (1865-1949)

Lucy Connely (1889-1957) as Queen Bivalve c. 1913

A. W. Connely (1847-1915)

198 | HARD SCRABBLE TO HALLELUJAH

On the immediate southern edge of Houma on the left descending bank of Bayou Terrebonne is Homestead, once a large sugar and potato farm. Clara Himel Connely, wife of Arthur Warren Connely, bought the land and the existing house for $3,000 on December 24, 1890.

The land upon which Homestead stands is part of the original claim by Pierre Menoux registered with the U.S. government on November 20, 1816 and confirmed on May 11, 1820.[1] It was purchased in 1852 by Aubin Bourg, who built his family's house on the property. Bourg served as Deputy Clerk of Court from 1852 to 1855, Houma Postmaster in 1853, as Sheriff 1855-1861, Clerk of Court 1874-1876, and Parish Treasurer 1888-1898.[2] His cattle brand was registered as number 482 dated December 23, 1863 in Terrebonne Parish records.

A.W. Connely first owned Mulberry plantation, which his father "Gilmore bought... on Bayou Dularge in 1843."[3] The Connely family owned Mulberry for about half a century.[3] Arthur Warren Connely's 1884-85 sugar production totaled 325 hogsheads according to annual planter production reports.[4]

During the Civil War, according to Miss Ruth Connely in a *Houma Courier* newspaper article, A.W. "ran the blockade in 1863 to join the Confederate Army in northern Louisiana. Arthur was aided by one of his Indiana cousins whose company was occupying Houma."

In a document dated June 4, 1865, he was given by Union authorities a parole and pass to return to his home unimpeded by U.S. authorities, "so long as he observes his parole and the laws in force where he may reside."[5]

Connely also served as Terrebonne Parish Sheriff from 1892 to 1915. From Aubin Bourg he purchased 100 acres and the existing residence at Homestead, the bill of sale dated December 24, 1890. One of the Connely daughters said decades later that the house was already an old house then.

Connely enlarged the first floor of the house in 1892 and added the second floor in the first decade of the 1900s. This enlargement was no doubt prompted by the fact that he and his wife, *née* Clara Himel, had 11 children to house and rear.

Their names ring true to that era's fashion in the naming of children after friends and family: Edmund McCollam, Clara Himel, Lucy Ella, Georgia Mallard, Arthur Warren, William Alexander, Flora Bowdoin, Lavinia May, Clerville Himel, Ruth, and Katherine Easton. Oldest son Edmund's name was given to him in honor of Edmund Slattery McCollam of Ellendale plantation. Edmund Connely became a medical doctor and married Frances Bisland, a daughter of the Aragon plantation Bislands.

The Civil War veteran and his family lived in a house that from its origin was set on cypress blocks, not bricks. Its construction was of hewn timbers with mortise and tenon joinery.[6] It is an outstanding example of a house that was continually modified from the mid-1800s to the present to accommodate family needs. Even today, evidence remains of the adjusted rooms

Account of Aubin Bourg funeral, The Weekly Thibodaux Sentinel and Journal of the 8th Senatorial District *March 5, 1898*

Aubin Bourg cattle brand December 23, 1863

Dr. Edmund McCollam Connely (1885-1957) at Louisiana State University c. 1903

Frances "Fannie" Bisland (1887-1976) c. 1905

Tenant house Homestead 2015

*Arthur W. Connely's cattle brand
March 5, 1869*

of the original house, in the flooring that was not changed in the 1892 enlargement; the rear porch that was enclosed when the young men's room was relocated from the side yard to make an in-house kitchen; and the bathroom that was relocated inside the house after electrification and running water were available, allowing for the outdoor privy to be converted into a cesspool.

The old house that once boasted of a dairy in the rear of the property remains today at 101 Oak Street, Houma, Louisiana, south of the junction of Bayou Terrebonne with the Gulf Intracoastal Waterway. It was dug through Homestead plantation in the 1930s.

Among other official records, A.W. Connely is included in brand registrations, number 530 dated March 5, 1869 in Terrebonne Parish records.

SOURCES
1. A Century of Lawmaking for a New Nation: U.S. Congressional Documents and Debates, 1774-1875, American State Papers, Senate
2. Terrebonne Parish officials' records
3. Helen Wurzlow, *I Dug Up Houma Terrebonne*, 1984
4. P.A. Champomier, Statement of the Sugar Crop Made in Louisiana
5. Document provided by William A. Ostheimer, a Connely descendant of Houma

Homestead 2015

Left and middle, bed and dresser purchased from the estate of Dr. Berwick Duval by Lavinia Connely; right, bed of Georgia M. Connely

HOMESTEAD | 201

JOSEPH A. GAGNÉ HOUSE

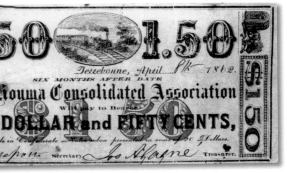

Notice of Joseph A. Gagné's appointment as Interim Sheriff by governor *The Houma Courier* July 15, 1854

Civil War scrip signed by Joseph A. Gagné

Cattle brand of Joseph A. Gagné, Sr. April 12, 1853

Ella K. Hooper (1886-1967), founder MacDonell French Mission School

Laura Morgan White (1869-1950), co-founder MacDonell French Mission School

On the near south side of the Gulf Intracoastal Waterway where it intersects Bayou Terrebonne across the bayou from Homestead in Houma stands a house built c. 1835[1] by Joseph Auguste Gagné, Sr., a Quebec, Canada native. The house is one of the oldest surviving homes in Houma.

It is situated on a claim by Pierre Menoux, registered on November 20, 1816 and confirmed on May 11, 1820.[2]

Gagné is listed in the 1860 U.S. Census as a 34-year-old Houma merchant with real estate valued at $1,200 and personal property valued at $10,700. He married Sarah Anne Dunn in Houma in 1851. Gagné was appointed Interim Sheriff by Governor Robert C. Wickliffe after the death of John H. Fields in 1855. It is for his wife's family that Dunn Street in Houma was named, because it was the longtime site of her father's brick yard. Her father, J.B. Dunn, was a native of Wilmington, Delaware. The Gagné couple had ten children together.

Joseph A. Gagné was treasurer of The Houma Consolidated Association which issued scrip during the Civil War. His signature accompanies that of A.J. Delaporte, the Association's secretary, on paper notes from that period. His signature also appears in the livestock brand records of the parish when he registered brand number 387 on April 12, 1853.

The Gagné house and grounds in 1919 were placed under the care of Miss Ella Keener Hooper, a Methodist Episcopal South Deaconess who first founded a Methodist mission school for girls in Terrebonne Parish. Laura Morgan White, a widow, was her co-worker in this endeavor.

Miss Hooper and Mrs. White moved to Houma in 1917 to do "the French work." Miss Hooper admired "The imposing Gagné house on East Main Street and its beautiful property along the Intracoastal Canal—all eighteen acres... and found it to be an ideal place for a school. When she and her co-worker, Mrs. Laura M. White, found out that it was for sale in 1919, they asked the Board [Methodist Woman's Board of Home Missions] to buy it."[3]

The Board did not do so at the time because of lack of funds, but "A ten thousand dollar gift from Centenary Methodist College was sent 'for the French work,' and the house and property were purchased....Wesley Houses were homes sponsored by the Methodists where young girls who needed help could live in a Christian atmosphere."[3]

Three years later, $24,000 was allocated to construct new buildings on the campus, which then accommodated both boys and girls. Wesley House became the MacDonell French Mission School. Miss Ella's sister, Wilhemina Hooper, taught at the school.[3] The MacDonell Wesley Community House was dedicated in the fall of 1921.[4]

MacDonell was named for Tochie MacDonell, who was the general secretary of the Methodist Woman's Council at the time. Downs Hall, a kitchen and dining hall, was added to the MacDonell Home in 1940,

MacDonell Wesley Community (Joseph A. Gagné) house c. 1925

Joseph A. Gagné house 2016

JOSEPH A. GAGNÉ HOUSE | **203**

Miss Ella K. Hooper and first six female students; temporary residence c. 1919

1924-25 school session MacDonell United Methodist Children's Home

Staff, MacDonell United Methodist Children's Home; Ella K. Hooper third from left, Laura M. White far right, 1924-25 school session

MacDonell dining hall c.1925

JOSEPH A. GAGNÉ HOUSE

206 HARD SCRABBLE TO HALLELUJAH

Bible class at MacDonell United Methodist Children's Home 1924-25 school session

JOSEPH A. GAGNÉ HOUSE

Visiting Wesley House c. 1920

Miss Ella K. Hooper's Model "T" Ford was frequently seen around Houma

named for Mrs. J.W. Downs, the administrative secretary for the Woman's Missionary Council at that time.[4]

One history of Methodism in Terrebonne Parish states, "From 1949 to 1953, the function of MacDonell changed from that of a boarding school to a school for Indians. MacDonell ceased operations as a school in 1953 and became a home for dependent children. The children who stayed there attended the public schools."[4] Miss Hooper left MacDonell School in 1949, the history continues, to help her sister, Wilhemina, at the Indian School in Dulac.

Miss Ella K. Hooper retired at the age of 65 in August 1951 and moved back to her home town, Rosedale, in Iberville Parish, Louisiana. Co-foundress Laura M. White had left a few years earlier. Miss Hooper died on June 28, 1967 at age 85.[2]

In February 1946 a fire threatened the lives of 14 school students in an upstairs dormitory, but their teacher, a Miss Gandy, was able to get them all to safety (including Zippy, a "long-eared, soot-black cocker spaniel pup," which was barely visible in the "smoke-filled upstairs apartment," according to *The Houma Courier* newspaper account).

McCoy Dining Hall c. 1920

The property surrounding the house evolved into the MacDonell United Methodist Children's Services, an agency dedicated to caring for children from broken families and single parents, and today operates as a residential treatment facility for children removed from abusive or unsafe domestic settings while waiting for court placement in safe, permanent family environments.

The home, now called Wesley House, was named to the National Register of Historic Places in 1982. It is a small, frame, story and a half, Greek Revival raised plantation house built partially under the influence of the French heritage hall-less plan.[1] It is surrounded by ancient oaks and by numerous structures including administrative building and housing units, with the names Keener Cottage, Downs Hall, Hooper Cottage, New Hope Cottage, Laskey Cottage, Staff Cottage, McGowen Hall, and Thomas House, as well as a workshop and an equipment barn. MacDonell's private grounds are located at 8326 East Main Street, Houma, Louisiana.

1937 graduation: Pat Lawford, Nettie Thibodaux, Bertha Martin, Inez Vicknair, Lydia Blanchard, and Harry Barrios

SOURCES

1. State Historic Preservation Office, Louisiana Comprehensive Statewide Survey, Terrebonne Parish
2. A Century of Lawmaking for a New Nation: U.S. Documents and Debates, 1774-1875, American State Papers, Senate
3. *Anatole's Story* by Polly Broussard Martin, Pelican Publishing Company, Gretna 2004
4. *Methodism Along the Bayou* Chapter 4: "Methodism 'Returns' to Terrebonne Parish," Louisiana Conference of the United Methodist Church, www.la-umc.org/chapter4

Interior of Joseph A. Gagné house, first two photos; top right and below, MacDonell United Methodist administration building 2016

JOSEPH A. GAGNÉ HOUSE | **209**

Emile A. Daigle home, 600 block of Main Street late 1800s

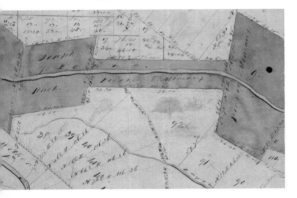

Pierre Menoux Claim State Land Grant Map December 24, 1831

Emile A. Daigle (1842-1913)

DAIGLEVILLE

The community of Daigleville below Houma on Bayou Terrebonne was not named so until just before the turn of the 20th century. In the year 1866, after the end of the War Between the States, a Houma newspaper described the "thick settlement extending some two miles from the town [Houma] on both sides of the Terrebonne" as one traveled south.[1] At the time, the southern boundary of the little town of Houma was around Roussell Street.

The article added, "These are small places, little homesteads, that have, in many instances, descended from father to son, from generation to generation. These little places are cultivated nicely, and although consisting only of a few acres, generally furnish the families living upon them with the necessaries of life."[1]

All of that land was part of the original claim of Pierre Menoux, who registered it with the U.S. government on November 20, 1816 and received confirmation of the claim on May 11, 1820.[2]

Businessman Emile Auguste Daigle began to enlarge the community in 1897. At the time of his death on June 14, 1913, Daigle was "one of the wealthiest and most prominent citizens of Houma," according to an article in the New Orleans *Times Picayune* with a Houma dateline of June 16. The obituary went on to briefly summarize his business life. "He led an active business life and was the owner of the Daigle Barge Line, Ltd. for a number of years. About five years ago, he sold his barge line and interests in sugar plantations and oyster packeries and since has lived quietly with his family in the old home in Main Street." The home had been built by Herman H. Loevenstein in the 600 block of Main Street, on a lot 336 feet wide with a depth to Belanger Street. (The house was torn down in the 1940s.)[3]

Emile A. Daigle started his life at Côte Malbrough in the northernmost part of Terrebonne Parish, and became one of Houma's largest property owners. Daigle's wife, Antoniata Cecilia Porche, (1846-1935) donated a new bell for St. Francis de Sales Catholic Church after the August 26, 1926 storm blew down the church's spire and destroyed the older bell. Mrs. Daigle also donated the St. Joseph stained glass window and altar to the old St. Francis Church, and her husband donated the St. Isidore window. These windows from the old church are now part of a church in Texas.[3]

Before he decided to phase out his participation in business enterprises, Daigle "owned all the houses in Daigleville... and more than 200 houses here in Houma," according to Helen Wurzlow in her book *I Dug Up Houma Terrebonne*.[3] After their construction, he rented or sold the homes. Even though he may have intended otherwise, earliest mention of the "village" in print, c. 1897 while the community was still under construction, is of Daigleville.

Through the years the name Daigleville persists in describing that area just across the Gulf Intracoastal Waterway (completed August 30, 1924), along East Main Street and across Bayou Terrebonne along East Park Avenue (although no map has been found designating the area as such officially). The recognized boundaries of Daigleville are the Gulf Intracoastal Waterway on the north, and on the south, Dixie Shipyard located on the left descending bank of Bayou Terrebonne. Both banks within that area are considered part of Daigleville.

A wooden bridge connected the community on both banks for decades. An article in the *Houma Courier* newspaper of July 8, 1897, contained the following mention of the community and the bridge:

"Mr. E. Daigle's steamer 'Harry' took an excursion party down to Bayou LaGresse Sunday. The excursion was given for the purpose of raising funds for the Daigleville bridge.... Mr. Daigle is making extensive improvements at his village 'Daigleville' and keeps several workmen busy building and repairing."

A new bridge that replaced the old wooden bridge crosses Bayou Terrebonne to the immediate left on East Main Street where Dug Road (Grand Caillou Road) intersects East Main on the southern side of the current twin spans over the Intracoastal Canal.

One of the landmarks of the community was the Daigleville School on the right side of East Main Street several blocks from the Intracoastal. In 1959 it became the first local school for Native Americans. The Terrebonne Parish School Board donated the property and building on July 21, 2015 for Native Americans to use, and it is currently being developed into a community center.

Just a short distance farther south is Our Lady of the Most Holy Rosary Church which has served Roman Catholics of the

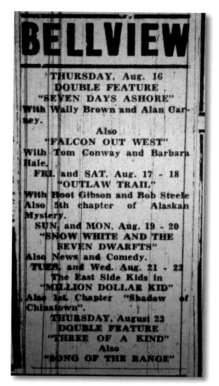

Bellview Theatre ad, The Houma Courier August 16, 1945

Daigle Barge Line, Ltd. invoices May 1, 1909 and January 21, 1924

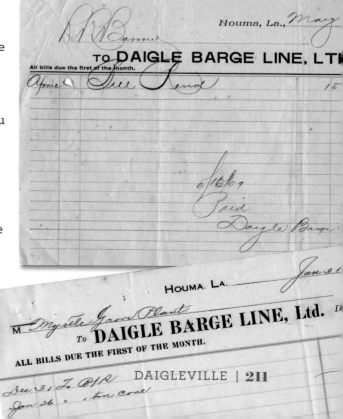

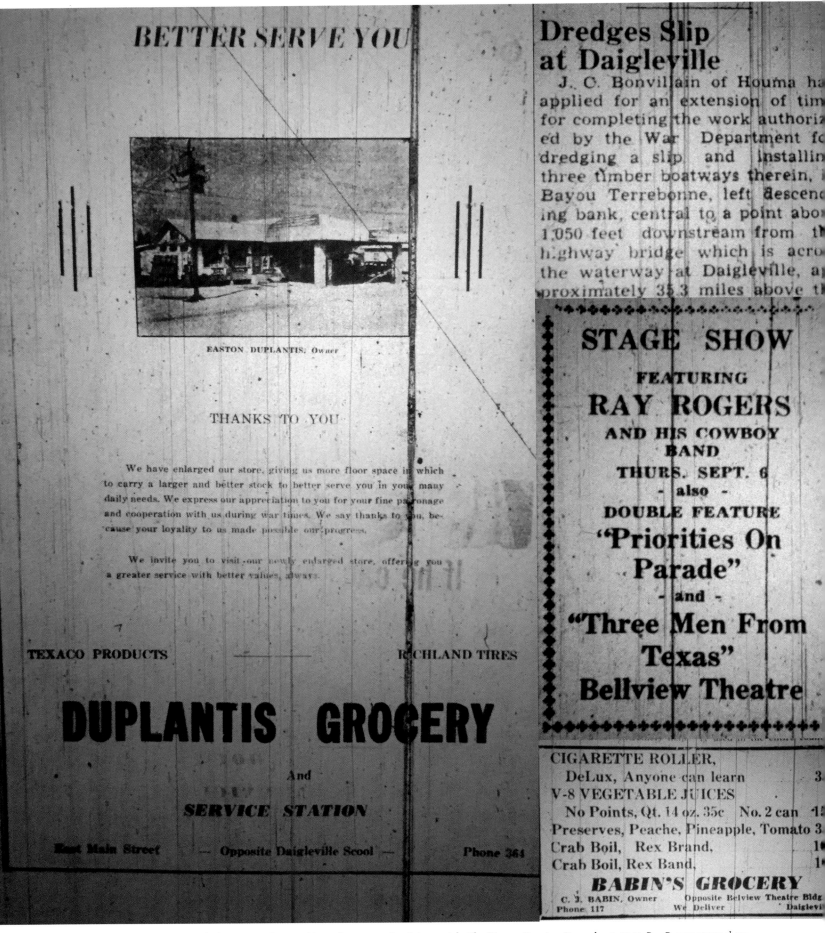

Duplantis Grocery Store ad, The Houma Courier, *November 1, 1945; Dredging article,* The Houma Courier, *December 2, 1943; Roy Rogers stage show ad,* The Houma Courier *August 30, 1945; Babin's Grocery Store ad,* The Houma Courier, *August 9, 1945.*

DAIGLEVILLE BAPTIST CHURCH
East Park Ave., Houma, La.

INDEPENDANT—FUNDAMENTAL—MISSIONARY

Prayer Meeting, Thursday	8:00 P. M.
Sunday School	10:00 A. M.
Preaching Service	11:00 A. M.
Evangelistic Service	8:00 P. M.

Missionary, E. N. Landry, Pastor
No Literature Except The Bible.

WE WANT TO EXPRESS

Our Appreciation to You for the Good Will and Patience Shown Us in these Days of Ration and Help Shortage.

GUIDRY'S GROC. & SERVICE STA.
WE DELIVER
E. Main St.   At Daigleville   Phone 700

Announcement!

WE HAVE PURCHASED

The

Bonvillian & Boudreaux

MACHINE AND WELDING WORKS

East Park Avenue at Daigleville

We take pleasure in making this announcement because we feel that experience and equipment qualify us to render the boat and oil field industry a genuine service.

Combined with the facilities of a modern machine shop, is the convenient weigh on the bank of Bayou Tererbonne, a fact that assures rapid boat repair work.

This is your invitation to conduct a personal inspection of the facilities which guarantee better boat service in a community where water transportation plays such an important part.

BAILES And SON
Machine And Welding Works

EAST PARK AT DAIGLEVILLE   Phone 2090

Be adaptable—
This year see what we've got before you decide what you want!

LET THIS FRIENDLY
STORE SERVE YOU IN
YOUR DAILY
KITCHEN NEEDS

BONVILLAIN'S GROCERY

Capt. Ernest Bonvillain Mgr.
E. Park At Daigleville

ALL WORK GUARANTEED HERE AND PROMPTLY ATTENDED TO.

Hig Grade Blacksmith and Wheelwright Work and Horseshoeing

Conveniently Located, on Main Street between Houma and Daigleville.

WILLIE HAMILTON, Proprietor.

Newly Remodeled To
BETTER SERVE YOU
GROCERIES — MEATS

DUPLANTIS GROCERY AND SERVICE STATION

TEXACO PRODUCTS
Daigleville
RICHLAND TIRES
Phone 364

Daigleville Baptist Church ad, The Houma Courier, *March 22, 1945; Bailes & Son ad,* The Houma Courier, *November 1, 1945; Guidry's Grocery ad,* The Houma Courier, *October 4, 1945; Bonvillain's Grocery ad,* The Houma Courier, *September 20, 1945; Willie Hamilton blacksmith ad,* The Houma Courier, *June 11, 1910; Duplantis Grocery and Service Station ad,* The Houma Courier, *June 13, 1946.*

214 | HARD SCRABBLE TO HALLELUJAH

Theriot Lumber Yard Daigleville 1943

DAIGLEVILLE | 215

Seated left to right, Paul E. Gaidry, Agelee Cadiere Gaidry, Arnold Gaidry, Adolphe J. Gaidry, Wilson J. Gaidry, Lowell Gaidry; standing, Serule Gaidry, Laura V. Gaidry, J. Wilfred Gaidry, Adoiskia Gaidry Theriot 1800s

area since the first Mass there was celebrated on September 19, 1948 in the community hall by the first pastor, the Reverend Anthony J. Wegmann.[4]

The new parish was requested by the pastor of St. Francis de Sales Church in Houma, the Rt. Reverend (later Bishop) Maurice Schexnayder. Terrebonne and surrounding parishes were still part of the Archdiocese of New Orleans at the time. Msgr. Schexnayder purchased the property in East Houma on behalf of St. Francis de Sales from Laura V. Gaidry for $13,000 on December 12, 1947. She retained the rights to live in her house at the time. On January 6, 1948, she sold the bayou side to the Rt. Rev. M. Schexnayder acting on behalf of St. Francis de Sales, for the sum of $3,000.

Archbishop Joseph Francis Rummel of the New Orleans Archdiocese canonically established the requested parish and appointed Fr. Wegmann on September 8, 1948. The church began as a chapel building from an unused Army base being dismantled near Gulfport, Mississippi. A committee of parishioners procured the building and had it moved by barge to Houma. Although the structure was damaged while being transported, the newly completed church building was blessed and dedicated by Archbishop Rummel on March 20, 1949.

Local businessman and parishioner "Hurry Hurry" Guidry demolished an old store building and barn on the church site, and used the material from those two structures to erect a community hall.[4]

The current address of the church site is 8594 East Main Street, Houma, Louisiana.

Another church of the area was the Daigleville Baptist Church advertised in the *Houma Courier* issue of June 21,

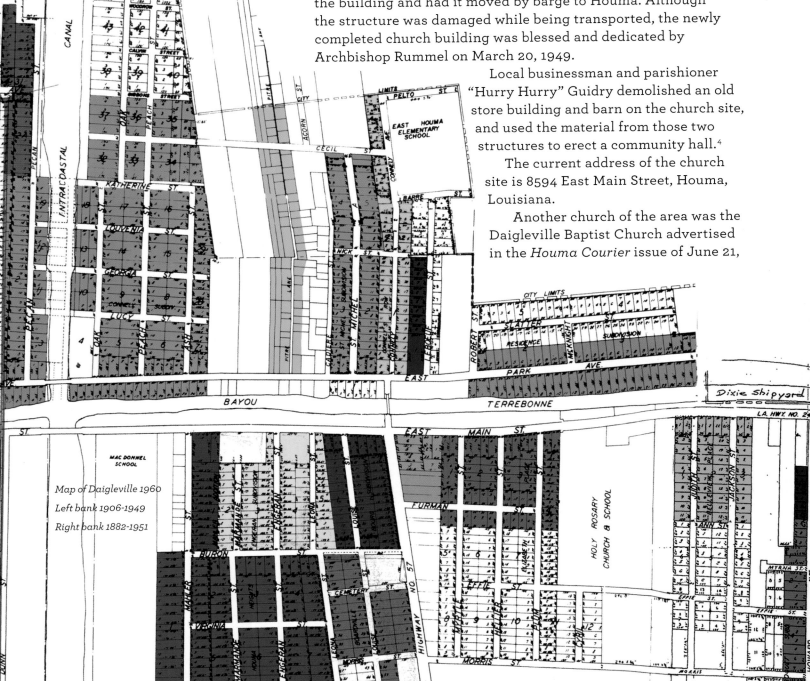

Map of Daigleville 1960
Left bank 1906-1949
Right bank 1882-1951

Original cornerstone Little Zion Baptist Church built c. 1860s photo 2016

Cornerstone New Zion Baptist Church, remodeled 1912

Former Army base chapel from Gulfport, Mississippi c. 1948 became the first Holy Rosary Catholic Church

1945 as independent, fundamental, and missionary. Pastor was missionary E.N. Landry.

Robert Ruffin Barrow, Sr. in 1859 donated the land for the earliest Baptist church for people of color in the area, the Little Zion Baptist Church on Grand Caillou Road.

The general area called Daigleville includes many subdivisions developed by different landowners through the years. On the right bank of Bayou Terrebonne are subdivisions Boardville, developed on January 10, 1882, and others developed by J.H. Hellier (March 31, 1902), J. Harry Hellier (Elizabeth Place, April 3, 1935), Alidore J. Mahler (March 7, 1938), Houma Heights (June 6, 1938), William Voss (Bellview Place, June 9, 1938), Theogene J. Engeron (April 24, 1939), Elphege Theriot (Roselawn, April 6, 1940) Harry J. Bourg (August 23, 1945), Horace J. Authement (November 28, 1948), G. P. Boquet (March 14, 1949 and Addition Number 1 April 28, 1951). Luke Subdivision, Voisin Place, and Saadi Sites were all developed in the 1950s. On the left bank of Bayou Terrebonne are the subdivisions developed by Dr. L.H. Jastremski (Connely's Row Subdivision January 3, 1906), A.W. Connely Estate (Connely Subdivision March 15, 1926), Mrs. W.J. Gaidry (Residence Subdivision July 29, 1930), N.J. Arceneaux (St. Michel Subdivision April 23, 1942), L.C. LeBoeuf (LeBoeuf Subdivision June 28, 1949) and George Pitre Lane.

[Note: Further information about the Daigle Barge Line can be found in the text on Pecan Tree plantation.]

SOURCES
1. *The Civic Guard* newspaper, Houma, Terrebonne, Louisiana, June 16, 1866
2. A Century of Lawmaking for a New Nation: U.S. Congressional Documents and Debates, 1774-1865, American State Papers, Senate
3. Helen Wurzlow, *I Dug Up Houma Terrebonne*, 1984
4. History of Holy Rosary Church compiled by Douglas Maier, 1995

Holy Rosary Catholic Church, The Houma Courier, November 6, 1997

Harry J. Bourg (1888-1963)

Alidore J. Mahler (1896-1987)

Theogene J. Engeron (1874-1947)

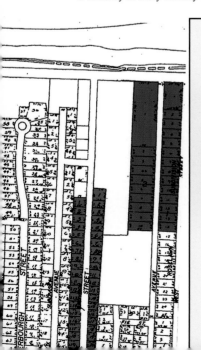

DAIGLEVILLE SUBDIVISIONS

LEFT BANK	RIGHT BANK
Connely Row-01/03/1906	Boardville-01/10/1882
Connely Subdivision-03/15/1926	J. H. Hellier-03/31/1902
Residence Subdivision-07/29/1930	Elizabeth Place -04/03/1935
Connely Subdivision addition 04/12/1937	Alidore J. Mahler Addition-03/07/1938
Connely Subdivision addition 01/20/1939	Houma Heights-06/06/1938
St. Michel Subdivision-04/23/1942	Bellview Place-06/09/1938
Lebouef Subdivision-06/28/1949	Theogene Engeron Subdivision-04/24/1939
George Pitre Lane	Roselawn Subdivision-04/06/1940
	Harry Bourg Subdivision-08/23/1945
	Authement Subdivision-11/28/1948
	G. P. Boquet Subdivision-03/14/1949
	Boquet Subdivision Addendum # 1-04/28/1951

Good Wine, Good Friends, *1986*,
Oil on canvas, 30 x 40 inches, George G. Rodrigue

Residence 2015

*Below, Pierre Menoux Claim
State Land Grant Map
December 24, 1831*

220 | HARD SCRABBLE TO HALLELUJAH

RESIDENCE
HOME PLACE

The first claimant of the 640 acres later known as Residence, four miles below Houma on the left descending bank of Bayou Terrebonne, was Pierre Menoux. He registered his claim with the U.S. government on November 20, 1816 and it was confirmed on May 11, 1820.[1] His succession of December 11, 1822 specified those 640 acres. James Bowie next laid claim to the land in 1828.

However, the person with whom Residence is most associated is Robert Ruffin Barrow, Sr. He arrived in Terrebonne Parish around 1828 and soon began to develop a veritable land empire in this and other parishes. Attracted by sugar land opportunities, he and his brother William entered Terrebonne from St. Francisville in West Feliciana Parish. His North Carolinian father Bartholomew lived at Afton Villa plantation, and gave his two sons $1,800 and two slaves, Daniel and Jennie Lyons (but in some sources referred to as General and Washington) and their children.[2] R.R. was described by his contemporaries as a striking figure atop his black steed named Tom Bennett.

With those resources William and R.R., as Robert Ruffin was known, managed to build an expansive network of agricultural holdings in almost every section of Terrebonne Parish, as well as in surrounding parishes. Family records and journals indicate that the first property the Barrows bought in Terrebonne was Pointe Farm along the extreme southern reaches of Bayou Terrebonne, around the year 1828.[3] According to a newspaper article of October 8, 1972, "Through hard work and strict management Robert's ventures began to be rewarding. With his profits, he bought more lands.... He also bought a piece of wooded land near Houma and named it Residence—for here he would build his home....Robert Barrow looked amid the trees and grasses across from the busy Bayou Terrebonne and visualized a home. A Residence."[4]

The two Barrow bachelors continued to acquire holdings for the next two decades, purchasing more and more land with profits from their earlier property investments.[5] The first house R.R. built on Residence land is thought to have been in use by 1832. The exact date of Barrow's purchase of the Menoux-Bowie land is unknown.

William B. Barrow died in 1842 without having married, and thus had no children as heirs. R.R. became the recipient of William's part of their assets. Eight years later, at age 52, R.R. married 25-year-old Volumnia Washington Hunley on Thursday, February 7, 1850 in an Episcopal ceremony at Christ Church Cathedral on Canal Street in New Orleans.[6] Volumnia was the sister of Horace L. Hunley, who created early hand-powered submarines including the *Pioneer* and the *Hunley*[7]

James Bowie (1796-1836)

Residence, May 23, 1887

William B. Barrow (1802-1842)

Robert Ruffin Barrow, Sr. (1798-1875)

Volumnia Washington Hunley Barrow (1825-1868)

Residence plantation's National Register of Historic Places plaque 2001

Top to bottom, left to right, Edmond J. Foolkes house; underground cistern; outhouse; chicken coop; manager's house; pigeon house; part of front lawn view from house 2016

Top to bottom, left to right, living room; Barrow family china; stairway; J. Faiore French piano 1855; Barrow family child's crib; bed (2016)

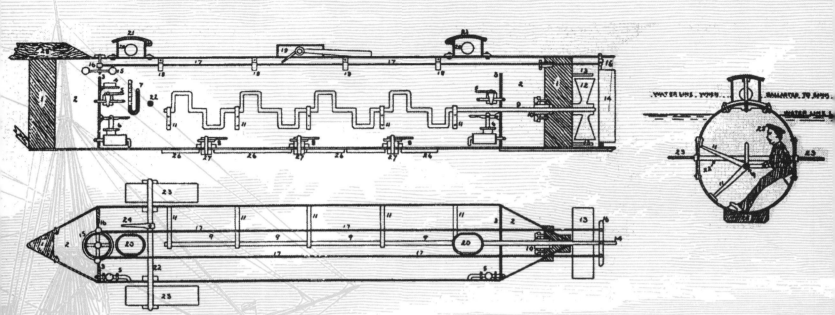

Schematics of the Hunley Submarine 1863

H.L. Hunley stencil hanging in Residence house 2016

224 | HARD SCRABBLE TO HALLELUJAH

for Civil War use. The latter sank the *USS Housatonic* in an 1864 battle and in the same battle the submarine itself sank in Charleston Harbor, South Carolina, killing everyone on board, including Volumnia's brother, Captain Hunley.

R.R. brought his bride home to Residence. The Barrows had two children, Volumnia Roberta, known as Roberta (1854-1900), and Robert Ruffin Barrow, Jr. (1858-1926). Robert, Jr.'s godfather was Episcopal Bishop Leonidas Polk.

Robert Ruffin, Sr., before the Civil War amassed a fortune in sugar lands. In 1860 he held all or part of 20 plantations both in Terrebonne and other area parishes totaling 21,256 acres valued at $1,062,000.[8] In Terrebonne he owned Residence, Mulberry, Point Farm, Roberta Grove, Myrtle Grove (where harvested cane crops from Residence went for grinding), Batey, Carothers, Honduras, Front Lawn, and had an interest in Cedar Grove, Crescent, Caillou Grove, Flora, Magnolia Grove, Ridgeland, Forest Grove, Idlewild, and Dularge. He owned Locust Grove in Assumption Parish and Oak Grove in Lafourche Parish, and also had land in Texas.

His plantations on Bayou Terrebonne were Batey in Schriever, Residence, Roberta Grove, Myrtle Grove, Front Lawn, and Point Farm. On Bayou Grand Caillou, his properties included Honduras and Caillou Grove. On Bayou Black, Barrow had interests in Carothers, Crescent, and Flora. On Bayou Dularge, R.R. and/or his partners worked Mulberry, Ridgeland, Idlewild, and Dularge. In Chacahoula Barrow's holdings were Cedar Grove and Forest Grove, and on Bayou Little Caillou, Magnolia Grove.

The Barrow-Hunley marriage unraveled, and Volumnia and Robert were legally separated in June 1866, with Volumnia retaining custody of their children. The courts ordered Robert to release to Volumnia the titles to Myrtle Grove, Roberta Grove, and Residence plantations, all south of Houma along Bayou Terrebonne. She died in November 1868 at age 43 without a divorce from Robert. Roberta was 14 years old, and Robert Ruffin, Jr. was ten.[8]

The two children, according to Thomas A. Becnel's book *The Barrow Family and the Barataria and Lafourche Canal*, eventually "clamored for their share of her estate. Despite the economic setbacks caused by the Civil War, Barrow and his children still owned extensive holdings. In 1871 property taxes in Terrebonne Parish alone came to $5,455, for Caillou Grove, Residence, Myrtle Grove, Point Farm, lots near Tigerville on Bayou Black, and various other tracts."

But the Barrows lost much of their lands bit by bit. Roberta Grove and Myrtle Grove were seized for taxes due on

Horace Lawson Hunley (1823-1863)

Robert Ruffin Barrow, Jr. (1858-1926)

226 | HARD SCRABBLE TO HALLELUJAH

Robert Ruffin Barrow, Jr. (1858-1926)

Volumnia Hunley Barrow's estate. Caillou Grove plantation, 3,300 acres on Grand Caillou valued at $150,000, was lost in a tax sale and subsequently bought for just $1,859.14 in back taxes on November 28, 1873.[8] By 1873 Robert Ruffin, Sr. was destitute and living in New Orleans. He died of cholera there in 1875 at his home in New Orleans at 219 Magazine Street.[6] Although his empire had crumbled around him, R.R. Barrow left his mark on Terrebonne Parish.

Barrow Street in Houma is named for Robert Ruffin, Sr., who gave significant gifts in the parish before his fortunes reversed in the years after the Civil War. He donated property for St. Francis de Sales Catholic Church in 1847, St. Matthew's Episcopal Church in 1857, Houma Academy in 1858, First Presbyterian Church in the 1860s (all in Houma), and St. Patrick Catholic Church and its cemetery in Gibson. In 1859 Barrow donated land on Grand Caillou Road for the Little Zion Baptist Church, the oldest church

228 | HARD SCRABBLE TO HALLELUJAH

Opposite page, left, first three top to bottom, St. Francis de Sales Catholic parish's first church 1850s; St. Francis de Sales second church c. 1910; St. Francis de Sales Church damaged in storm of 1926. Right side, top, St. Francis de Sales Academy entrance c. 1950; right middle, Houma Academy, later St. Francis de Sales School 1890s, Bottom, from left, St. Matthew's Episcopal first church 1860s; First Presbyterian Church built 1895; St. Patrick Catholic Church and Rectory in Gibson 1925.

New Zion Baptist Church 1912-1968

New Zion Baptist Church 2016

Rev. Wilbert J. Hanks tomb (1907-1992) "Standing over his flock," 2016

of that denomination for African Americans in Terrebonne Parish. The Barrow family donated a bell named Suzanna Roberta in 1857 to St. Matthew's Episcopal Church. The Barrows also donated funds for the St. Francis de Sales Catholic Church bell and tower in 1847, but the tower was felled by the storm of August 26, 1926, and the bell cracked.[9] (Antoniata Cecilia Porche, Mrs. Emile A. Daigle, donated a new bell for the church.)

Much later, the Barrows donated property as a site of the Residence Baptist Church on Isaac Street in Mechanicsville on the right descending bank of Bayou Terrebonne, as well as the bell for the church. In the 1940s Miss Laura Gaidry, niece of W.J. Gaidry, Sr. of Residence, sold property for the location of Our Lady of the Most Holy Rosary Catholic Church on East Main Street near Daigleville, according to a history of the church.

R.R. Barrow constructed Dug Road beginning in what is now called East Houma by having a road dug with shovels by his

Bell at Residence Baptist Church 2016

Residence Baptist Church 2016

Cornerstone of Residence Baptist Church founded 1910, built 1912

RESIDENCE | **229**

William J. Slatter (1838-1917)

The Residence dance
The Houma Courier,
July 24, 1897

Volumnia W. Hunley Barrow (1825-1868) and her daughter Volumnia Roberta Barrow (1854-1900) c.1860

Opposite page: Residence gas pump and delivery wagon 2015

slaves between his two plantations Residence and Honduras, the roadbed formed by compacted dirt. Barrow's Dug Road began at Residence on the right descending bank of Bayou Terrebonne and made a gradual arc until it reached the Honduras plantation mill on the southern side of what is now Barrow Street crossed by Point Street. (Barrow Street ended at present-day Honduras Street, with a gate across Barrow accessing the boundary of Honduras plantation.) The road was Barrow's desire, but an inadvertent transportation boon to the community resulted from the project.

Barrow also had a vested interest (half of the stock) in Barataria-Lafourche Canal Company No. 2 as well as partial ownership of other canals that proved valuable to Terrebonne Parish interests other than Barrow's own. Thomas Becnel's book details the Barrows' role in improving water transportation between Terrebonne and various markets for parish goods.

R.R. and Volumnia's daughter Volumnia Roberta Barrow eloped with William J. Slatter (1838-1917) in 1871. They were married 16 years and had five children: Clara, born 1872; Volumnia Louise, born 1874; Aannis "Annie," born 1875; Bessie, born 1877; and David, born 1879. Roberta and Slatter divorced and she married her second husband T. Albert Woods (b. 1833) in 1888. Woods died in 1889, only a year after their marriage. William J. Slatter, the father of Roberta's children, died in 1917.

While Roberta lived in Tennessee when married to William J. Slatter, she and her brother Robert in 1882 divided Barrow property they held jointly on Bayou Terrebonne. She became the owner of Residence, and Robert became the owner of Roberta Grove. When Roberta was widowed in 1889, she made her home at Residence.

The house at Residence had become dilapidated by 1895, so Volumnia Roberta, then Mrs. Woods, had it dismantled in 1895 and rebuilt the home, using structural beams and lumber from the original house to complete construction by 1898.[3] During construction, the family lived in a house adjacent to Residence on its south at present-day 9031 East Park Avenue. The "new" house at Residence is in the Queen Anne Revival style, with the interior including two mantels in the Eastlake style.[10]

Robert, Jr. married the daughter of his business partner, Charles Tennent, in 1879. Jennie Lodiski Tennent was a descendant of Don Manuel Gayoso de Lemos, the last Spanish governor of Louisiana.[8] Robert and Jennie became the parents of six children, two of whom died in childhood. Volumnia Hunley Barrow was born in 1881 and died in 1882. Robert Ruffin Barrow II lived to only six months old in 1885. The rest of the children were Irene, born 1883; Zoe, born 1886; Jennie, born 1888; and Hallette, born 1892.

After Volumnia Roberta's death on September 17, 1900, her children Aannis "Annie" Slatter Hickman and Clara Katherine Slatter Gaidry were the next inheritors of Residence.[7] Their siblings Volumnia Louise, Bessie, and David had all died in infancy.

THE RESIDENCE DANCE.

Many of our society folks are determined not to let the intemperate heat of this summer frighten them out of having a good time. It is not the fault of the young people if the elements combined to produce a summer of unprecedented heat and disastrous humidity. They laid their triggers and formulated their plans for a gay social season, this year long before summer donned its swaddling clothes, and therefore they claim that they are entitled to the floor and if any change in the schedule must be made the principal concessions should be made by the Weather Bureau rather than by them.

Milk container from Residence Dairy c. 1940

Trespass notice The Houma Courier *October 17, 1917*

Roberta Louise Gaidry Piel (1912-2000) c. 1922

Residence cane wagon c. 1930s

Annie married Thomas Smith Hickman in 1902, and Clara became the wife of Wilson J. Gaidry, Sr. Annie and Thomas had no children, and therefore no heirs. Clara, her husband, and their three children Harold Langdon (born 1898), W.J. Gaidry, Jr. (born 1901) and Roberta L. (born 1913) made Residence their home.

Harold Langdon Gaidry died in 1952. Wilson J. Gaidry, Jr. became the new operator of Residence following his mother's death in 1955. Wilson, Jr.'s son Wilson J. Gaidry III inherited his father's share of the property upon Wilson Jr.'s death in 1973 and the share of his aunt Roberta L. (Mrs. Peter Piel) in 2000. Wilson III's mother was Lillie Lea McKnight Gaidry, who died in 2011 at the age of 103 years.[11]

The house that R.R. Barrow's daughter built was named to the National Register of Historic Places in September 2001. The property continues to be used as a working farm, with the cultivation of crops and livestock. Original outbuildings that still exist at Residence include a blacksmith shop, manager's home, chicken coop, outhouse, and pigeonniere.

Residence was early on the site of a sawmill, as well, which in 1875 was deconstructed and rebuilt at another Barrow plantation, Myrtle Grove. Residence opened a dairy in 1909. Fresh, raw, and pasteurized milk came from the plantation's 700 acres at the time, with about 300 head of cattle.[3]

Lillie Lea Gaidry, the wife of Wilson Joseph Gaidry, Jr., (married 1930) said in an oral interview commissioned by the Terrebonne Parish Consolidated Government in 1982 that her husband operated the dairy after his father died in 1927. She remembered that when she first lived at Residence, milk was delivered "with a little cart pulled by two little mules called Pete and Jenny....Milk would sell for say five cents a pint, ten cents a quart." She recalled that in 1930 the cows, Jerseys, Holsteins, and Guernseys, were still milked by hand, at noon and midnight. Mrs. Gaidry also said that at one time within her recall, there were

Top, left to right, Wilson J. Gaidry Sr. (1868-1925); Harold Langdon Gaidry (1898-1952); Wilson J. Gaidry Jr. (1901-1973); Wilson J. "Doc" Gaidry III (1939-)

Wilson J. Gaidry, Sr. on Residence Dairy Wagon c. 1910

about 30 small dairies in the parish. The Gaidrys bought a truck c.1935 to replace the horse-drawn delivery wagon and workers did house-to-house delivery. Residence Dairy closed in 1965.

The historic home's current address is 8951 East Park Avenue, Houma, Louisiana on the left descending bank of Bayou Terrebonne. Wilson J. "Doc" Gaidry III and his wife Wanda Faye Claire Ledet are currently working to prepare the home and remaining grounds and structures as a tourist attraction.

R.R. Barrow, Sr. registered livestock brands number 141 dated March 30, 1832, and number 346 dated September 13, 1848, according to Terrebonne Parish records. Robert Ruffin Barrow, Jr.'s brand number was number 791 dated August 19, 1914.

Aurelie Ledet and Residence Dairy truck c. 1940

SOURCES

1. A Century of Lawmaking for a New Nation: U.S. Congressional Documents and Debates, 1774-1875, American State Papers, Senate
2. Powell, William S., *Dictionary of North Carolina Biography*, University of North Carolina Press, 1979-200, Volumes 1-6. Robert Ruffin Barrow entry written by Claiborne T. Smith, 1979
3. Interviews with Barrow descendant W.J.Gaidry III, current occupant of Residence, 2016
4. Coats, Sherri, "Street's Namesake Left Lasting Legacy" article, *Houma Courier* October 8, 1972
5. Floyd, William Barrow, *The Barrow Family of Old Louisiana*, published by the author, Lexington, Kentucky, March 1963
6. Barrow and Cole Families Timeline by Angie Trahan
7. Pizzuto, Greg, *The Life and Death of H.L. Hunley*, published in *The Charger*, Fall 2001
8. Thomas A. Becnel, *The Barrow Family and the Barataria and Lafourche Canal: The Transportation Revolution in Louisiana, 1829-1925*, Louisiana State University Press, Baton Rouge and London, 1989
9. Helen Wurzlow, *I Dug Up Houma Terrebonne*, copyright 1985
10. National Division of Historic Preservation, Standing Structures Survey of Terrebonne Parish
11. www.findagrave.com and *Terrebonne Life Lines*, Vol. II Number 4

Residence Dairy milk bottle and cap c. 1940

RRB

Cattle brands of Robert Ruffin Barrow, Sr., September 13, 1848; Robert Ruffin Barrow, Jr., August 19, 1914; Robert Ruffin Barrow, Sr., March 30, 1832

MECHANICSVILLE BARROWTOWN

The community of Mechanicsville received its name because Roberta Grove mechanics, who worked on the plantation's farm implements, lived there first.

Mechanicsville now consists of Acklen, Banks, King, and Samuel streets in East Houma. It is on the right descending bank of Bayou Terrebonne in what would have been the northern reaches of Roberta Grove plantation. It was created in December 1883, after Robert Ruffin Barrow, Jr. and his sister Volumnia Roberta Barrow Slatter Woods divided the Barrow estate in 1882.

Earliest ownership records note that Citizens Bank of Louisiana from New Orleans first held title to the land with Houman J.C. Bourg acting as its agent.

Gabriel Montegut and Aubin Bourg are known to have divided the land and sold lots in December 1883, the land surveyed by A. Jolet, Jr. The four streets of the community were created at that time. Although the map creating the subdivision dated 1883, and the map of 1914 confirming the earlier survey, both identify the community as Mechanicsville, through the years the area has transformed into Mechanicville in local parlance.

Joseph Thompson, Sr., known as Kingfish, was also called the "Mayor of Mechanicville" at one time. East Street School nearby was the site of the Roberta Grove plantation mule barn and stables. The small community's boundaries along East Main Street are East Street and Rosemary Street. During one stretch of time, the Bellview Dairy owned by the W.J. Whitney family occupied the rectangular stretch of property bounded by East Main Street at front, East Street at its western extreme, Senator Street at the south, and the back boundaries of Mechanicsville residences along Acklen Avenue on the east.

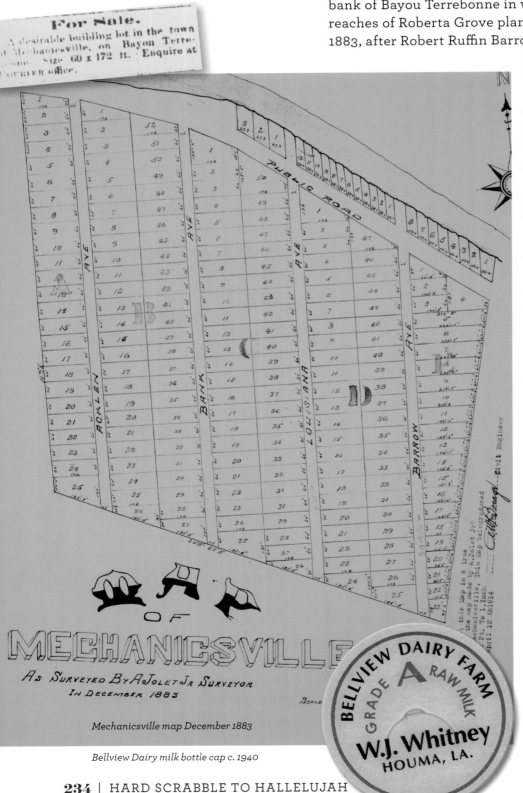

Ad lot for sale The Houma Courier, October 13, 1894

For Sale.
A desirable building lot in the town of Mechanicsville, on Bayou Terrebonne. Size 60 x 172 ft. Enquire at COURIER office.

Mechanicsville map December 1883

Bellview Dairy milk bottle cap c. 1940

Across East Street from Mechanicsville and the dairy property lies another subdivision created by the largest landowners in the area, the Barrows, on March 12, 1924. Barrowtown Subdivision fronts on East Main Street, with its Houma-side boundary on West Street and its eastern boundary at East Street. It extends back from the main thoroughfare of East Main seven blocks, with the following streets connecting East and West streets today (front to back): Bryant, Isaac, Isabel, Larry, Madge, Truman, and St. Joseph.

The Barrows, besides creating these small communities, donated land in the vicinity for Residence Baptist Church, Little Zion Baptist Church, and New Mount Pilgrim Baptist Church.

An illustrious African American poet and historian was born at Mechanicsville on March 8, 1900. Marcus Bruce Christian was the son of Emanuel Banks Christian and Rebecca Harris, and received his education from the Houma Academy and evening classes at a New Orleans public school. Ebed Christian, Marcus' grandfather, was a former slave who acted as director of Lafourche Parish Public Schools during Reconstruction, and his father was also a teacher.[1]

Marcus joined the Federal Writers' Project in 1936 and worked with the acclaimed author Lyle Saxon. Christian's work with "a circle of black artists and writers" both at Dillard University and beyond inspired his research into Louisiana history that became part of Lyle Saxon's 1945 publication *Gumbo Ya-Ya*. He became special lecturer and writer-in-residence at the University of New Orleans from 1969 until 1976. He contributed poetry and prose to many publications before his death on November 21, 1976 in New Orleans. He left his papers to the University of New Orleans.[1]

Marcus B. Christian, poet and teacher (1900-1976)

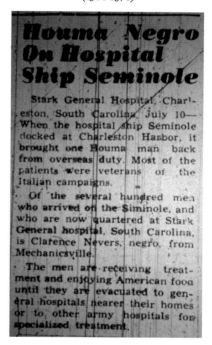

Article on Clarence Nevers from Mechanicsville during World War II
The Houma Courier, *July 27, 1944*

SOURCES
1. *Dictionary C, Louisiana Historical Association* http://lahistory.org/site20.php

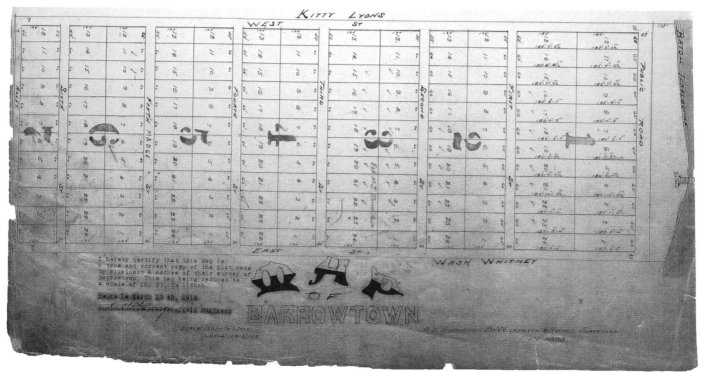

Map of Barrowtown March 12, 1924

ROBERTA GROVE

The same land of Pierre Menoux, whose 1820-certified claim included Residence, was also the foundation of Roberta Grove plantation situated on both banks of Bayou Terrebonne about five miles below Houma.[1]

Holden E. Wright purchased the land and built the first house there in 1829. The Wrightsville, Pennsylvania native married Mississippi-born Nancy Sarah Griffin in 1822[2] before settling in Terrebonne Parish.

Robert Ruffin (known as R.R.) Barrow, Sr. of Residence plantation next owned Roberta Grove, named for his daughter Volumnia Roberta Barrow, who went by the name Roberta rather than her first name. Roberta Grove and Myrtle Grove plantations were "2,800 acres of fertile land six miles below Houma," which typically produced 450 tons per day, processed at the mill at Myrtle Grove.[3] After R.R. and his wife Volumnia Washington Hunley Barrow separated in 1866, the courts ordered him to release to her Myrtle Grove, Roberta Grove, and Residence plantations.[4]

Mrs. Barrow, Sr. died two years later in 1868. Her estranged husband was declared destitute in 1873, brought to bankruptcy in part by the Civil War and its aftermath. John Bradford Pittman, his business partner, nephew, and tutor of his children, bought Roberta Grove and Myrtle Grove at a Sheriff's Sale for $4,277.54 in 1874. The next year R.R. died of cholera on July 27, 1875 at his home in New Orleans at 219 Magazine Street.[2]

In 1876 his son Robert Ruffin Barrow, Jr. purchased Roberta Grove for $22,000. Robert Ruffin and Roberta divided the Barrow estate in 1882, with Roberta assuming sole ownership of

Roberta Grove c. 1886

Robert Ruffin Barrow, Sr. (1798-1875)

Volumnia Roberta Barrow (1854-1900)

Robert Ruffin Barrow, Jr. (1858-1926)

Roberta Grove second house built 1891

Left to right: Hallette Barrow Cole wedding photo June 17, 1914; Robert Ruffin Barrow, Jr. and Jennie Lodiski Tennent Barrow, Venice, Italy 1900; Clockwise from left, Zoe Gayoso Barrow; Robert Ruffin Barrow, Jr.; Irene Felicie Barrow; Jennie Lodiski Tennent Barrow (sitting) c.1910

Jennie Lodiski Tennent Barrow (1859-1942) c. 1900s

Residence, and Robert taking title of Roberta Grove. The siblings agreed to co-own Myrtle Grove plantation.

The original home at Roberta Grove burned on May 15, 1889, and Robert Ruffin, Jr. completed construction of the second house, a duplicate of the original plantation home, in 1891. The Barrows' daughter Hallette Mary Barrow received Roberta Grove as a wedding gift in 1914.[2]

R.R. Barrow, Inc. next took possession of 1,800 acres at Roberta Grove on August 20, 1925. Author Thomas A. Becnel, in his book *The Barrow Family and the Barataria and Lafourche Canal*, described the structure of the corporation. Robert Jr. owned 400 shares of stock valued at $100 each; his wife from whom he was separated, Jennie Lodiski Tennent Barrow, owned 110 shares; and their daughters Irene Felicie, Zoe Gayoso, Jennie Tennent, and Hallette Mary, each owned 60 shares. The charter provided for the election of officers at an annual meeting, and to keep his controlling interest "he had to be present to vote at the meeting." However, when he was too ill to attend the meeting, his estranged wife and daughters elected the mother, Jennie Lodiski Tennent Barrow, president. Robert Ruffin Jr. filed suit but died on March 24, 1926 before the suit was settled.[4]

Both R.R., Sr. and Robert Ruffin, Jr. registered livestock brands in Terrebonne Parish records, the father's first brand numbered 141 dated March 30, 1832 and his second numbered 346 dated September 13, 1848. The son's brand was registered as number 791 dated August 18, 1914.

Hallette purchased 566 acres from the corporation in 1927. Hallette and her husband Christian Grenes Cole, M.D. remodeled the house in 1954 and owned Roberta Grove until her death in 1976 and his in 1977.[2] The hurricane of 1926 had destroyed the two smaller windows on each side of the larger stained glass window at the front of St. Matthew's Episcopal Church in Houma. Mrs. Cole and her sister, Mrs. Harris P. Dawson, donated two stained glass windows to the church in memory of their grandparents Robert Ruffin and Volumnia Washington Hunley Barrow, and their parents Robert Ruffin Barrow, Jr. and Jennie Lodiski Tennent Barrow.[3] The Roberta Grove house, which stood at 9317 East Main Street, had its contents go to auction in 1981.

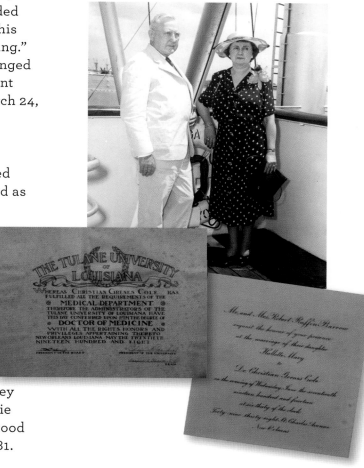

Wedding invitation of Hallette Mary to Dr. Christian Grenes Cole June 17, 1914; Dr. Christian Grenes Cole (1880-1977) and Mrs. Hallette Barrow Cole (1892-1976) sailing on Queen Elizabeth August 5, 1962; Christian Grenes Cole, Doctor of Medicine diploma May 20, 1908 Tulane Medical School

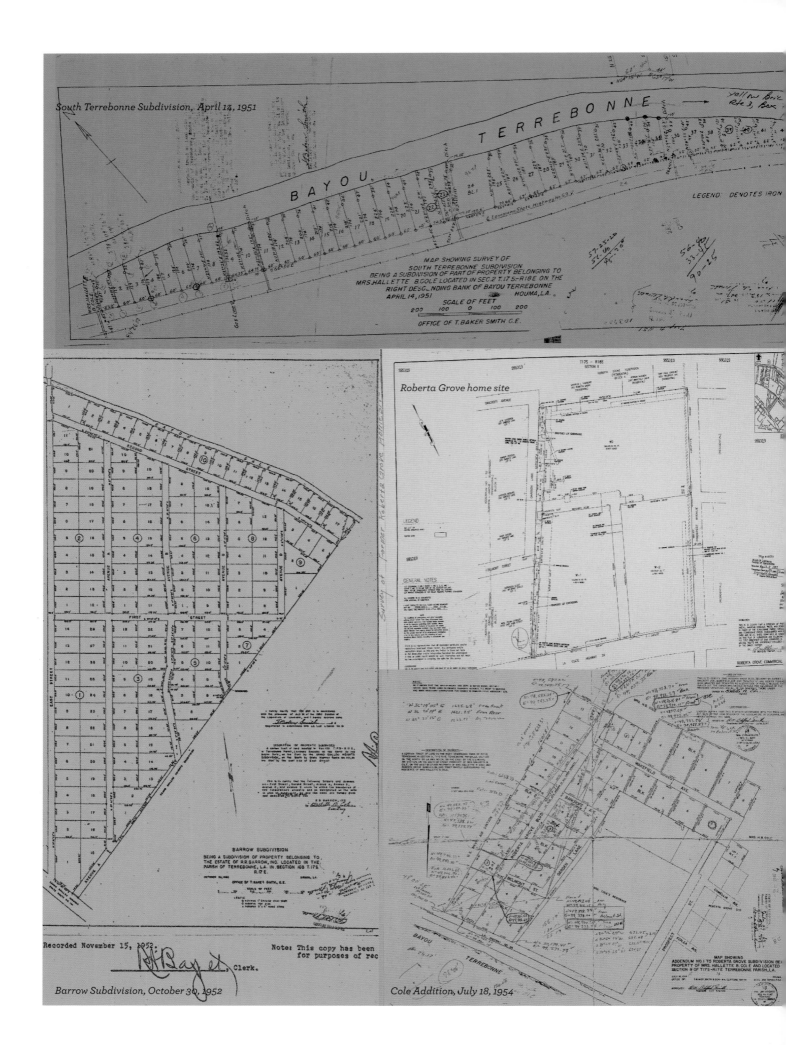

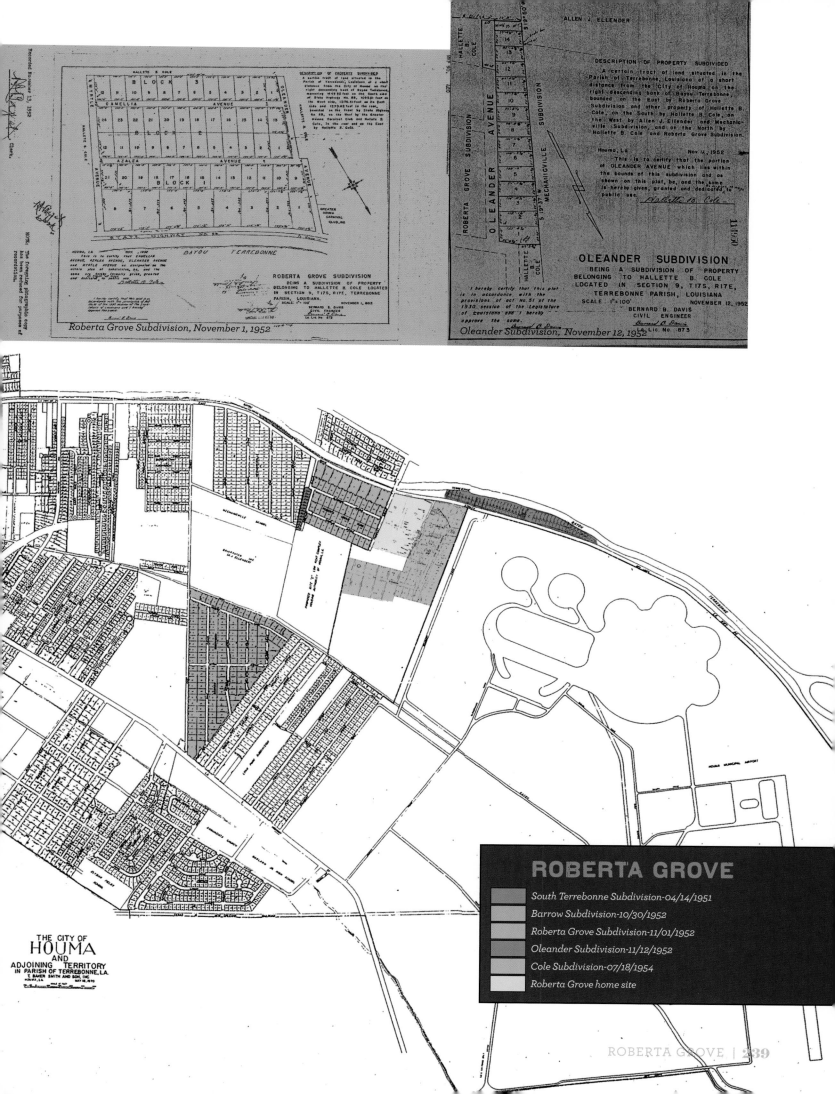

Roberta Grove chicken coop and yard c. 1940

Roberta Grove front porch c. 1940

R. R. Barrow, Jr. living room in home at 4938 St. Charles Avenue, New Orleans

Garden Society, 1968: standing, Mrs. Christian Grenes (Hallette) Cole and Mrs. Randolph A. (May) Bazet, Sr.; seated, Mrs. Charles R. (Ruth) Patterson, Sr., Mrs. Ashby W. (Anita) Pettigrew, Jr., and Mrs. Claude J. (Thelma) Ellender

The house itself succumbed to fire caused by arson in 1987.[2]

During the course of its history, Roberta Grove lands underwent many divisions. As early as 1882 when post-Civil War conditions left the family in financial difficulty, Citizens Bank of Louisiana held a mortgage on Roberta Grove. Its agent J.C. Bourg directed Gabriel Montegut and Aubin Bourg's real estate concern to have a part of Roberta Grove surveyed, which was done by A. Jolet, Jr.[5] That part of Roberta Grove, known thereafter as Mechanicsville because plantation mechanics lived there, was divided into lots and sold off beginning December 1883. Boundaries were East Street to Rosemary Street in East Houma. The streets of the community created by Jolet's survey were Acklen, Bank, Barrow, and Louisiana streets. (One native of Mechanicsville was Marcus Bruce Christian, a poet and historian born in 1900. He was the grandson of a former slave.)

Other communities to which Roberta Grove lands gave rise were Barrowtown, established on March 12, 1924, extending from West Street to East Street; South Terrebonne Subdivision, its 42 lots created April 14, 1951; Barrow Subdivision, created October 30, 1952; Roberta Grove Subdivision, its 57 lots created November 1, 1952, and its addition Cole Subdivision, created July 18, 1954; and Oleander Subdivision, its 14 lots created November 12, 1952.[2]

In 1943, the U.S. Government appropriated 300 acres for the U.S. Naval Air Station (LTA) Houma.[2]

Main Street entrance Roberta Grove c. 1960

Aerial photo of Roberta Grove, June 10, 1947

SOURCES

1. A Century of Lawmaking for a New Nation: U.S. Congressional Documents and Debates, 1774-1875, American State Papers, Senate
2. Barrow and Cole Families Timeline by Angie Trahan
3. Helen Wurzlow, *I Dug Up Houma Terrebonne*, 1984
4. Thomas Becnel, *The Barrow Family and the Barataria and Lafourche Canal: The Transportation Revolution in Louisiana, 1829-1925*, Louisiana State University Press, 1989
5. Terrebonne Parish public records

U.S. Naval Air Station (LTA) Houma 1944

Roberta Grove burned December 8, 1987

Governor Don Manuel Gayoso de Lemos bed

Birth certificate of Veranese Evans (Douglas), born on Roberta Grove plantation July 23, 1939

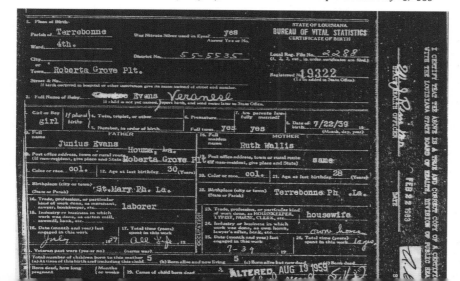

Roberta Grove Boehm Bird collection 1971

ROBERTA GROVE | 241

U.S. NAVAL AIR STATION (LTA) HOUMA

Memorial service for nine men killed in crash of K-133 over the Gulf of Mexico April 29, 1944

Right, highlights from Blimp Base newspaper 1944 and article in The Houma Courier, December 2, 1943

All The News From The
HOUMA NAVAL AIR STATION

U.S. NAVAL AIR STATION LTA HOUMA | 243

Above, Texaco plane Keystone Loening 1936 model at U.S. Naval Air Station (LTA) Houma during WWII

Article about Arab dignitaries visiting U.S. Naval Air Station (LTA) The Houma Courier, April 26, 1945

Arabs Stop Over at NAS

In their traditional dress, members of the royal family of Saudi Arabia, stopped over at the Houma Naval Air Station one day last week. They stopped long enough for their plane to be refueled and to chat a little with the local "natives" at the station.

One member of the group spoke English well and all members of the group took advantage of the stop-over to stretch their legs a little, drink cokes and talk a little.

They are members of a group of delegates to the San Francisco conference to open on April 25. One member of the group, Prince Amir Nawaf Ibn Abdul Aziz, is the nine year old son of the king of Saudi Arabia.

Members of the group stated that they had enjoyed the sights in New York and other cities and that they had attended and enjoyed a circus in New York. They plan to tour the country after the conference and that they hoped to have a tour of Washington performed on one of the Princes before returning to Arabia.

Terrebonne Parish's coast, its important oil interests on land and principally in tankers transporting oil in the Gulf of Mexico, prompted the U.S. military during World War II to locate an air station near Houma on the right bank of Bayou Terrebonne for lighter-than-air ships (K-class blimps, in this case) as well as airplanes.

The installation was important to defense against German U-boats, which had been wreaking havoc upon shipping in the Gulf. On May 4, 1942, German sub U-507 sank U.S. tankers *Norlindo*, *Munger T. Ball*, and *Joseph M. Cudahy* off the Florida Keys, and on May 6 sank the U.S. freighter *Alcoa Puritan* south of Mobile, Alabama. The U-507 on May 6 torpedoed the U.S. tanker *Sun* and sank the *William C. McTarnahan* on May 15, about 25 miles east of Ship Shoal Lighthouse. The U-boat sank the U.S. tanker *Guilfoil* in Gulf waters 50 kilometers south of New Orleans. On July 9 a U-boat sank the *S.S. Benjamin Brewster* just two and a half miles off the coast of Grand Isle, Louisiana. The steamer *R.M. Parker, Jr.* was sunk not far from Ship Shoal Lighthouse on August 13, 1942.[1]

The Civilian Air Patrol chose Grand Isle south of Lafourche Parish (but part of Jefferson Parish) to have the Louisiana Highway Department construct a 900-foot runway for launching air strikes in defense against the U-boats and to rescue survivors of U-boat attacks.[2]

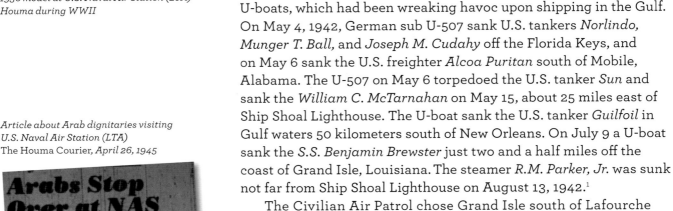

Arab dignitaries visiting Houma NAS and Texaco's Lake Pelto Field in 1945 included (front row, second from left) nine-year-old Prince Amir Nawaf Ibn Abdulaziz Al Saud (1932-2015)

In response to the crisis, the Texas Company (Texaco) at the time developed "Port Texaco," an innovative floating port terminal for storing and discharging crude oil to large tankers[3] offshore of Terrebonne Parish. The tankers stored oil offloaded from barges that transported it from tank batteries located in the fields (salt domes) at Bay Junop, Four Isle, Dog Lake, Bay St. Elaine, Lake Pelto, Lake Barré, Caillou Island, Golden Meadow, and Leeville. Tugs dispatched from a dock complex at the end of the road in Montegut took crews back and forth to Port Texaco. Large tankers onloaded oil that had been stored in three World War I era tankers at Port Texaco and transported the oil to the Texas Company's refinery at Port Arthur, Texas.[3] Those and other oil tankers were the primary target for U-boats in the Gulf.

In the area the U.S. Navy called the Gulf Sea Frontier—from the eastern coast of Florida to Brownsville, Texas—41 ships were sunk in May, 1942.[3] The losses suffered by Allied shipping included 93 attacks which sank 57 ships and damaged 17 in a four-month period.[4] The Germans referred to this period as the "happy time."[5] Two dozen German U-boats were operating in the Gulf.[3]

The place locals referred to as the "Blimp Base" rose in open land in 1943 about three miles south of Houma off the upper extremes of Grand Caillou Road as one boundary. The station extended all the way to what is now Highway 56, "Little Caillou road," its southern extreme situated along Bayou Terrebonne after it forks at Presqu'ile. The station continued as a U.S. Navy installation until 1947.[6] The term blimp has its roots in British experiments with the lighter-than-air craft during World War I; a name given to one prototype was the B-limp (referencing its non-rigid structure).

According to USS KIDD Veterans Memorial documents on Louisiana's Military Heritage, "This installation saw its origins in July of 1942 when the U.S. Coast Guard stationed a detachment of J4F *Widgeon* aircraft here to help patrol the Gulf coast against the threat of German U-boats. The U.S. Navy later purchased 1,743 acres of land from Terrebonne Parish and the City of Houma, the South Coast Corporation, R.R. Barrow, Inc., and several small landowners for the establishment of a Naval Air Station. Included in this land purchase was the 613-acre Houma Airport. Construction was begun on August 6, 1942, of an air station that would house an LTA (Lighter Than Air) squadron of blimps."

Above, S.S. Benjamin Brewster *sank July 9, 1942*

Port Texaco located on Terrebonne Coast, 1943

Sun, *sank May 6, 1942*

William C. McTarnahan, *sank May 15, 1942*

R. M. Parker, Jr., *sank August 13, 1942*

U.S. NAVAL AIR STATION LTA HOUMA | 245

Texaco Hangar, U.S. Naval Air Station (LTA) Houma, La., Bernard B. Davis Aerial Photographic Service c. 1940

First commanding officer of the Houma Naval Air base, Commander Bernard F. Jenkins, USN, 1943

Combat Air Crew arrives at Houma Navy Base The Houma Courier *August 3, 1944*

During clearing of the land, the Myrtle Grove sugar mill had to be demolished, and the small Texas Company hangar had to be relocated; that hangar remains to this day on the spot to which it was moved. The W. Horace Williams Co. of New Orleans was the prime contractor on the Naval Air Station.[7] Later in the war, a "shell plant" operated by Leach's Machine and Boilerworks manufactured artillery shells in Houma.[7]

The Louisiana's Military Heritage documents continue, "The 1,000-foot wooden blimp hangars at NAS Houma were unique. Other such hangars had sliding, sectional doors that were moved on overhead tracks in the hangar's doorsills. Due to the soft, shifting soil of south Louisiana, the hangar doors at NAS Houma were built in a clamshell design that moved on tracks and rolled outward away from the entrances. A landing mat 2,000 feet in diameter accommodated the blimps while two (2) concrete runways and several hangars were used by the HTA (Heavier Than Air) aircraft of the Navy, Coast Guard, and the Texas Company.... NAS Houma was commissioned into service on May 01, 1943, under the command of Commander Bernard F. Jenkins, USN."[8] The original cost of the hangar's construction was $3,800,000.

Although the above references more than one hangar, only one was built, since by the end of 1943 the U-boat threat was gone from the Gulf of Mexico. The hangar was 300 feet wide and 225 feet high. According to an article (number 23) by C.J. Christ entitled "History of NAS Houma Filled with Interesting Stories," "The men and materials [for the projected second hangar at the Houma base] were diverted to Trinidad instead."

Naval Airship Association, Inc., records from the World War II Era described the hangars at major airship bases as being built of timber to preserve steel, including that at Houma. The association made the statement that the hangars were possibly the largest timber structures ever built, adding that only Houma's hangar featured office-building sized movable doors. The hangar accommodated six inflated blimps, and was six football fields in length.

A total of 17 hangars were built, at Richmond, Florida; Brunswick, Georgia; Hitchcock, Texas; Tillamook, Oregon; Santa Ana, California; and Houma. A total of 11 blimp squadrons served at these facilities; Houma had only one of those squadrons.

Houma Airport postcard c. 1940, NAS Houma matchbook cover, Bayou Bomber patch, envelope bearing NAS Houma Dedication, postmarked May 1, 1943

U.S. NAVAL AIR STATION LTA HOUMA | 247

U. S. Naval Air Station (LTA) Houma c. 1944

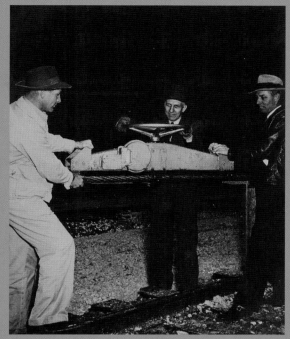
Closing the hangar door: Ernest Slade and R. J. Druealing

Opening of hangar doors c. 1944

U.S. NAVAL AIR STATION LTA HOUMA | **249**

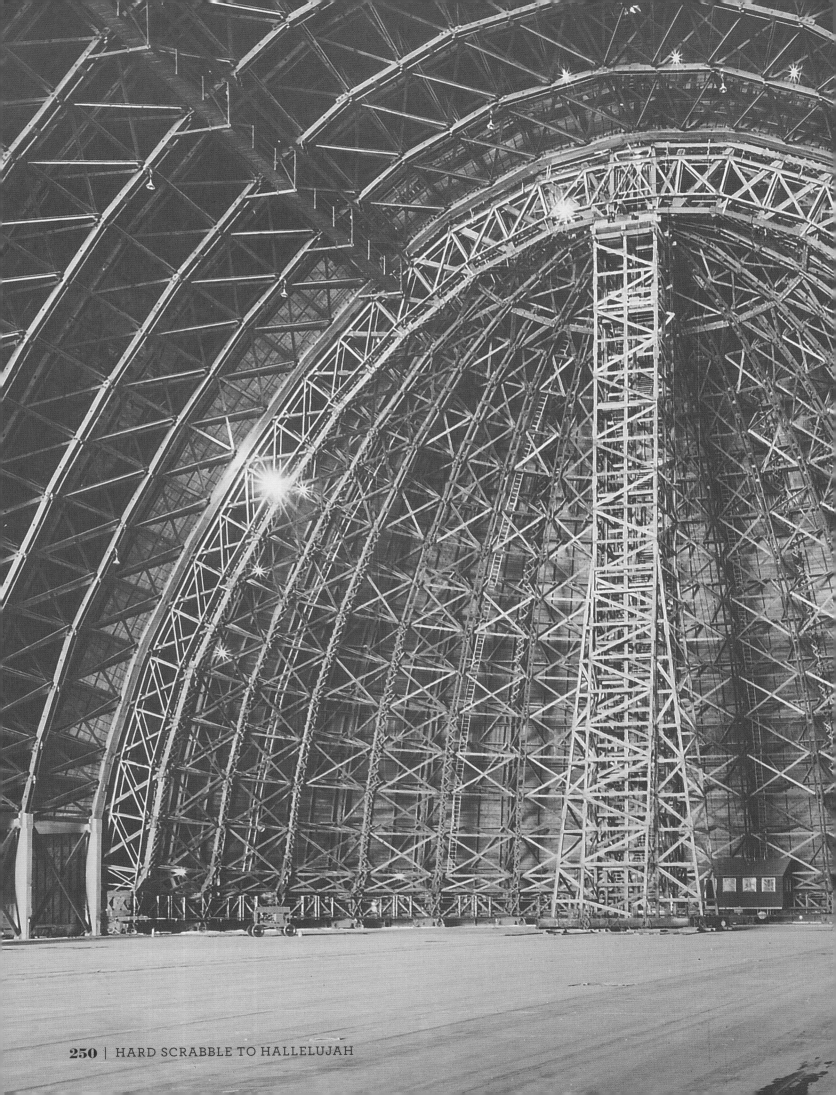

North door, U. S. Naval Air Station (LTA)
Houma November 10, 1944

Inset shows a man standing
inside the hangar, bottom right

Continue Search For Men Killed In Blimp Wreck

Investigate Accidents in Which Nine Men Killed. Three Blimps Destroyed, Another Damaged.

A search continues for the bodies of the nine missing men who died in the disastrous accident of a U. S. Navy blimp which was destroyed during a tornado Tuesday of last week. An investigation is being made of the accident as well as the losses Thursday night in which two other blimps were destroyed and a third badly damaged during a freak wind which ripped them from their moorings.

Monday the public relations office of the Eighth Naval District released the names of

Article on search after 1944 blimp crash in Gulf of Mexico, Terrebonne Press, April 28, 1944

U. S. Navy blimp K-62 burns in Houma on April 21, 1944 after being blown by strong winds from its hangar

Tribute Paid To Crew of 9

A memorial service in memory of the nine men who gave their lives in the crash of the K133 navy blimp, was held at the Houma Naval Air station Saturday morning, April 29.

The service took place on the "take off surface" of the mat where the squadron, in dress whites, lined up in formation.

Nine wreaths, brought to the ship by nine men of the squadron, were given to Lieutenant Commander H. M. Harris, commanding officer of the squadron, who delivered them to Chaplain W. L. Lancey aboard ship.

At the conclusion of the service aboard the mat, the ship took off out over the Gulf. Prayers were offered for the souls of the

Article on Navy paying tribute to victims of April 19, 1944 blimp crash, The Houma Courier, April 27, 1944

Above, Ensign William "Billy" Thewes on bottom left, after rescue from blimp crash

Sign at U.S. Naval Air Station entrance Grand Caillou Road, 1943

252 | HARD SCRABBLE TO HALLELUJAH

Documents from the KIDD Veterans Memorial gave further details about the base. "LTA Squadron ZP-22 was commissioned into service at NAS Houma on May 15, 1943, and began conducting anti-submarine patrols of the Gulf coast. Operations continued until the unit was decommissioned on September 12, 1944. The base ceased to be an LTA facility on September 21, 1944, and began serving as an HTA training facility with Coast Guard air/sea rescue flight operations continuing. In 1945, NAS Houma was redesignated as a Naval Air Facility (NAF). Following the cessation of hostilities, the base served as an aircraft storage facility until October of 1947. . . .The field serves today as the Houma-Terrebonne Regional Airport." The Navy discontinued the use of blimps as of March 19, 1959 when it retired K-43, the last K-ship in service.[9]

Names given to the blimps were Dopey for K-17, Bayou Bomber for K-22, Maurauder for K-40, Acey-Deucy for K-53, Jean Lafitte for K-54, and Lafitte-related names Dominique You for K-57, Nez Coupe for another, and Gambi for yet another.[10] No submarine sightings were recorded from blimps based in Houma.

"Blimp Drops Bomb on Golf Course" was the headline for a Terrebonne Press news item on Friday, April 28, 1944: "A bomb was accidentally dropped on the Houma golf course Wednesday afternoon, according to an official statement by Lieut. Commander H.M. Harris, commander of the blimp squadron at the Houma Naval Air Station." The Eighth Naval District said the missile, embedded in the golf course, was harmless in its present state and that bomb disposal officers had been sent to remove it.

Incidents involving LTA ships from the Houma station occurred just two days apart in April 1944. A list of airship accidents chronicled that the U.S. Navy airship *K-133*, of Airship Patrol Squadron 22 was caught in a thunderstorm on April 19 while patrolling over the Gulf of Mexico. Nine of ten crew members were lost, and the sole survivor spent 21 hours in the water before being recovered. He was the young pilot Ensign William Thewes, who suffered from exposure but recovered quickly when found. The tragedy was called the worst World War II airship accident on record.

Two days later, April 21, the southeast door of the local blimp hangar was chained open, when a gust of wind carried three K-class blimps out into the night. The airship K-56 traveled 4.5 miles before crashing into trees. The K-57 blimp traveled four miles from the station, where it exploded and burned. The third, K-62, hit a high-tension power line a quarter mile away on the base and burned. The K-56 was salvaged and sent to Goodyear Tire and Rubber Company in Akron, Ohio, for repair, and returned to service.[9]

The Houma air station was staffed not only by military personnel, but also by a whole cadre of local support personnel.

Associated Press article on blimp accidentally dropping bomb on Houma Golf Club links

Houma Country Club March 17, 1942

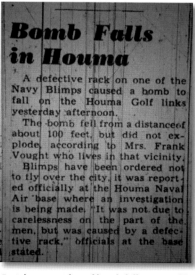

Article on accidental bomb falling on golf course The Houma Courier, April 26, 1944

Article on blimp dropping bomb on golf course Terrebonne Press, April 28, 1944

254 | HARD SCRABBLE TO HALLELUJAH

Both pages top row, first left, Ship's Company marching to Recreation Hall for memorial service for Secretary of the Navy Frank Knox, on April 30, 1944; left foreground, the Rec. Hall and left center, the Mess Hall, with the main gate in the far right background. Next photo, memorial service for three officers and six enlisted men lost in K-133 which went down in the Gulf during a storm. The blimp shown flew over the Gulf for services conducted by chaplain. Above, Mess Hall. Top right, Ship's Company in front of Administration Building. Both pages center row, first left, base infirmary. Next photo, Administration Building. At right, Recreation Hall contained an auditorium, library, pool room, service store and canteen, post office, barber shop, several offices, restrooms, and later a bowling alley. Both pages bottom row, first left, executive officers' quarters on left, and bachelor officers' quarters on right. Next photo, Barracks B. Below, commissioning of NAS Houma on May 1, 1943. Bottom right, the first blimp to arrive at NAS of the first squadron ZP 22. It arrived at 3 a.m. on May 13, 1943, circled the base, and moored at 6:05 a.m. The blimp then circled the base and surrounding area all day.

Some military men married Terrebonne Parish girls and settled in the parish after the war.

The air station included a base chapel for all denominations, a 100-bed hospital, four tennis courts, bachelor officers' quarters, post exchange and gymnasium, enlisted men's barracks, WAVE/nurses/female officers' quarters, railroad to hangar and fuel farm, explosives bunker, and necessary administrative and other facilities having to do with the workings of the base. A prisoner of war camp for Germans was at the southern end of the air station and across from Woodlawn plantation house, now Woodlawn Ranch Road.

The local squadron was deactivated on September 12, 1944. The Terrebonne Parish School Board leased two barracks buildings to use as the first high school for African Americans of the parish after the war's end.[6]

Beginning in 1946, requests by local governmental officials, and U.S. Navy decisions, determined the fate of Houma air station facilities' use as follows:

2/15/46 Sale of NAS surplus

10/7/46 Mayor Leon Gary of Houma requested acquisition of NAS facilities.

12/15/46 Portion of NAS opened for civilian aviation use, Veterans Flying School, and development of air transport service to serve the community.

1/31/47 Rear Admiral J.J. Clark announced that the U.S. Navy would continue to use the previously-requested facilities.

6/11/47 Houma City Council and Terrebonne Parish Police Jury agree to proceed with design and construction of a 100-bed community hospital and bond issue.

10/15/47 NAS Houma was deactivated. Valuation: $13 million. Hospital (100-bed), airport facilities, workshops, offices and furnishings, bachelor officers' quarters, cooking equipment and tableware. The base consisted of 1743.18 acres purchased by the Navy for $251,190. Improvements valued at $13,725,930. Two hundred fifty acres in buildings, hangars, concrete and asphalt runways, 1,500 acres of agricultural land. All available for lease of one year, renewable without notice, with stipulation of 30 days' notice of termination of lease by the government in the event of a national emergency. Maintenance of the property as a public airport by the parish and city as ample consideration.

1/27/48 Terrebonne Parish Police Jury and City of Houma create Airport Commission. Everette Richaud, chairman, signs lease after U.S. Navy agrees to amendments concerning insurance and fire protection requirements.

3/10/48 Texas Company bid $800 per month for ten buildings and $75 per month for commanding officer's residence.

10/7/48 Terrebonne Parish Medical Society agreed to put in motion plans for the establishment of a hospital in Houma "as soon as possible." Urged public officials to proceed with an election to ratify a constitutional amendment which would permit public ownership of a hospital to be operated on a lease by a private or

Top to bottom, sailors swimming at Legion Pool The Houma Courier *August 17, 1944; sailors swimming at Grand Isle* The Houma Courier *July 1, 1944; article on USO shows on blimp base* The Houma Courier *April 27, 1944; article on Waves arrival* The Houma Courier *August 17, 1944; Marines cutting sugar cane to prepare for Naval Air Station* The Houma Courier *November 11, 1943; below, Ship's Service Restaurant (Mess Hall) c. 1944; above left, first road names on U.S. Naval Air Station were Moffet and Berry, for deceased officers* The Houma Courier *July 14, 1944*

religious organization, financed by a bond issue. The amendment would allow the operators to make legitimate changes sufficient to make the facility self-sustaining at no cost to the taxpayers.

7/1/54 Terrebonne General opened as a 76-bed hospital with 80 employees and nine doctors. In 2014, it celebrated its 60th anniversary. TGMC in 2016 has 321 beds, 1,400 employees, and 300 physicians. It is the largest medical facility in the tri-parish area of Terrebonne, Lafourche and St. Mary.

When the enormous blimp hangar became a safety concern by standing two years past its initial projected duration of five years, the Cleveland Demolition Company strategically set dynamite on February 16, 1949, and in two explosions set six seconds apart, the hangar came down. *Courier* reporter Jonathan Edwards described it at the time. "A twist of the wrist, a huge flash of flame, a deafening roar, and what had become a landmark to Terrebonne was no more."[11]

Superintendent of the demolition, Arthur Caya, estimated that about 3 million board feet of lumber remained from the structure after the blast.[11] In newspaper articles and advertisements preceding the actual demolition, Cleveland Demolition Co. representatives said they would set up a small sawmill at the base, to prepare the salvaged wood for sale locally. The company projected that whatever didn't sell locally would be shipped to California. Wood from the hangar was sold to build any number of structures in and near Terrebonne Parish.[12]

SOURCES

1. World War II Database: Caribbean and Gulf of Mexico Campaigns, 16 Feb 1942-1 Jan 1944 and Regional Military Museum, Houma, Louisiana
2. *The Lafourche Gazette*, August 31, 2014
3. Jason P. Theriot, "German U-Boats and the Oil War Off Louisiana's Coast, 1942-1943," in *The Lafourche Country III*, edited by John P. Doucet and Stephen S. Michot, Lafourche Heritage Society, 2010
4. C.J. Christ, Article 55: "Blimps: How They Work"
5. C.J. Christ, Article 15: "War Activity in Gulf Centered on Oil"
6. C.J. Christ, "Blimp Wrap-Up," *The Courier*, February 3, 2008
7. C.J. Christ, "Highlighting the Civilian Side of the War Effort in Houma," *The Courier* newspaper, January 29, 2006
8. USS KIDD Veterans Memorial, "Louisiana's Military Heritage"
9. http://www.joebaugher.com/navy serials/third series4.html:U.S. Navy and U.S. Marine Corps BuNos Third Series
10. C.J. Christ, Article 22: "Houma Blimp Base Was Busy During WWII"
11. "Dynamite Blasts Bring Down Houma Blimp Hangar Wed.," *The Houma Courier*, February 18, 1949
12. *The Terrebonne Press* newspaper, September 26, 1948

Top to bottom, article on dynamiting Houma blimp hangar, The Houma Courier, February 18, 1949; second from top, ad for sale of salvaged hangar lumber, 1949; third from top, ad for blimp hangar lumber and equipment, Terrebonne Press, September 26, 1948; arrival of demolition crew, Terrebonne Press, September 26, 1948; blimp hangar dynamited, February 16, 1949

Louisiana license plate made out of sugar cane fiber board (bagasse) 1944

U.S. NAVAL AIR STATION LTA HOUMA | 259

PRISONER OF WAR CAMPS

Terrebonne Parish was the site of three POW camps during World War II. The first, called Houma #1, was established in 1943, adjacent to and north of the Woodlawn plantation house on present-day Woodlawn Ranch Road. The second, established in 1944 and called Houma #2, was in Houma on West Main Street between Boykin Street and Wolfe Parkway near Terrebonne High School and the current site of the Spanish Christian Pentecostal church. The third was established in Montegut in 1945.

In the United States at war's end, 175 branch camps served 511 area camps (one source placed the number at 700 camps during the war[1]) housing over 425,000 prisoners of war, most of them Germans and most of the prisoners located in the South, because of the high cost of heating the barracks in other areas.[2] Eventually every state but three had POW camps.[2] The POW camps in the U.S. were a result of a request by the British to alleviate POW housing problems in Great Britain.[1]

As many as 40,000 prisoners were housed in Louisiana camps from July 1943 until 1946, many of them from Rommel's Afrika Corps. The New Orleans port was the first U.S. soil most of the prisoners touched before being transferred. Four main posts in Louisiana were Camps Livingston near Alexandria, Polk near Leesville, Ruston near the city of that name, and Plauché, located near the Huey P. Long Bridge, where Elmwood Industrial Park is now. All Terrebonne camps were branch camps under Camp Livingston.

The camps in Houma were an anomaly, because War Department regulations prevented building POW camps within 150 miles of any coast. The regulation was changed at the request of Louisiana's senators at the urging of the American Sugar Cane League and Valentine Sugars to allow prisoners to be used as field laborers.[3] World War II was disruptive to worldwide production and manufacturing of sugar, when there were increased industrial and economic needs. The disruption was felt directly in Japanese-held Philippines and Indonesia, and in Europe, to beet fields. Indirectly affected were Hawaii, India, Puerto Rico, and Cuba. In the United States, Louisiana and Florida sugar cane production suffered from the labor shortage brought on by the war.[3]

A solution to the problem was presented to the Southern Defense Command on February 25, 1942 by the American Sugar Cane League to employ the prisoners in agriculture. The community, business or agricultural interests would carry the cost to establish a camp while the government contracted the labor. The Army charged $1.50 to $1.80 per hour of prisoner labor, while 90 cents went toward prisoner pay.[4] The first Louisiana POW camp rose in Franklin in 1943.

The South Coast Company was the initial employer of the

Honolulu Plantation Company 10 pound sugar sack, c. 1920

Narcisa Sugar Company token, 50 cents, Havana, Cuba, c. 1915

Cuba sugar cane carts Carret a Manzanillo c. 1914
Indonesia sugar cane factory c. 1910

260 | HARD SCRABBLE TO HALLELUJAH

1943

No War Prisoner Labor Available For Cane Harvest

Appeal to Local People to Assist in Working Out Labor Shortage Problem

Since announcement that war-prisoner labor will not be available to harvest this year's sugar cane crop, County Agent M. J. Andrepont has attacked the labor shortage problem with renewed vigor. Contacts are being made in various areas to obtain help and an effort is being put forth to induce labor from non-essential lines of endeavor to take part.

Ask Farmers to Register

Mr. Andrepont has requested all farmers to register with his department, stating the amount of labor they will need and how much they already have available. Non-essential labor sources in other areas will be tapped to take care of each planter according to his requirements as an allocation plan is inaugurated.

Articles about Prisoners of War Working During Sugar Cane Harvests, 1943

The Houma Courier
September 23, 1943

The Houma Courier
September 2, 1943

Terrebonne Press
August 13, 1943

The Houma Courier
October 21, 1943

The Houma Courier
September 23, 1943

The Houma Courier
November 11, 1943

Terrebonne Press
November 12, 1943

The Houma Courier
December 24, 1943

Times Picayune
October 9, 1943

Prisoners of War to Remain Here for Sugar Cane Season

May Use Country Club

George Harmount was appointed to call upon the officials of the Houma Country club to find out if that building can be used to house the officers and guards, if necessary.

Commander Percy Porche of the local American Legion post agreed to call a meeting of his post to see if they would allow the use of the Legion grand stand for housing the prisoners if it could be remodeled in time for this purpose, but it has been found that it would be too costly and two or three camps will have to be established.

Application for one of these camps has already been sent in to C. E. Kimberley, Jr., state director of emergency labor in the office of the extension service, Baton Rouge, Mr. Andrepont said today.

It was explained by the chairman that housing and transportation for the prisoners are to be furnished by the community. A six-foot barbed wire fence must surround the camp, which must be fully equipped with flood lights and a guard will be furnished by the army for every four prisoners.

Regulations

According to government regulations, each man must have 40 square feet of space, and lights, water, screening and refrigeration must be available. The government will provide cots for sleeping, it was announced.

Pay

The prisoners are to be paid from 50 to 75 per cent of the prevailing wage rate in the community, and their pay must be at least $1.50 per work day, it was explained.

Quarters and mess hall, sanitation and bath facilities for the prisoners must be separate from those of the guards, it was pointed out, and employers must provide transportation to and from the camp and the tools for working.

Prisoner labor must not supplant free labor and the prisoners must not work longer than civilians in the vicinity, it was explained. The county agent cautioned all farmers not to relax their efforts to secure all available labor for the harvesting season.

Present at the meeting were Mayor Elward Wright, Dr. M. V. Marmande, Sheriff Peter Bourgeois, Lieutenant Henry O'Connor, E. R. Theriot, Commander Percy Porche and Earl Boudreaux of the American Legion, Numa Olivier, Elliott Jones, M. Brien, John Daigle, Temus Bonnette, Superintendent H. L. Bourgeois, George Harmount, Ashby Pettigrew, Jr., A. D. O'Neil, George Keen, James Chauvin, Ivy Redmond, Ensign Lehman, Mr. Andrepont nad Roy Hebert, assistant county agent.

German War Prisoners Will Arrive At Ashland Plantation Tomorrow

County Agent Milton J. Andrepont announced yesterday that the German war prisoners and their guards will arrive tomorrow and will be taken directly to Ashland plantation in special railroad cars. The group will be a part of a larger number scheduled to arrive at Schriever, one allotment going to Lafourche and the other to Ashland.

Precautions Taken

Mr. Andrepont said the tent village to house the men will be adequately guarded with an armed guard at each fence corner. Stakes have been placed within the fences and no prisoner will be allowed beyond them. The lighting which is provided by REA is controlled by one large switch outside the enclosure but the wiring is such that it takes more than one switch inside to turn off lights, a precaution that makes it possible to keep the enclosure brightly lighted regardless of possible maneuvering by prisoners.

Check System Inaugurated

An accurate system of checking the prisoners and their cane knives [illegible]. All knives will be laid out on the ground and checked in by number before the truck leaves the field. If a knife is missing it will be found before the crew returns to quarters. The knives will then be placed in a locked box out of reach of prisoners who will be under constant observation by armed guards on the cab of the truck and by another guard who will follow in another vehicle.

Open Space Required

Mr. Andrepont stated that all prisoners will work in open places and that an opening at least 12 feet square is required at all times. The number of guards in each instance will be governed by the type of field and type of cane being cut. The men will be under constant surveillance during rest and lunch periods. They will be away from the camp an average of ten hours daily.

Heated Tents

Mr. Andrepont said the men will live in tents and that each will be adequately heated by a stove of sufficient capacity to keep the quarters comfortable. No tent or any other obstruction will be within [illegible] feet of the fence.

Prisoners Here for Harvest

Two hundred and fifty German prisoners of war arrived here by train Saturday morning to aid in harvesting Terrebonne's 22,000-acre cane crop. They were taken by truck to the camp at Woodlawn plantation where the advance guard had set up tents.

The prisoners are to be transported to and from the cane fields in trucks where they must be seated on benches during the time they are riding, said County Agent Milton J. Andrepont. Guards will follow the trucks carrying the prisoners in another truck or will be seated on a specially constructed platform on the cab of the prisoners' trucks.

Upon reaching the field the prisoners are to file out and cane knives will be issued, but before cutting cane in a particular square, a strip of cane of not less than twelve feet wide must be cleared so that the guards may have a clear view of the patch from all angles, the agent said.

At the conclusion of their day's work, each prisoner is checked out by turning in his cane knife before leaving the field, the agent pointed out, and until all knives are accounted for, the trucks cannot be moved.

When the prisoners return to camp, they are released and checked in. The whole camp is brightly lighted, and guards are stationed around it. These guards are specially trained men in the armed services, Mr. Andrepont pointed out.

From all reports, he added, the prisoners are doing a good job of cutting cane.

The camp at Woodlawn is a private camp belonging to the Southcoast corporation and put up at their expense, it was explained.

Fifty War Prisoners Arrive Mon.

Fifty additional German prisoners of war arrived here Monday to help save the approximately 3300 acres of sugar cane left in the fields when the hard freeze hit here last week, County Agent Milton Andrepont said today.

This makes a total of 330 German prisoners now at work harvesting cane in Terrebonne, the agent stated. From all reports, he said, the prisoners in this parish are well disciplined and are perfoming as excellent a service in the cane fields here as can be expected.

The weather is turning warmer and in ten days' time the cane will be sour and the sugar lost, Mr. Andrepont stated. Still more help is needed to save the entire crop in the short time left to do

and in ten days' time the cane will be sour and the sugar lost, Mr. Andrepont stated. Still more help is needed to save the entire crop in the short time left to do
so, and an official request has been made to the army through the extension service of the war manpower commission for 150 soldiers, the agent stated. Had the weather remained cold, it was explained, there would have been more time to harvest the cane. About 70,000 to 80,000 tons of cane were left in the fields when the temperature dropped to 24 degrees last week, he said.

The agent expressed the hope that farmers and plantation owners, as they complete their harvesting, will assist those who have cane remaining in the fields.

Felix Delatte of Southdown reported a drop in Houma and Thibodaux to 24 degrees, with a white frost Wednesday of last week. At Schriever, a temperature of 22 degeres was reported.

The regular winter vegetable gardens were not affected to any great extent by the freeze, Mr. Andrepont stated. Cabbage and carrots and other hardy winter crops will survive, he said.

PAGE SIXTEEN

EIGHT CAMPS SET FOR SUGAR AREAS

War Prisoners Soon to Begin Cutting Cane

(The Associated Press)

Dallas, Oct. 8.—German war prisoners of war housed in temporary camps will be harvesting rice this month in Southern Texas and cutting sugar cane in Louisiana.

Three temporary camps in the rice counties of Southern Texas already have been established, the Eighth Service Command prisoner of war branch announced today. They are located at Wharton, Liberty and Rosenberg. Three others will be opened within a few days, in Alvin, Bay City and Beaumont.

In the sugar cane areas of Louisiana, eight temporary camps have been authorized and it is probable that others will be established. Most of them will be in operation within two weeks.

Among the communities in Louisiana where camps will be set up to provide workers for nearby cane fields are Donaldsonville, Franklin, Napoleonville, Raceland, Port Allen, Hahnville, Mathews and Schriever.

Typical of the camps will be that at Donaldsonville, where fair grounds buildings will be used to house the prisoners and the guard company. Prospective employers of prisoners of war labor have arranged with the South Louisiana fair for use of the buildings, and will bear the cost of sheltering and transporting the prisoners.

Prisoners will be moved into the temporary camps from the permanent stockades of Huntsville and Camp Swift in Texas and Camp Livingston and Camp Polk in Louisiana. They will leave the temporary camps in the morning and be returned to them at night. While at work, as at all other times, they will be under guard.

Numbers of the prisoners at the camps will vary from 150 to more than 1000. The program calls for numbers sufficient to meet the local needs, the announcement said.

The war man power commission has certified that there is a need for agricultural workers in these areas, that hiring of prisoners of war will not be in competition with free labor, and will not impair wages, working conditions and employment opportunities, or displace employed workers.

Prevailing wages as certified by the war man power commission will be paid by the employer, with the prisoner of war receiving 80 cents per day and the rest going into the United States treasury, where it serves to reimburse the government in part for the prisoner's keep.

Seek Recruits In Sugar Cane Crop

Need 3,000 Additional Laborers To Harvest Terrebonne's 21,000-Acre Sugar Cane Crop This Season

With the harvesting of sugar cane ready to start in the parish about October 5, County Agent Milton J. Andrepont finds that the securing of labor for the harvest season is going to be extremely difficult for many reasons, because of the number of men in the armed services, the defense industries and processing plants.

The responsibility of recruiting and placement of agricultural labor on farms has been placed on the shoulders of the extension service, according to public law 45, and county agents in all sections of the sugar area are tapping all possible sources of labor for the cane fields.

Terrebonne will need approximately 3,000 additional laborers for the harvesting of this year's promising crop of approximately 21,000 acres of sugar cane, the county agent stated.

Prison Camp Site Ready

The advance group, guards [illegible] soldiers who will set up tents [illegible] the camp for the German prisoners of war which is being constructed at Woodlawn are scheduled to arrive here today, according to County Agent Milton Andrepont.

The camp, which will ho[illegible] about 250 German prisoners [illegible] will assist in the harvesting [illegible] the sugar cane crop, is priva[illegible] owned. It has been approved [illegible] the war department and is ow[illegible] and constructed at the expens[illegible] the Southcoast company, [illegible] made the contract for the emp[illegible] ment of the prisoners, the a[illegible]

1,000 German Prisoners and 250 Guards May Be Here for 90-day Harvest Season

In an effort to solve the labor problem connected with the harvesting of Terrebonne's 21,000 acre sugar cane crop [illegible] the war department has agreed that prisoners may be employed in the parish for the 90-day harvest season, it was revealed at a meeting of the Terrebonne parish labor committee held here Friday afternoon, attended by parish and city officials, representatives from the [illegible], Chamber of Commerce, school board and several [illegible] attorneys.

The meeting was conducted in the police jury room of the parish courthouse by County Agent Milton J. Andrepont, who pointed out that similar action is being taken throughout the sugar and rice sections of the state in order to harvest the season's crops.

To Be Well Guarded

It was emphasized by the group that, if ways and means can be found to establish a prison camp in the community, this camp will be well guarded with floodlights and high surrounding fences, and that there will be a guard for every four prisoners.

It was estimated that about 1,000 German prisoners will be needed here for 90 days to assist in the harvesting of this year's large crop and 250 guards will be required.

Mr. Andrepont appointed the following committee to make a survey of all available buildings in the parish to house the prisoners:

E. R. Theriot, Superintendent H. L. Bourgeois, Elliott Jones, George Keen and Dr. M. V. Marmande.

May be Named to Higher Post

Margaret Alice (Peggy) Toups Davis Wurzlow painting by German prisoner of war Lehmann c. 1944

Painting by German prisoner of war Otto Weber c. 1944

Alfred J. (Freddie) Ruiz painting by German prisoner of war Otto Weber c. 1944

Painting by German prisoner of war Otto Weber c. 1944

Aerial photo of Woodlawn POW camp Woodlawn plantation 1944

PRISONER OF WAR CAMPS, TERREBONNE | 263

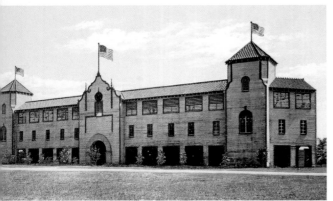

Lenox Hotard Post No. 31 American Legion Baseball Grandstand 1930

Prisoner of war camp site in Montegut, La. 2015

Prisoners of war cutting cane at Camp Livingston, near Alexandria, La., Wilkerson plantation October 1944

POWs in Terrebonne. The Houma Indians Baseball stadium and subsequently the clubhouse of the Houma Golf Course, Booty's Southerner, were discussed as possible sites of the first camp, but South Coast determined those locations to be too costly, and built at the Woodlawn site in 1943 instead. South Coast opened the Montegut camp in 1945.

Realty Operators (Southdown) began participation in 1944 and constructed Houma #2 camp on West Main Street. Realty Operators also established the Hollywood facility on Naquin Street in 1945 as a staging area, and a second staging facility at 7064 Main Street, the site of the old Mayfair Club.

The second staging area had one large tent set back about 200 feet off the street. Prosper Toups, Jr. and his sister Priscilla remembered riding with their father, Prosper Sr., who was overseer at Greenwood plantation near Gibson, to pick up the prisoners and take them to work in the morning and back in the afternoon. The prisoners began work at Greenwood at seven in the morning and were returned to the staging area at 4:30 in the afternoon. They were transported in a stake body truck with an armed guard seated atop the cab, and two guards in the rear of the truck. There were 20 prisoners seated ten on each side for the trip.

Another staging area was constructed near the Naval Air Station at the site of the current Louis Miller Terrebonne Career and Technical High School.

A March 16, 1945 document lists "Preliminary Recommendation for Utilization of Prisoners of War as Agricultural Workers" in Louisiana. The document recorded a 1945 allotment of 300 prisoners to the Houma camp, which then had a capacity of 300; Woodlawn's allotment was 350 with a capacity listed as 350 at the time.[5] Enlisted soldiers working in the fields were paid a minimum of 80 cents a day in scrip, roughly the equivalent to the pay of an American private.[6] Sugar cane farmers for whom prisoners labored would pay 80 cents a day up to $1.20 a day depending on the individual's productivity. High producers also received privileges such as a shorter work day once the specific task that day was performed.

Prisoners in Louisiana harvested rice, sugar cane and cotton. They worked in oyster packing houses, performed levee work, cleared land and broke corn in the spring and summer. Area camps were located close to where prisoners worked.

The Articles of the Geneva Convention of 1929, of which the U.S. was a signatory, restricted prisoners from producing or transporting arms and munitions or materials for combat units or in any degrading, unhealthful or hazardous work. The POWs could work on farms or elsewhere only if they were also paid for their labor, and officers could not be compelled to work, according to Geneva Convention mandates.

Weekly reports on prisoners of war from the Office of the Provost Marshal General in Washington, D.C. record the numbers of POWs at Houma camps as 656 as of November 1, 1944. Sixteen days later, as of November 16, Houma #1 housed 347 prisoners and #2 had 336, for a total of 683. By October 15, 1945, the Woodlawn camp had 352 and the

1944

Elijah "Lule" Babin, bridge tender at Presquile who shot prisoner escapee, article in The Houma Courier *May 11, 1944*

Articles about Prisoners of War Working During Sugar Cane Harvest, 1944

The Houma Courier
March 30, 1944

The Houma Courier
January 27, 1944

The Houma Courier
May 11, 1944

Terrebonne Press
April 28, 1944

The Houma Courier
April 27, 1944

The Houma Courier
August 17, 1944

The Houma Courier
July 15, 1944

Times Picayune
August 18, 1944

PRISONER OF WAR CAMPS, TERREBONNE | 265

Above, Valentine Prisoner of War Camp 1943
Below, North Thibodaux Prisoner of War Camp 1945

266 | HARD SCRABBLE TO HALLELUJAH

Above, North Thibodaux Prisoner of War Camp 1945
Below, Valentine Prisoner of War Camp 1943

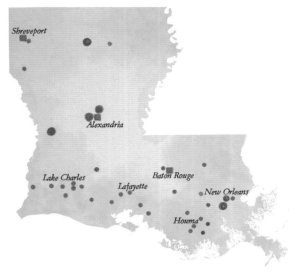

Prisoner of war camps in Louisiana 1943-1946

Houma #2 had 352, for a total of 746 in Terrebonne Parish.[7]

That date's official records tallied a total of 373,501 prisoners of war in the United States, 353,050 of them Germans, 15,521 Italians, and 4,930 Japanese. A total of 12,825 prisoners were located in the state of Louisiana.[7]

The semi-monthly report on POWs of May 1, 1945 listed the "Terrebonne" camp as having 276 prisoners working in agriculture, and the Woodlawn plantation camp accommodating 337 prisoners, also working in agriculture.[7] The prisoners at both camps labored in sugar cane fields, transported each day by truck or bus, under guard.

On November 1, 1945 the report listed 351 at the Woodlawn camp and 412 in the other Terrebonne camp. The number rose to a total of 819, the highest number recorded locally, as of November 16, 1945, and had dwindled only slightly by December 15 that year, to 815. By January 15, 1946, only one Terrebonne camp is listed, with only 266 prisoners remaining.[7]

Carolyn Portier Gorman wrote that her father, Clovis Portier, was 17 when "he worked for Sarah plantation near Chauvin and was put in charge of a busload of POWs. Dad was given a rifle, but he would laugh at that and point out that the men all had machetes to cut the cane. But they were not interested in escaping."

Grand Caillou native Gerald Collins was staying at his grandfather Joseph Fryou's overseer's house at Sarah plantation during the war. The pre-teenager drove a water cart to hydrate the field workers, and brought them lunches; they'd often reciprocate by giving him Hershey bars. Collins remembered fondly that in the afternoons the Germans stopped at the Fryou house, before being transported back to camp, to play a volleyball game against the children who lived with the Fryous, using a fence as a net.

Dr. Herman Walker, Jr. of Bourg related good memories of Sunday musicales by German prisoners at the home of his parents, Herman Ernest Walker, Sr. and Ellen Mary Ellender Walker. Mr. Walker farmed Front Lawn plantation (on the left descending bank of Bayou Terrebonne), and had 16 Camp Woodlawn POWs working his land. On Sundays, Mrs. Walker prepared a large lunch for her family and a number of prisoners who were all musicians. Because the Walkers were a musical family, they had instruments to lend the Germans, who performed concerts in the Walkers' living room--all under guard, of course. In an oral history conducted by Mr. Walker, Sr.'s granddaughter Angela Feyerabend, he recalled prisoners Herbert Heffermann, a jeweler named Willy Gartwohl, a concert violinist, a wrestler, and a painter.

In the same interview, Feyerabend discovered that neighboring farmer Wallace R. Ellender, Jr. also had 16 German prisoners working on his farm, among them Willy Kamus of Frankfurt, Irvin Georg of Berlin, a bricklayer, and a pottery and china maker named Emil Abendziller. Abendziller sent the Ellenders china he had made once back in Germany.

Good relations with the POWs are reflected also in the fact that Ellender and his wife Clara Crossman Ellender, after the war, received letters (which they kept) from German POWs who had worked the

1945

THE HOUMA COURIER
A GREATER NEWSPAPER FOR A GREATER HOUMA
Houma, Louisiana, Thursday, August 30, 1945

PRISONERS OF WAR TO STAY FOR TERREBONNE SUGAR CANE HARVESTING

Contract for PW Labor on Lily Machine

A contract for the employment of prisoners of war to work on the water lily machine in the parish has been signed and is in effect, it was reported to the parish police jury at its meeting here last week by members of the water lily committee.

Those serving on this committee are Ben Long, chairman, Rivers F. Breaux, Norbert Voisin and Thaddeus Pellegrin.

Farmers To Discuss Labor Problem

Farmers of Terrebonne parish have been invited to attend a meeting in the County Agent's office tonight, Thursday, April 5, for the purpose of discussing the local farm labor problem, according to a letter addressed to them by M. J. Andrepont, county agent.

Mr. Andrepont states in his letter that prisoners of war have been taken from the Houma camp to work on the Mississippi river.

The county agent states further that plans are being made for organizing school children as they did last year. Farmers are being urged to attend this meeting and Mr. Andrepont is soliciting their advice, he states.

Assurance that prisoners of war will be available to help harvest the sugar cane, rice and other crops in Louisiana this year have been received here by Senator Allen J. Ellender from headquarters of the Eight Service Command. The assurance has been received by Senator Ellender in response to numerous letters and telegrams that have been sent to the Provost Marshall of the Eight Service Command.

In a recent telegram to Paul V. McNutt, War Manpower Commission, Senator Ellender stated that unless the prisoners were available for the harvest season, sugar and rice interests and the pulpwood industry would be gravely handicapped.

Senator Ellender has received a letter from Eight Service Command Headquarters in Dallas, Texas, giving assurance that the prisoners would remain in Louisiana for the harvest season. The letter, signed by Waton H. Walker, Lieut. General, U. S. A. follows:

ARMY SERVICE FORCES
Headquarters, Eight Service Command
Dallas 2, Texas
20 August 1945

Honorable Allen J. Ellender
Member of Congress
Houma, Louisiana

Dear Mr. Ellender:

Reference is made to your letter of August 13, 1945, relative to a proposed reduction in prisoner of war labor to be made available in Louisiana during the next two or three months.

The allotment of prisoners of war to the state of Louisiana for agriculture work is at the present considerably above what it has been during previous months of the 1945 crop season. The allotment will be further increased during the months of November and December.

MORE ABOUT PRISONER OF WAR

recent events makes an adequate supply of free labor available.

The distribution of prisoners of war within the state of Louisiana is furnished this headquarters by the State Supervisor of the Emergency Farm Labor as concurred in by the Louisiana Farm Council, to meet the harvesting requirements of the state as a whole.

The number of prisoners of war available for industrial work in the entire Service Command for the remainder of the year has been reduced. This headquarters is keeping in very close contact with the War Manpower Commission in the allotment of prisoners of war to industrial work in order that such labor is utilized in areas of the most critical labor shortage and in the production of essential products.

The effect of recent events is not as yet fully apparent. You may rest assured that every effort will be exerted to effect a close coordination with all agencies involved and yet not create a condition whereby prisoner of war labor is placed in competition with free labor and remain within the War Department's policy in the repatriation of prisoners of war.

Faithfully yours,
Walton H. Walker
Lieutenant General, USA
Commanding

Articles on prisoners of war working in various capacities in Terrebonne Parish, 1945

The Houma Courier
March 22, 1945

The Houma Courier
February 15, 1945

The Houma Courier
August 30, 1945

Prisoner of war scrip for Camp Livingston, Alexandria, La.

PRISONER OF WAR CAMPS, TERREBONNE | 269

Letters to Wallace R. Ellender, Jr. and Clara Crossman Ellender after WWII from former prisoners of war

270 | HARD SCRABBLE TO HALLELUJAH

PRISONER OF WAR CAMPS, TERREBONNE | 271

Prisoner of war camp sketch, Main Street, Houma by Joe Boudreaux 1944

Prisoners of war loading cane near Alexandria, La., Westover plantation 1944

German prisoner of war passport 1944

Ellenders' Hope Farm in Bourg.

Keneth L. Rembert recalled prisoners of war working at the oyster shop of his grandfather Theo Engeron, Lake Fish & Oyster Company on East Main Street.

Gerald J. Voisin had vivid memories from his boyhood on Bayou Dularge of watching the German prisoners of war cutting sugar cane by hand in their white uniforms on Mulberry plantation, with armed guards mounted on horseback.

Another local, Joe Boudreaux, remembered the Germans at the Houma camp on West Main Street. A local newspaper quoted Boudreaux as stating that four guard shacks were located at each corner of the campground, but only two were manned at a time. However, the Germans never knew which ones were manned. Boudreaux could not remember any stories of attempted escapes.

Boudreaux said one of his vivid memories was that a prisoner painted onto one of the barracks artwork depicting the tools of their trade in the fields--rakes, shovels, picks and cane knives. German artists among the prisoners left a legacy of a number of paintings with families in the local area when they returned home.

Besides working in the fields, local POWs helped military personnel at the Naval Air Station plant native trees and shrubs there.[8]

Woodlawn camp prisoners were transported to the field every morning by a cane truck with a canvas top. That camp's prisoners were overwhelmingly compliant, but four Germans did break loose one night, only to be caught a short time later by authorities, but not before one of them had been shot in the leg by Presqu'ile bridge tender Elijah "Lule" Babin.

Terrebonne Parish provided a place in its sugar industry for German field workers to do hard but non-hazardous work, in the open air. Though it was spoken by a German prisoner of war at some other location, POWs here could probably have agreed with having been well fed: "When I was captured I weighed 128 pounds. After two years as an American POW I weighed 185. I had gotten so fat you could no longer see my eyes."[7]

All POW camps in the U.S. were closed and the last prisoners transported to Europe via France and England by June 1, 1946.

SOURCES

1. Michael Bowman, "World War II Prisoner of War Camps" http://www.encyclopediaofarkansas.net, *The Encyclopedia of Arkansas History and Culture*
2. "List of World War II prisoner-of-war camps in the United States," Wikipedia online encyclopedia
3. Joseph T. Butler, Jr. "Prisoner of War Labor in the Sugar Cane Fields of Lafourche Parish, Louisiana: 1943-1944," *Louisiana History*, published by the Louisiana Historical Association, in cooperation with the Center for Louisiana Studies of The University of Southwestern Louisiana, Summer, 1973
4. Dr. Matthew Schott, quoted in Sandra Corday's article "The Lives of German POWs: An Interesting Chapter in Louisiana History," Loyola University of the South publication
5. Nicholls State University Archives, Ellender Collection, Box 140, Folder 5
6. Antonio Thompson, *Men in German Uniform: POWs in America during World War II*, University of Tennessee Press, 2010 and Glenn Sytko, "German POWs in North America: The Journey to Prison Camps," (http://www.uboat.net/men/pow/pow-in-america-transport.htm),Uboat.net
7. Headquarters Army Service Forces, Office of the Provost Marshal General, Washington 25, D.C. (Restricted Classification Removed Per Executive Order 10501)
8. USS KIDD Veterans Memorial, Louisiana's Military Heritage: Forts, Camps, and Bases http://www.usskidd.com/heritage-bases.html

1946

SPECIAL ATTENTION!

The Houma Motor Co., Inc. Announces The Removal of Their Motor Co., Into a Temporary Building on the site of their New Modern Home, to be Constructed in the Early Future-Opposite POW Camp.

THANK YOU! for your patience and cooperation extended us in our previous crowded quarters at 510 Lafayette St. Our temporary structure is a prefabricated building on the site of our new location, with greater floor space enabling us to better serve you, until such time we are able to get materials to construct our new and modern designed building.

We are just outside the city limits opposite the POW Camps, on West Main St. With our larger building, more parking space, and a skilled staff, we know that you will find highly satisfactory service at our International and Packard agency.

Houma Motor Co. Inc.

INTERNATIONAL TRUCKS—PACKARD AUTOMOBILES
SALES and SERVICE
H. T. Lee, Gen. Mgr.
West Main St.—Opposite P. O. W. Camp—Phone 420

Ad for Houma Motor Co., which was located opposite the POW Camp on West Main St. Houma 1946

The "Big House" at Myrtle Grove plantation c. 1890

Inset: Trespass notice The Houma Courier *April 10, 1926*

MYRTLE GROVE

Claims by Luis Duma and Pierre Cazeau[x] were the roots of Myrtle Grove plantation on both banks of Bayou Terrebonne about five miles south of Houma. Their claims were registered with the U.S. government on November 20, 1816 and confirmed on May 11, 1820.[1]

Wrightsville, Pennsylvania native Holden E. Wright arrived in Terrebonne Parish in 1829 and soon purchased Roberta Grove plantation. He also owned Myrtle Grove[2] on both sides of Bayou Terrebonne below Residence and Roberta Grove. Wright's wife was Nancy Sarah Griffin of Woodville, Mississippi; the couple married on September 22, 1822.

The next owner of record was Robert Ruffin Barrow, Sr., but as early as 1836 Thomas R. Shields purchased the property. Cane crops of Myrtle Grove, Roberta Grove and Residence plantations were all processed at the Myrtle Grove mill, since all three farms were connected by both proximity and ownership of the Barrow family. Residence and Myrtle Grove together produced 35,000 gallons of molasses at the mill in 1844.³

Robert Ruffin Barrow, Sr. again owned Myrtle Grove beginning 1852, but he was forced to cede Myrtle Grove, Residence, and Roberta Grove plantations to his estranged wife, Volumnia Washington Hunley Barrow, in 1866.⁴

Following Volumnia's death in 1868, her two children Volumnia Roberta (known as Roberta) and Robert Ruffin, Jr. had joint ownership of Myrtle Grove. John Bradford Pittman, confidante to the Barrow family, bought Myrtle Grove at a tax sale in 1874, one year before R.R. Barrow, Sr. died destitute. In 1875 the sawmill at Residence was moved to Myrtle Grove.³ The Barrow siblings assumed Myrtle Grove ownership again in 1876. Thomas Cage and James Lunny held the property in 1878.

However, in 1882 Myrtle Grove and all other Barrow properties were separated from Residence, which Roberta had in sole possession, and Roberta Grove, for which Robert Ruffin received sole ownership. Roberta and Robert had joint ownership of all other Barrow properties beginning 1882.

Duncan S. Cage of Woodlawn plantation took ownership of Myrtle Grove in 1887. Robert Ruffin Barrow, Jr. purchased his sister Volumnia Roberta Barrow Slatter's half-interest in Myrtle Grove at a Sheriff's Sale for $9,000 in 1902.⁴ Lessee under Barrow, Jr.'s ownership was Barrow & Duplantis, under the managership of Henry Clay Duplantis, overseer of the farm at Myrtle Grove. As manager, Duplantis received a salary of half the profits of the plantation.⁴ No owner ever built a "plantation master's house" at Myrtle Grove. Duplantis held a lease there from 1888 until 1919.

Property owned by Myrtle Grove at the time included ten sugar barges, two sugar cane loaders on the plantation, seven sugar cane hoists at Company Canal, a sugar cane hoist at Bayou Terrebonne, a sugar cane scale at Bayou Little Caillou and another sugar cane scale at Bayou Blue.⁴

The *1897 Directory of the Parish of Terrebonne* by E.C. Wurzlow described Myrtle Grove as the main farm of Robert Ruffin Barrow, Jr., totaling 885.64 acres, 550 planted in cane and 350 in corn. As many plantations did, Myrtle Grove at the time had a general store on site, with clerks Joseph A. Gagné and Omer Duplantis. Merchant was D.F. Gray.

Myrtle Grove, Inc. with principals Robert Ruffin Barrow, Jr. and his daughter Zoe Gayoso Barrow were the holders of 1800 acres beginning 1923. Three Myrtle Grove entities took form in the late 1920s. Myrtle Grove Syrup Co., Inc. was formed in 1926.³ First president of the company was A.P. Cantrelle, whose employ was terminated in 1928. Mrs. Hallette Barrow Cole of Roberta Grove plantation, another of Barrow, Jr.'s daughters, then became president. The company declared bankruptcy on August 4, 1933 in U.S. District Court of New Orleans.³ Myrtle Grove plantation in 1928 was under direction of Charles L. Chauvin, keeper, and

Holden Wright (1792-1868) on his horse Boston

Henry Clay Duplantis (1846-1910)

Frank Campbell, 128 years old, former slave, c. 1941

Next spread page 276, Corporation organization for the Myrtle Grove Syrup Co., Inc., October 18, 1926; page 277 left top to bottom, C. Cenac & Co. invoice to Myrtle Grove Syrup Co., Inc., December 11, 1930; Myrtle Grove Sugars Inc. stock certificate; Bourg State Bank & Trust Co. checks; Barrow & Bonvillain business card; right, top to bottom Myrtle Grove Syrup Co., Inc. Incorporation by Louisiana Secretary of State December 16, 1926; Myrtle Grove Sugars Inc. letterhead; Myrtle Grove Syrup Co., Inc. invoice October 23, 1930; Myrtle Grove Syrup Co., Inc. envelope

ALBERT P. CANTRELL, President   HALLETTE B. COLE, Secty.   C. GRENES COLE, M.D., Treas.

MYRTLE GROVE SYRUP CO., INC.

HOUMA, LA.

SHEET NO (ONE)

THE MYRTLE GROVE SYRUP CO INC
================================

A corporation organized on the Eighteenth day of the month of October, Nineteen Hundred Twenty Six under the laws of incorporation of the State of Louisiana.

The object and purpose for which this corporation is organized are to engage in the cultivation of Sugar Cane and other crops, to manufacture refine, buy, sell, same, to operate railroads, Barge Lines and in connection to engae in other entrepriseto furthur the success of this Corporation, as evidence by copy of charter filed in the office of the Clerk of Court of the Parish of Terrebonne, under date of October 18th 1926.

The Capital Stock authorize being fixed @ $20,000.00 which may be increased to $50,000.00.
Divided into 200 shares of the par value of One Hundred Dollars each, which shall be paid for in Cash.

The following are subscribers to Stock.

Mr A.P. Cantrelle	100 Shares	Houma La,
Dr C.G. Cole	20 "	17 Newcomb Blvd N.O.
Mrs Zoe B. Topping	20 "	4938 St Charles St N.O.
Mrs Hallette Cole	20 "	17 Newcomb Blvd N.O.
Mrs Jennie B Dawson	20 "	20 Clarilon Ave Montgomery ALA,
Mr Harris Gagne	20 "	Houma La,

The following shows ASSETS and LIABILITIES as of Date

September 1st 1927.

Assets

Fixed Assets.

Live Stock	7,970.00	
Farming Implements,	8,933.15	
Barges and Boats,	5,000.00	21,903.15

Improvments and Repairs.

Repairs to Factory	11,792.17	
" to Buildings	1,406.97	
" to Railroad	1,827.79	
" to Barges	184.82	15,211.75

HOUMA, LA., Dec 11 192_

To MYRTLE GROVE SYRUP CO., Inc., Dr.

State of Louisiana

I, THE UNDERSIGNED **SECRETARY OF STATE**, OF THE STATE OF LOUISIANA, DO HEREBY CERTIFY THAT

a certified copy of Articles of Incorporation of the

MYRTLE GROVE SYRUP COMPANY, INC.,

domiciled at _____ in the _____ Parish of Terrebonne, State of Louisiana, a corporation organized under the provisions of Act No. 267 of the Session Acts of the General Assembly of this State for the year 1914, approved July 9th, 1914, being by act before Herbert Clarence Wurzlow, Chief Deputy Clerk of Court and Ex-officio the said Parish and State, of date the eighteenth day of October 1926, and recorded in _____ book No. 2 Folio 171 of the records of the Recorder of the Parish of Terrebonne, has been duly filed and recorded in this office, in book "Record of Charters" No. 117 folio _____ et seq. on the sixteenth day of December 1926; the Incorporation Tax of Ten and no/100 is $ 10.00 Dollars, which is one twentieth of one per cent of the capital stock, and other fees having been paid as required by law, the said corporation is now authorized to transact business in this State subject to the restrictions imposed by law, and especially the provisions of Act No. 267 of the General Assembly of 1914, approved July 9th, 1914.

INCORPORATED UNDER THE LAWS OF LOUISIANA

NUMBER 7   SHARES

MYRTLE GROVE SUGARS, INC.
HOUMA, LOUISIANA
Capital Stock $10,000.00

This Certifies that _____ is the owner of _____ Shares of the Capital Stock of **MYRTLE GROVE SUGARS, INC.** transferable only on the books of this Corporation in person or by Attorney upon surrender of this Certificate properly endorsed.

IN WITNESS WHEREOF, the said Corporation has caused this Certificate to be signed by its duly authorized officers and its Corporate Seal to be hereunto affixed this _____ day of _____ A.D. 19__

SECRETARY   PRESIDENT

SHARES $100.00 EACH

DR. C. G. COLE, President   LOUIS M. DILL, Sec'y.   HALLETTE B. COLE, Treas.

MYRTLE GROVE SUGARS, INC.
HOUMA, LA.

Mr A R Dizurie
Montegut, La

Dear Sir

Yours received in regards to the cane you have down the St Louis Canal will say that it is to far and out of the way for us to handle, but in case we can spare the Barges when some of our sellers get through, and you can arrange to deliver enough to justify us making ...

HOUMA, LA., _____
BOURG ST[ATE BANK & TRUST CO.]
PAY TO THE ORDER OF _____
_____ DOLLARS
MYRTLE GROVE SYRUP CO., Inc.
President _____ Secretary

HOUMA, LA., _____ 192__ No. _____
BOURG STATE BANK & TRUST CO.
84-340
PAY TO THE ORDER OF _____ $ _____
_____ DOLLARS
MYRTLE GROVE SYRUP CO., Inc.

R. R. BARROW
F. A. BONVILLAIN

BARROW & BONVILLAIN
MYRTLE GROVE PLANTATION
Houma, La., Dec 11th

HOUMA, LA. Oct 23 1930

M _____ Hutching

To MYRTLE GROVE SYRUP CO., Inc., Dr.

Freeman Williams
4x1-35	Compass	1 acre @ 2.00	2.00
4x1-29	Compass	3/4 acre @ 2.00	1.50
4x1-35	Compass	1 acre @ 2.00	2-
4x1-35	Compass	1 acre @ 2.00	2-
3x1-17	Compass	1/2 acre @ 1.50	.75
3x1-19	Compass	1/2 acre @ 1.50	.75

Peter White
4x1-8	Compass	1/4 acre @ 2.00	.50
3x1-17	Compass	1/2 acre @ 1.50	.75
3x1-8	Compass	1/4 acre @ 1.50	.40

AFTER FIVE DAYS RETURN TO
MYRTLE GROVE SYRUP CO., Inc.
HOUMA, LA.

Notice to Trespassers.

The undersigned would hereby notify the public from and after this date February 5th 1887, he will prosecute to the fullest extent of the law any one found hunting or tresspassing on the premises of the Myrtle Grove plantation this parish.

DUNCAN S CAGE.

Feb-5 3m.

Miss Edna Daspit, after a pleasant stay of two weeks at Myrtle Grove is at home to her many friends again.

Hunting, fishing, trapping, moss picking and otherwise tresspassing on Myrtle Grove Plantation and other lands adjoining, belonging to me, is strictly forbidden according to law.

R. R. BARROW.

Mr. Duncan S. Cage has removed from Ashland plantation, and taken up his home on Myrtle Grove.

R. V. Davis, colored, a resident of Beattyville, had two of his ribs broken last Tuesday morning. He was employed on Myrtle Grove plantation, near Houma, and fell under a hand car. The car ran over him.

THE COURIER received its first samples of sugar this week. Mr. Winslow Hatch who is taking off the crop at Myrtlegrove plantation sent us samples of the kinds of sugar he is making. Mr. Ernest Lirette another one of our Houma young men also sent us a sample showing the quality of sugar he is making on Greenwood plantation.

Top to bottom, Duncan S. Cage Myrtle Grove trespass notice, The Houma Courier *February 5, 1887; Edna Daspit home from visiting Myrtle Grove,* The Houma Courier *August 21, 1897; R.R. Barrow trespass notice,* The Houma Courier *March 9, 1912; Duncan S. Cage announcement relocating to Myrtle Grove,* The Houma Courier *November 14, 1903; Myrtle Grove employee R.V. Davis injured on job,* The Houma Courier *December 4, 1897; Winslow Hatch harvesting cane at Myrtle Grove,* The Houma Courier *November 14, 1903*

Myrtle Grove, loading cane c. 1910

On Myrtle Grove plantation: Carlos family left in photo; Vincent and Eva Defelice family at right. Sixth from left, Kiss Defelice, Gertie Pellegrin in white dress, Viona Lapeyrouse Chabert, small girl far right back row c. 1930

Tenant house near Myrtle Grove sugar mill 1910

Above, Myrtle Grove Refinery on Bayou Terrebonne c. 1930

Left, Henry Clay Duplantis, far left, and Myrtle Grove plantation house c. 1900

Evelida Daspit Duplantis (1853-1895)

Salvador Daspit, assistant keeper. Myrtle Grove Sugars, Inc. was formed in 1928 with the leadership of president, Christian Grenes Cole, M.D. (husband of Hallette Barrow Cole); secretary, Charles J. Chauvin; and treasurer, Hallette Barrow Cole.[3]

The Myrtle Grove Mill closed in the late 1930s.[2] It was torn down by the U.S. Government to use the land for the U.S. Naval Air Station (LTA) Houma in 1942. LTA signifies Lighter Than Air, and referred to blimps.

The home of overseer Henry Clay Duplantis and his family was on the left descending bank of Bayou Terrebonne at bayou side, facing the bayou, a half mile south of the Prospect Street Bridge on East Park Avenue.[5]

SOURCES
1. A Century of Lawmaking for a New Nation: U.S. Congressional Documents and Debates, 1774-1875, American State Papers, Senate
2. Helen Wurzlow, *I Dug Up Houma Terrebonne*, 1984
3. H.C. Minor Estate Partnership Papers, Louisiana State University Hill Memorial Library, Special Collections, Baton Rouge, Louisiana
4. Thomas Becnel, *The Barrow Family and the Barataria and Lafourche Canal: The Transportation Revolution in Louisiana, 1829-1925*
5. Executive Director Patty Whitney of the Bayou History Center, Inc., and Dolly Domangue Duplantis

Clemence Neze Guidry (1828-1893), Henry Clay Duplantis's mother-in-law

PRESQU'ILE
(PRESQUILLE)

The northern half of Presqu'ile lands belonged to early settlers and first claimants Pierre Gueno and Joseph Gueno as early as 1816, the year they registered the claim with the U.S. government on November 20; it was confirmed on May 11, 1820.[1]

The southern half of Presqu'ile was first claimed by Charles and Michel Billiot, for which they registered with the U.S. government on November 20, 1816, and which was confirmed May 11, 1820,[1] at the same time as the Guenos' claim.

The name of the plantation that developed there comes from the French expression meaning "almost an island" because it is located where Bayou Terrebonne divides, veers left (east) and continues to the coast at Seabreeze. Bayou Little Caillou has its origin at that point, and becomes the western border of the property, thereby creating a near-island between the two bayous. However, the plantation straddled both banks of both bayous about five miles below Houma.

The next owners of Presqu'ile (not as yet named as such) were Joseph Gueno, Delmas Gueno and Antoine Ramel, who purchased the property on December 16, 1822 for $13,850. Ramel died in 1822, and his succession was curated for Ramel's absent heirs by Henry Schuyler Thibodaux. Dated August 26, 1822, the inventory

Aerial photo of origin of Bayou Little Caillou off Bayou Terrebonne at Presqui'le plantation 1953

Aerial photo of Presqui'le mill between Bayou Terrebonne and Bayou Little Caillou 1953

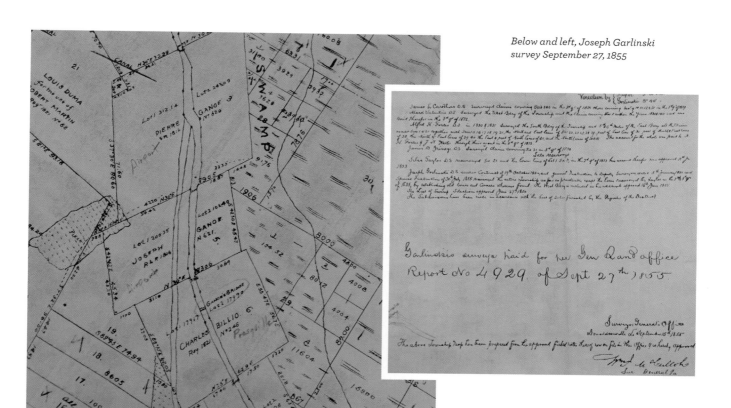

Below and left, Joseph Garlinski survey September 27, 1855

Below, Joseph P. Gueno ad in sugar report of 1876-1877

describes the partnership, and states that his heirs' plantation was valued at $4,500.00. An interesting point of his property inventory is that it detailed the origin of some of his slaves: Pierre (Houma Nation); Alexis (Senegal), Lundy and Julie (Congo), Adelaide (Congo Nation) from a list of 19 slaves including those with exotic names such as Mandinga and Figaro.[2]

One record describes the public sale on December 16, 1822 of the plantation of "Pierre Cazeau[x], Joseph Gueno, Delmas Gueno and Antoine Ramel deceased (his land purchased by the 3 remaining partners. The plantation sold for $13,850.00 ($150.00 loss), described as 27 arpents and 36 links fronting each side of Bayou Terrebonne, to depth certificate calls for, bounded above by land of Pierre Cazeau[x] and below by land of Joseph Gueno. . . ."[3]

After the death of Joseph Gueno, an inventory of the slaves of the partnership (Pierre Cazeau[x], Delmas Gueno, Joseph Gueno) dated June 27, 1829 records the sale of a number of slaves to Delmas Gueno for $2,390, and one dated August 24, 1829 records the sale of seven slaves to Pierre Cazeau[x] for $1,740.[2]

Land baron Robert Ruffin Barrow, Sr. was the subsequent holder of Presqu'ile land until it was purchased by Verret & Company in 1840.

The Gueno family once more became associated with Presqu'ile when Joachin Gueno became owner in 1846. Helping him to work his plantation in 1850 were overseer Armogene F. Aycock, carpenter Lucien Savoie and coopers Joachin and Zerphirin Boudraux. In 1858 the partnership of Bush & Gueno

Second from top to bottom, trespass notice, Cambon and Champagne, The Houma Courier *September 8, 1900; trespass notice, Gueno Brothers* The Houma Courier *September 17, 1903; Trespass notice, A.R. Viguerie* The Houma Courier *October 17, 1917*

Presqu'ile Plantation Mill c. 1900

Sugar Properties Pass to Whitney-Central Bank, Mortgagee

Houma, La., Feb. 21.—Two of the largest sugar plantations in Terrebonne parish changed hands here Saturday when the property of the Terrebonne Sugar Company and Ashland Planting Company were sold by the sheriff. Both were bid in by the Whitney-Central Trust and Savings Bank of New Orleans. This bank held mortgages on both plantations.

The Terrebonne Sugar Company was sold for $180,000 and Ashland for $175,000. Neither of these plantations was cultivated during the past year.

J. K. Wright, local attorney, has been appointed liquidator of the affairs of the People's Bank and Trust Company here. The bank closed voluntarily January 7.

Jackson Breaux, charged with breaking and entering in the nighttime and grand larceny, broke out of jail here by digging a hole through the brick wall of his cell. A slit in the screen over the window of his cell indicates that Breaux received help from the outside.

Breaux is an ex-convict, having been sent to the state penitentiary about two years ago for stealing fur in Lafourche parish. He and three other men were arrested here two months ago as suspicious characters. He later confessed to stealing $2500.

Above, Albert Robert Viguerie (1871-1954); cattle brand of Joachim Gueno April 14, 1849; notice of sugar properties passing to mortgagee Whitney-Central Trust and Savings Bank, The Times Picayune December 1927

were listed as Presqu'ile landholders with Joseph P. Gueno acting as manager.

Joseph P. Gueno was the son of Joachim and Amelesie Robichaux Gueno. The 1892 *Biographical and Historical Memoirs of Louisiana* includes a biography of Joseph P. His father Joachim died in 1880, and is credited with clearing the plantation on which his son Joseph P. then resided—1,120 acres. Joachim also sired seven sons and one daughter. He had a cattle brand registered as number 352 dated April 14, 1849. Joseph P. was educated at home and in public schools, served in the Second Louisiana Calvary in the Civil War, and returned home to manage the Presqu'ile plantation.[4]

The *Memoirs* recorded that he and his brothers A.O., J.O. and A.E. built their refinery in 1883 and in 1889 he and his brothers purchased the entire Presqu'ile estate. In 1890 they made a million pounds of sugar from their 1,020 acres at Presqu'ile. Joseph P. also owned the 1,700-acre Collins plantation which he renamed Sarah for his wife Sarah E. Davis, and had a sugar house there as well.

Gueno Brothers was the official owner from 1897 to 1904, with lessees the Cambon Brothers and Gueno & Smith (1904-05). In 1907 H.F. Posey, part owner and manager of the plantation, wrote a letter to the federal government encouraging the widening and deepening of Bayou Terrebonne. Dated June 18, 1907, Posey's letter described the total sugar output of "Presquille's" plantation as about eight million pounds of sugar annually, "not counting molasses."[5] Gueno and Posey sold the 650-acre Presqu'ile plantation and a 400-ton factory in 1909 for $75,000 to the Lower

Terrebonne Refining and Manufacturing Company.[6]

Terrebonne Sugar Company under the presidency of Albert R. Viguerie purchased Presqu'ile in 1912 and held it until December 5, 1927, when Whitney-Central Trust & Savings Bank foreclosed. At a sheriff's sale, the refinery and plantation were valued at $180,000. Carl F. Dahlberg & Co. purchased the properties on January 19, 1928.

South Coast Company, with Emile Maier president of the Terrebonne Division, obtained the land on September 27, 1935 when Presqu'ile acreage was still mostly in sugar cane.

The Jim Walter Corporation purchased the South Coast Company on June 19, 1968. Walter Land Company in 1970 with its agent Logan H. Babin, Inc. began to develop Presqu'ile and other previous agricultural lands throughout Terrebonne Parish as residential properties.

The Presqu'ile Mill burned in 1945. The store near the Presqu'ile Bridge was managed by Adney Aubin Adams, Sr. from c. 1945 until 1962, and closed when Mr. Adams went to Ashland plantation. The present address of the store location is 10627 East Main Street, Houma, Louisiana.

The main house site at Presqu'ile has not been identified. One overseer's house, that of a Mr. Plaisance from 1930 until 1943, then of Raymond Rodrigue from 1943 until closure of the plantation, subsequently purchased by John Peter Lirette, is located at 3796 Hwy. 24, Bourg, Louisiana. A second overseer's house owned by Stanley Joseph Poche (died 1999) has a current address of 3803 Hwy. 24, Bourg, Louisiana.

SOURCES

1. *A Century of Lawmaking for a New Nation: U.S. Documents and Debates, 1774-1875,* American State Papers, Senate
2. *Terrebonne Life Lines, Volume 22 Number 2,* Summer 2003
3. *Terrebonne Life Lines, Volume 15 Number 2,* Summer 1996
4. *Biographical and Historical Memoirs of Louisiana, Volume 1,* The Goodspeed Publishing Company, Chicago, 1892
5. Letter entered into the records of the 60th U.S. Congress, 1907
6. *The American Sugar Industry and Beet Sugar Gazette, Volume XI,* Chicago, January 1909

Adney and Rubella Bergeron Adams in store at Ashland plantation c. 1965

Presqui'le plantation overseer house 2015

South Coast tokens, left to right, 1 cent; one meal; 50 cents

South Coast Presqui'le Store c. 1930s

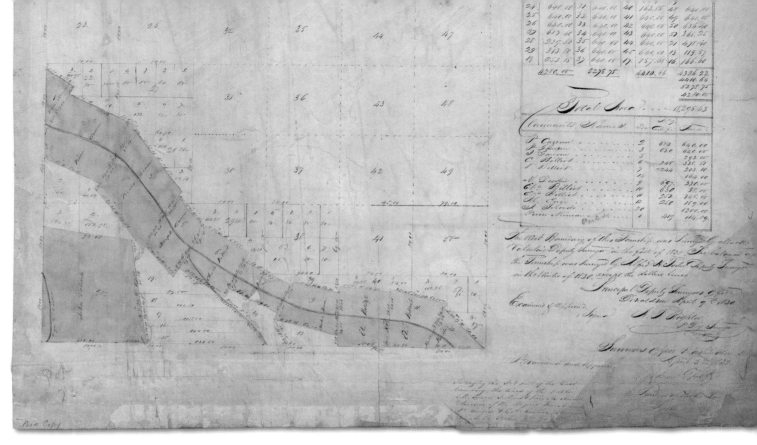

Michael Dardar Claim
State Land Grant Map September 28, 1856

FRONT LAWN
FRONTLAWN

Seven miles south of Houma on Bayou Terrebonne, Michael Dardar claimed land which he registered with the U.S. government on November 20, 1816 and which was confirmed on May 11, 1820.[1] He was the first claimant of land that would become known as Frontlawn plantation on both banks of Bayou Terrebonne in Bourg.

Dr. John W. Danks laid claim to the property in 1846. The Mississippi native, married to Catherine P. Gwin of Tennessee, died in 1859 in Terrebonne Parish.[2] The next owner of record was Robert Ruffin Barrow, Sr., who held extensive lands in Terrebonne and other parishes. His Frontlawn plantation in 1850 was overseen by W. Benedict.

In 1858 Dr. Francis Epes Robertson of Virginia bought Frontlawn for $69,362.00.[3] He was married to Kentucky native Mary Frances Moore in Houma in 1847. By the time of the 1860 U.S. Census, F.E. Robertson age 42, planter, owned $90,000 in real estate and $82,950 in personal possessions, according to census records. His eldest son John is listed in the 1880 U.S. Census as store manager, presumably of the plantation store, along with his brother and five sisters. John P. Robertson became owner of Frontlawn in 1878.

After the Robertson era at Frontlawn, the Bonvillain Brothers purchased the property in 1879. Alcide, Felix and Martial Bonvillain also owned Argyle, Ridgeland, Mulberry, and Crescent plantations elsewhere in Terrebonne Parish.[4] W.W. Pugh for a

Left to right, Martial J. Bonvillain (1868-1918), Felix A. Bonvillain (1865-1946), and Senator Alcide J. Bonvillain (1873-1937)

short time was the holder of Frontlawn before Alfred Boudreaux purchased it in 1886, and worked the land until 1902. Boudreaux was from Chenier in Jefferson, Louisiana, married to Louisiana Aimee Fields of Terrebonne Parish.[2]

Alfred Boudreaux is profiled as "among the prominent planters" of Terrebonne in the 1892 *Biographical and Historical Memoirs of Louisiana*. The magazine describes Boudreaux as living on "one of the finest plantations on the bayou" seven miles southwest of Houma. "It consists of 1,500 acres of land, with 450 acres under cultivation, and it is being improved and extended every year," the profile continues. "In 1890 and 1891 he made 665,000 pounds of sugar, and refined at Gueno Bros' refinery. He will soon erect a refinery on his plantation."[5] An 1897 Terrebonne Parish publication described the plantation as consisting of 1,100 acres of land on both sides of Bayou Terrebonne, 365 acres in cane and 175 in corn.[6] In 1890 the overseer of the mill was F.A. Aycock; Basile Breaux was assistant overseer, and Theodule Boudreaux was blacksmith.

The heirs of Alfred Boudreaux continued in their fathers' enterprise from 1904 until 1908, when the Lower Terrebonne Refining & Mfg. Co., Ltd. purchased Frontlawn in 1908. The company held the property until it was purchased by Terrebonne Sugar Company. That company was Frontlawn's owner beginning in 1916 until Herman Ernest Walker, Sr., purchased 80 arpents of Frontlawn on both sides of Bayou Terrebonne on June 9, 1926. Parts of Frontlawn have been under the ownership of the Walker family and heirs since that time.

The Frontlawn plantation home, on the left bank of Bayou Terrebonne, burned on July 5, 1958. World War II veteran Eugene Louis Guidroz died fighting the blaze. The 36-year-old was the first fireman of the Houma Fire Department to die in the line of duty.[7] The current address of that location is 4009 Hwy. 24, Bourg, Louisiana.

Herman E. Walker, Sr. and his wife Ellen Mary Ellender had five sons and four daughters. He bought lots 9 and 10 in 1938, which currently encompasses the homesites of Dr. Herman E. Walker, Jr., Linda Walker Russel, and Dr. Leslie T. Walker, Sr. (deceased). In 1944 he purchased lots 1 and 2, upon which daughter Mary Ellen Walker Feyerabend Dugas situated her home, now owned by Jack and Myra Porche.[8]

SOURCES
1. A Century of Lawmaking for a New Nation: U.S. Congressional Documents and Debates, 1774-1875, American State Papers, Senate
2. Ancestry.com
3. *The Weekly Messenger* newspaper, St. Martinville, Louisiana, June 24, 1893
4. Helen Wurzlow, *I Dug Up Houma Terrebonne*, 1984
5. *Biographical and Historical Memoirs of Louisiana*, Goodspeed Publishing Company, Chicago, 1892
6. *1897 Directory of the Parish of Terrebonne*, by E.C. Wurzlow
7. *Terrebonne Press* newspaper article, July 8, 1958
8. Information from the Walker family

Eugene L. Guidroz (1920-1958) Terrebonne Press July 8, 1958

Above, Eugene L. Guidroz, fireman who died of a heart attack at Front Lawn fire Terrebonne Press July 8, 1958

Right, article on Front Lawn fire, Terrebonne Press July 8, 1958

Herman E. Walker, Sr. (1904-1980) c. 1930s and Ellen Ellender Walker (1911-2000) c. 1930s

Adam Boquet family at Front Lawn c. 1915

Former Front Lawn home site 2015

EDMUND FANGUY PROPERTY

Cattle brand of Edmund Fanguy March 7, 1833

Charles and Vincent Fanguy Land Patent # 237 September 20, 1842

Edmund Fanguy Claim State Land Grant map June 4, 1832

Edmund Fanguy registered a land claim on Bayou Terrebonne with the U.S. government on November 20, 1816, which was confirmed on May 11, 1820. Fanguy's land was described as "a tract about a mile above Canal Bellanger, 6 arpents front on each side of Bayou Terrebonne," in probate records of Terrebonne Parish dated February 1, 1834. The plantation had been mortgaged for $700 to merchant banker Pierre Dubuys on January 1, 1823.

Pierre Billiot and Marie Jeanne, both free people of color, sold a piece of land to Fanguy on July 21, 1828, as recorded in Terrebonne Parish conveyance records. Fanguy's large farm eventually consisted of 200 arpents on both sides of Bayou Terrebonne.

Fanguy's first land ownership in Terrebonne Parish, however, was recorded as along Bayou Grand Caillou in the American State Papers documents. His registration reads, "Edmund Fanguy claims a tract of land situate in the county of Lafourche [Terrebonne Parish had not been established yet], containing ten arpens [sic] front on both sides of the Bayou Grand Caillou, and twenty arpens [sic] in depth, joining lands of George Toupes. The claimant proves possession and cultivation prior to the year 1800." His claim was registered with the U.S. government on November 20, 1816 and confirmed on May 11, 1820.[1]

A map by A.F. Rightor dated April 16, 1831, and approved in 1832 shows the Fanguy tract on both sides of Bayou Grand Caillou.[2] That tract later became Cane Brake plantation, to the south of LaCarpe plantation, which itself lay south of the large Cedar Grove plantation. Above Cedar Grove and its eastern

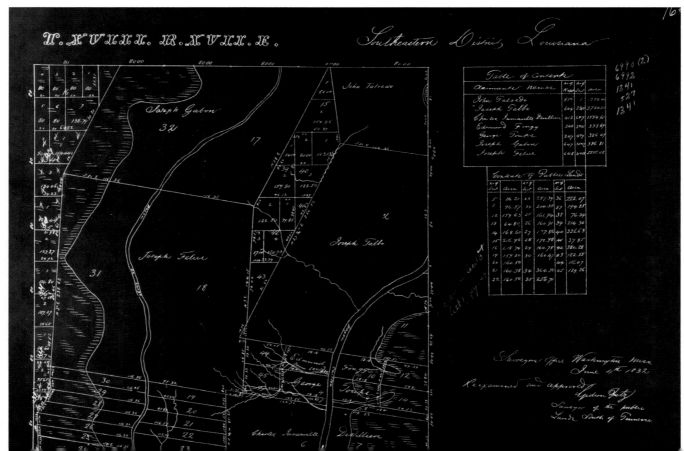

neighbor Caillou Grove was the extensive Ashland plantation; Woodlawn was situated north of Ashland.

Edmund Fanguy was descended from Pierre Fangui and Catherine Agoutte (Aoushe) of Boujean-Sur-Libron, France. The following two generations of his lineage lived in Montpellier, province of Languedoc, France. His great-grandfather Jean was a miller, and his grandfather Antoine was a china maker. Edmund's father Vincent (recorded as Vincente Fangui by Spanish authorities) immigrated to the United States and married Marie Magdeleine Laporte of New Orleans, Louisiana. He was the youngest of nine children, with his sisters and brothers seemingly immersed in the dominant Spanish culture of the time; one of his sisters married Bartholome Bosque, one married Vincent Odozgoyti, another married Don Jose de Vargas, and another married Don Philipe De Agesta.[3]

Edmund himself was a sailor, soldier, and farmer. He married Heloise Falgout in Plattenville, Louisiana in May 1814 and at some point moved to Terrebonne Parish. The couple had 11 children, the oldest of whom, Charles, was the father of Victorine-Aimée Fanguy. She became the wife of Jean-Pierre Cenac, a French immigrant from Barbazan-Debat in the Hautes Pyrénées, France.[3]

Jean-Pierre and Victorine were the parents of 14 children, grandparents of 75, and the progenitors of a large number of Cenacs and their kin from Terrebonne Parish.[3]

After Edmund's death in 1833 his wife Heloise Falgout Fanguy and heirs made no objection to a public sale of the Bayou Terrebonne land when notified in March 1843. The land was appraised at $1,200 on May 3, 1843.[4]

Edmond's sons Charles and Vincent Fanguy bought a patent for public lands along the right descending bank of Bayou Terrebonne on September 20, 1842 described as section 55 of Township 18S, Range 18E. Edmond's widow Heloise and her son Clodamire bought a patent for public lands adjacent to Charles and Vincent's purchase, section 54, T18S, R18E, also on September 20, 1842. Total area in each of their tracts was 162.90 acres.[5] Those two tracts later became part of Ranch plantation. Many of Edmond's descendants became residents of Grand Caillou.

SOURCES

1. *A Century of Lawmaking for a New Nation: U.S. Congressional Documents and Debates, 1774-1875*, American State Papers, Senate
2. Map by surveyor A.F. Rightor, April 16, 1831, of the Southeastern District, Louisiana, re-examined and approved June 4, 1832 by the Surveyor of the Public Lands South of Tennessee
3. *Eyes of an Eagle: Jean-Pierre Cenac, Patriarch*, by Christopher Everette Cenac, Sr. with Claire Domangue Joller, 2011
4. *Terrebonne Life Lines Volume 22 Number 4*, Fall 2003
5. U.S. General Land Office Documents, Charles and Vincent's certificate numbered 227; Heloise and Clodamire's certificate numbered 237

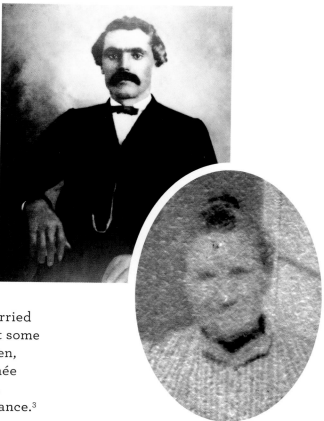

Jean-Pierre Cenac, Sr. (1838-1914) and Victorine-Aimée Fanguy Cenac (1844-1926)

Cattle brand of Jean-Pierre Cenac, Sr. January 31, 1865

Google map, highlighted sections showing Vincent and Charles Fanguy's, and Heloise and Clodamire Fanguy's, land purchases September 20, 1842

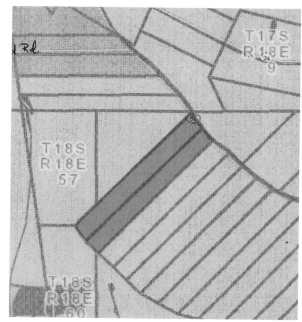

SEMPLE & SHIELDS
OAKWOOD

A triracial freewoman of color named Marianne Marie Nerisse Iris received a Spanish land grant in 1788 on both banks along lower Bayou Terrebonne that would later encompass Semple and Shields plantation.

Henry Schuyler Thibodaux's claim and Michael Dardar's claim to that land were registered with the U.S. government on November 20, 1816; the claims were confirmed on May 11, 1820.[1]

Joseph Semple and W.B. Shields in 1845 purchased the land and established their plantation there, the right bank property extending to Bayou Little Caillou. Joseph Semple is listed as sole owner in 1858.

Semple died on July 31, 1873, and in August an inventory of his property showed he owned land "a short distance below Houma, measuring 2 arpents front by depth of 10 arpents," and a sugar plantation "about 7 miles below Houma known as Plantation of Semple and Shields on both sides of the bayou. The first tract is bounded above by the land of J. Gueno and below on the left bank by land of E. Guidry and on the right bank by land of J.R. Saboular, with buildings and improvements such as sugar house, sugar mill, engine, and cabins, etc." The plantation house was on the left bank. Total value of the estate was given as $12,500.[2]

Lepine & Buron owned the plantation from 1868 until 1872. Citizens Bank of Louisiana in New Orleans acquired the properties at a Sheriff's Sale on December 1, 1877, for $9,560.00. R.H. Bagley bought the plantation in 1878.

The next year, on January 13, Semple's wife Eliza Davis Semple petitioned that "a portion of the Semple and Shields plantation has been sold, the first property was appraised at an excessive amount" and she requested a new inventory on behalf of herself and other heirs Emily Newell, Sarah Semple Rockwood heirs, Lydia Semple Paul heirs, and Mary Semple Carradine and children.[2]

In March of 1880, the property had been subdivided into 39 lots and appraised for $47,081.50.[2]

*William Bayard Shields Land Patent #1687
June 13, 1844*

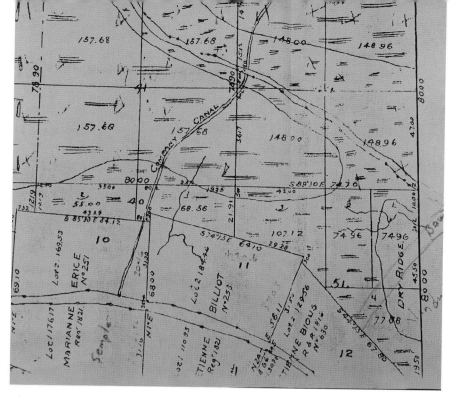

Above and below, Joseph Garlinski survey September 27, 1855

Mrs. Semple died on January 4, 1883. John R. Bisland, executor of the will of Mrs. Semple, and Duncan S. Cage, themselves owners of other plantations, put up a $1,600 surety bond on the property.[2]

In his later years prominent Terrebonnean Gabriel Montegut reminisced about his agency breaking up plantations and selling small tracts, among them "Semple Plantation adjoining Bourg."[3]

SOURCES

1. A Century of Lawmaking for a New Nation: U.S. Documents and Debates, 1774-1875, American State Papers, Senate
2. Terrebonne Parish Civil Records, Succession of Joseph Semple, submitted by Nancy Wright
3. Helen Wurzlow, *I Dug Up Houma Terrebonne*, 1984

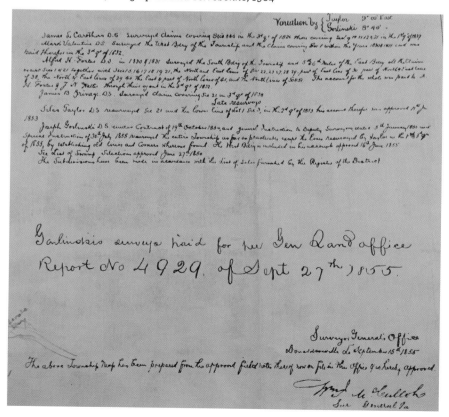

Robert H. Semple obituary Thibodaux Minerva December 16, 1854

Article on casting of Semple & Shields driving wheels Thibodaux Minerva June 11, 1853

Re-emp. Oct. 19th June 1834,

Exd 57

THE UNITED STATES OF AMERICA,

CERTIFICATE No. 2691

To all to whom these Presents shall come, Greeting:

WHEREAS Jesse Batey, assignee of François Lecompte, and Joseph Lecompte,

ha*s* deposited in the **GENERAL LAND OFFICE** of the United States, a Certificate of the REGISTER OF THE LAND OFFICE at New Orleans whereby it appears that full payment has been made by the said François Lecompte, and Joseph Lecompte, according to the provisions of the Act of Congress of the 24th of April, 1820, entitled "An Act making further provision for the sale of the Public Lands," for the Lot or Section twenty in Township seven (West of the Mississippi River) of Range nine East, in the District of Lands subject to Sale at New Orleans Louisiana Containing one hundred and eighty three acres and six hundredths of an acre,

according to the official plat of the survey of the said Lands, returned to the General Land Office by the **SURVEYOR GENERAL**, which said tract has been purchased by the said François Lecompte and Joseph Lecompte,

NOW KNOW YE, That the United States of America, in consideration of the Premises, and in conformity with the several acts of Congress, in such case made and provided, **HAVE GIVEN AND GRANTED**, and by these presents **DO GIVE AND GRANT**, unto the said Jesse Batey,

and to his heirs, the said tract above described: **TO HAVE AND TO HOLD** the same, together with all the rights, privileges, immunities, and appurtenances of whatsoever nature, thereunto belonging, unto the said Jesse Batey,

and to his heirs and assigns forever.

In Testimony Whereof, I, John Tyler PRESIDENT OF THE UNITED STATES OF AMERICA, have caused these Letters to be made **PATENT**, and the SEAL of the GENERAL LAND OFFICE to be hereunto affixed.

GIVEN under my hand, at the CITY OF WASHINGTON, the twentieth day of September in the year of our Lord one thousand eight hundred and forty two and of the **INDEPENDENCE OF THE UNITED STATES** the Sixty seventh.

[L.S.]

BY THE PRESIDENT: John Tyler

By J. Williamson
RECORDER of the General Land Office.

R. Tyler, Sec'y.

François and Joseph LeCompte Land Patent #2691 September 20, 1842

PECAN GROVE

Michael Dardar and Etienne Billiot along with Charles Billiot registered land claims on the left bank of Bayou Terrebonne with the U.S. government on November 20, 1816 and had the claims confirmed on May 11, 1820.[1]

Anatole Matherne and François LeCompte established a small holding on that land along Bayou Terrebonne in 1872.

Matherne was born July 3, 1818 in Plattenville, Assumption Parish. He married Celeste Savoie in Thibodaux on November 2, 1841. The couple had eight children.[2] LeCompte was born in Donaldsonville, Ascension Parish on April 2, 1808. He married Adelaide Gisclair on June 14, 1830 in Thibodaux. François and Adelaide had 12 children.[2] The LeComptes' son Joseph François was born in Bourg in 1833 and died on March 17, 1930.[2] The families called the property Pecan Grove, with the house located at what is now 4139 Highway 24, Bourg, Louisiana.

Locals still refer to the property as Pecan Grove. However, there is now no sugar cane acreage in the immediate area. Only the remnants of a grove of pecan trees mark the site.

SOURCES
1. A Century of Lawmaking for a New Nation: U.S. Congressional Documents and Debates, 1774-1875, American State Papers, Senate
2. Ancestry.com

Home site of Pecan Grove 2016

Spinning Cotton in Erath, 1977, Oil on canvas, 30 x 40 inches, George G. Rodrigue

Above, aerial photo of Bourg 1953
Below, Joseph Garlinski Survey September 27, 1855

294 | HARD SCRABBLE TO HALLELUJAH

Canal Belanger c. 1910

BOURG
HUBERT BELANGER AND CANAL BELANGER
NEWPORT AND COMPANY CANAL
MARIANNE MARIE NERISSE IRIS – SPANISH GRANT
JEAN BAPTISTE GUIDRY
MATHERNE DAIRY

Landowner Hubert Belanger donated a right of way on October 7, 1824 for a valuable canal thereafter named Canal Belanger, which began at the canal's intersection with Bayou Terrebonne in what is now Bourg. The land was 15 feet wide, entrusted to a citizens committee "so a canal can be made and cut away through the prairies of floating marshes, to communicate from the Bayou Terrebonne to the Bayou Lafourche. . . ." as reads the deed recorded in Terrebonne Parish, including the stipulation that "no building can be built on the property." The canal extended to Lockport on Bayou Lafourche 300 feet south of the Company Canal.

A year and a half later, on May 13, 1826, "The Canal Belanger Society" was formed "to dig a canal from the Plantation of Hubert Belanger to the canal of William Fields on Bayou Lafourche," according to a document in the Terrebonne Parish Clerk of Court Office's files. Signers were "Hubert Bellenger, P. Cazeaux, Jh. Gueno, Pre. Gautier, Bte. Duplanti, Laurent Pichof, Gerome Dupre, Pierre Brunet, Charles Purkins, Jean Jn. Tyson, J. Bte. Boudreaux,

Company Canal locks Westwego, La. 1918

Book E, No. 944, pp 283, 284, 285 3/596

State of Louisiana

Parish of Terrebonne

Today this second day of the month of June of the year of our Lord, one thousand eight hundred thirty, and the fifty-fourth of the Independence of the United States of America

Before me, Leufroy Barras, judge for the Parish of Terrebonne and ex-officio notary public in the said parish, was present Mr. Hubert Bellanger, inhabitant of this parish, who has in the presence of the undersigned witnesses declared that he had sold, ceded, and abandoned with all the warranties against all the troubles, mortgages, evictions, substitution or other impediment generally of whatsoever nature to Mr. Pierre Gautier, also inhabitant domiciled in this parish, here present in person and accepting six lots of land or pieces of ground situated in the town named "New Port", established on the upper part of land of the said Bellanger and adjoining Bellanger Canal, being Nos. 2,3,4,5,6, and 7 according to the plan placed in this office, made by James L. Carothers, dated May 1830, each piece of land having seventy feet front and eighty feet depth, being also on the land confirmed in the name of Etienne Billiot, who sold it to Mr. Henry S. Thibodaux, who sold it to the present vendor, Bellanger. And I, said judge, hereby certify that on the said land there does not exist any mortgage apparent on the records of my office.

This sale is made for and in consideration of the sum of two hundred dollars of which sum fifty-four dollars have been paid before the passing of the presents and for the one hundred and forty-six dollars the said Gautier furnished his obligation which has been signed "Ne Varietur" by me, said judge, in order to identify it with those presents and, in order to assure the payment of the said obligation, the said Gautier affects and mortgages in favor of the said Bellanger and all other legal owner or owners of the said obligation.

In faith whereof the parties have signed in the presence of and with Messrs. Hypolite Porche and Evariste Porche, domiciled witnesses, and me, said judge. The word "established" erased on the first page—approved.

E. Porche   Hypolite Porche   Hubert Bellanger

Witness

French document and English translation of Hubert Bellanger (Belanger) land sale to Pierre Gautier at Newport, Canal Belanger June 2, 1830

U.S. Delatour *dredging Bayou Terrebonne at Bourg c. 1920*

Pelegrain Boudreau, Isidore Boudreaux, Francois Boudreaux, Charles Dupre, Jean Dupre Sr., J. Charles Naquin, Widow [of] Jean Naquin, Alexandre Verdin, Jean M. Naquin, Jacques Billiot, Bt. Gregoire, Jean Charpentier, Ch. Bourque, Etienne Billiot, Pierre Billiot, Pierre Billiot, Michel Billiot, Louis Toups, Joseph P. Darce, Js. Pre. Ledet, Joseph Milome, Jean Baptis Henri, Gabriel LeBouef, J. Bt. Theodore Henry, Jean Dupre Jr., Valentin Cevin, Eugene Janvier, Solomon Verret, J. Monsan, Renaud Boudreaux, Guillaume Henry, Ch. M. Henry and Joseph Boudreaux."

When the Company Canal was dug (completed in 1836) as a leg of the Barataria & Lafourche Company No. 1 waterway system, Canal Belanger still existed 300 feet south of the newer canal, and parallel to it. Canal Belanger has dwindled to the size of a drainage ditch, the only evidence of the onetime important canal.

Company Canal locks Lockport, La. c. 1990s

Hubert Bellanger (hereinafter Belanger) and his wife, Sophie Marie Comeau Belanger were among the first pioneers to settle in what would become Terrebonne Parish. Belanger's ancestors were from Quebec, Canada, the men working as canal builders and engineers. His family of origin lived in Pointe Coupee Parish when Hubert was born in 1785, and moved to the Lafourche Interior about the time of the Louisiana Purchase in 1803. He fought in the War of 1812 as part of the DeClouet Regiment, 15th Division.[1]

As early as 1816 Hubert and Sophie lived above the current community of Schriever, where they conducted farming and ranching on their own land in the uppermost limits of Bayou Terrebonne. Hubert registered his land claim with the U.S. government on November 20, 1816 and received confirmation on May 11, 1820.[2]

This was similar timing to that when Hubert's brother-in-law Henry Schuyler Thibodaux began to amass property nearby along Bayou Terrebonne to form St. Brigitte plantation, named for Hubert's sister and Thibodaux's wife Brigitte Emelie Belanger. Another sister, Marie Agnes Celeste Belanger, was married to

> A PIONEER GONE.—Again does the painful duty devolve upon us to record the demise of another of those who were among the early adventurers in the settlement of this, now beautiful, portion of Louisiana, in the person of HUBERT BELLANGER, a brother of the late Mrs. Henry S. Thibodaux, who died of apoplexy, at his residence, parish of Terrebonne, on Wednesday the 28th ult.
>
> The deceased was a descendant in a direct line from the great French Navigator, Jacques Cartier, the discoverer of Canada, being a great-great-grand-son. He was born in the parish of West Baton Rouge, in this State, sometime during the year 1783, and consequently, at his death, had reached the seventy-second year of his age. He was one of the few remaining links which connect the present with the passing generation, and has descended to the grave honored with years, respected and cherished by his numerous relatives and friends. He was noted for his good sense, ready wit, and many amiable and agreeable social qualities. *Requiescat in pace.*

Article about legacy of Hubert Bellanger (Belanger) (1785-1855), Thibodaux Minerva *April 7, 1855*

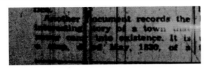

Article segments on Courthouse records 1822, The Houma Courier April 11, 1946

R.R. Barrow trespass notice for Canal Belanger, The Houma Courier March 14, 1903, and ad for Canal Belanger Fair The Houma Courier June 26, 1897; below, article about 1868 Great Western Railroad mail routes, including Canal Belanger, New Orleans Crescent May 28, 1868

Lemuel Tanner who founded Magnolia Grove/Waubun plantation.

U.S. Congressional Documents 1774-1875, Claims in Eastern Louisiana, gives some insight into Hubert's desire for land: "Hubert Bellanger claims a tract of land on Bayou Derbonne, containing six hundred and sixty-three and forty-eight-hundredths superficial acres. It is proved that this land was settled by permission of the proper Spanish officer, prior to the 20th December 1803; but as no settler by permission is entitled to more land than six hundred and forty acres, we are of opinion the present claimant cannot be entitled to a greater quantity."[2]

Sometime before 1820 Hubert and Sophie sold their land in northern Terrebonne and relocated to property they bought from Henry Schuyler Thibodaux below Montegut at present-day Madison Canal. By 1820 they were settled on land nine or so miles south of the present-day city of Houma along both banks of Bayou Terrebonne. At the final place of residence for the Belangers in what is now Bourg, Hubert was a sugar planter and tavern owner. His plantation there became a considerable estate which was worth over $30,000 in cash, "a fortune in today's money," according to an account by Edward Domangue in *Terrebonne Life Lines,* Vol. 20, No. 1.

Hubert's son, Hubert Madison Belanger (b. 1810), when he was age 21, divided land on the left descending bank of Bayou Terrebonne near his father's plantation into lots in 1830 and christened the potential community Newport. He named the streets Front, Main, Canal, Market, and Cypress.

In Terrebonne Parish records is a document dated June 2, 1830, which sheds light not only on Hubert Belanger's property ownership and sale of Newport lots, but also chronicles the earliest known land claimants of the land there. The sale was from Hubert Belanger (Bellanger in the document) to "Mr. Pierre Gautier...accepting six lots of land or pieces of ground situated in the town named 'New Port', established... according to the plan placed in this office, made by James L. Carothers, dated May, 1830, each piece of land having seventy feet front and eighty depth, being also on the land confirmed in the name of Etienne Billiot, who sold it to Me.[sic] Henry S. Thibodaux, who sold it to the present vender, Bellanger." Judge Leufroy Barras was the signer of the transaction.

A community had already centered around a tavern where Canal Belanger met the Belanger plantation. The name Newport was soon dropped.

Retired Houma Mayor Calvin Wurzlow was quoted in a *Houma Courier* Christmas 1927 edition of the newspaper, "The canal is known as Canal Belanger, the name of the village where the canal entered Bayou Terrebonne, having been changed to Bourg Post Office, in honor of the pioneer who assisted in the building of the community." The community itself continued to be referred to as Canal Belanger into the first two decades of the 1900s.

In 1894 the U.S. Postal Service changed the name of the post office from Canal Belanger to Bourg, with James P. Hotard as the first postmaster. A postal service list of Terrebonne Parish post offices records that the village of 150 in 1894 was named for Aubin Bourg, "who secured the post office." Bourg was an early Houma postmaster, in the 1850s.

The Belanger family is identified not only with the Bourg area, but also with the founding of the town of Houma. Hubert Madison Belanger (married to Marie Celine Mars Belanger) and his brother-in-law Richard H. Grinage (married to Celeste Belanger) in 1834 donated the property upon which the town of Houma grew as parish seat. The two men are credited with being Houma's founders.

When Hubert Madison Belanger died in 1843, his estate included land on Canal Belanger, a number of slaves, and four lots in Newport, valued at $13,765.

His father, Hubert Belanger, died in 1855. The land he had accumulated was described in the report of the estate sale on March 6, 1856 as "A certain tract of land, situated in this Parish, about twenty two miles below the Town of Houma, on both sides of Bayou Terrebonne, being the tract lately cultivated by Hubert Bellanger, deceased as a Sugar Plantation, having a front of twenty one arpents, more or less, on the left bank, descending, of Bayou Terrebonne, & of thirty four arpents more or less, on the right bank . . . ." The Belanger house was situated on the left bank.

The present address of what was once Hubert and Sophie Belanger's sugar plantation is 4365 Highway 24 at the intersection with Company Canal Road in Bourg, Louisiana.

Present-day Bourg is a sprawling area along both banks of Bayou Terrebonne, but its recognized town grew at the junction of Bayou Terrebonne and the Company Canal. The Company Canal, perpendicular to Bayou Terrebonne on the bayou's left bank, was dug as part of an extended canal system by the Barataria & Lafourche Canal Company No. 1 which opened on September 3, 1836.[3]

Thomas Becnel wrote, "The Company Canal was to begin where Westwego is today, on the Mississippi River opposite New Orleans. There the company [B&L No. 1] would dig a canal to Bayou Segnette, which flowed into Lake Salvador. From this lake the B&L would cut a canal to Bayou Lafourche near the settlement of Lockport. On the west bank of Bayou Lafourche the canal was to continue through Lake Field[s] to Lake Long and thence to Bayou Terrebonne near Bourg, along the old Belanger canal route. Then the route would proceed up Bayou Terrebonne to Houma, where a short canal through town would reach Bayou Black, which veered sharply west toward Morgan City. . . ."[3]

The short canal alluded to above was the Barataria Canal in Houma. The entirety of the Barataria and Lafourche Canal was often called the Company Canal, not just the section in the Bourg area.[3]

Marianne Marie Nerisse Iris, a free woman of color, was given a Spanish land grant in what became Bourg as long ago as 1787, the earliest grant recorded for Terrebonne Parish. Marianne was triracial; her father and mother were Jean Baptiste Iris and Françoise St. Therese. A native of Lafourche, Louisiana (born

Article about Aubin Bourg death, Times Picayune, 1898

Marie Celine Mars Belanger Hotard (1815-1872)

c. 1754), she married Jean Baptist Billiot in 1781 in Plaquemines Parish, Louisiana. Billiot's brother Louis Sauvage was the father of Rosalie Courteaux, a Native American woman who purchased swampland below Montegut in 1859, and who is important to the history of Native Americans locally.

Conveyance records show that Laurent Pichoff sold property on the left descending bank of Bayou Terrebonne on November 9, 1838, and that Gaubert Heirs sold a half arpent on the right descending bank on February 5, 1840. With the Belangers and Marianne Iris, these were among the earliest landowners in the vicinity of Bourg near Company Canal.

Jean Baptiste Guidry (1763-1874) bought the Marianne Iris property in the late 1700s or early 1800s. The next title holder of the land was St. James Parish native Jerome Guidry, (1817-1877), who married Thibodaux native Annette Rose Arcemond in Tensas Parish in 1839. Guidry was a carpenter and operated flatboats along Bayou Terrebonne.

Ernest Louis Guidry (1843-1914) next had the property near Bourg that had been the Marianne Iris lands. He was a native of Thibodaux married to Odelia Marie "Lucy" Champagne of Lafourche Crossing, Louisiana. Ernest was a mechanical engineer, a farmer, and ran a flatboat busines. His livestock brand was registered as number 592 on September 9, 1880 in Terrebonne Parish records. He enlisted in the Confederate States Army at the age of 16, and was a soldier in the Civil War from 1861 through 1865. Ernest served as justice of the peace in later years.

Cattle brand of Ernest L. Guidry September 9, 1880

Bourg home of Farquard P. Guidry; woman on porch Grace Aloysia Chauvin c. 1920

Ernest and Odelia's son Farquard P. Guidry (1871-1965) married Grace Aloysia Chauvin in 1897. When he was nine, he asked his parents' permission to leave school so that he could help his ill father run the small plantation his parents owned; his teacher insisted that he continue his studies under her tutelage on Saturdays. When he was ten, he rented a 12-acre plot of land, hiring friends to help him plant sweet potatoes for market. Later he purchased swampland in Bourg, and with his father's help built a sawmill to convert his raw timber into lumber. He also owned a substantial interest in the Lower Terrebonne Refining Company. At the age of 39, he founded the Bank of Terrebonne and Trust Company,[4] which he later served as president beginning in 1932. He was with the bank until his death on July 20, 1965.

When he moved to Houma in 1933, he made his home at 401 [currently 7905] East Park Avenue, the previous homesite of Jean Pierre and Victorine Cenac. Two of Farquard and Grace's daughters married two sons of John Joseph Guidry of Eloise plantation in lower Terrebonne. Loretta married Horace "Jack" Guidry and Edith married Robley "P.K." Guidry. Ray Anthony, one of Farquard's sons, registered his livestock brand as number 1,026 on May 26, 1939.

The town of Bourg is where St. Ann Catholic Church anchors the largely Roman Catholic community, and where a few retail and other businesses are located, as well. Families that have long been residents of Bourg include Hornsby, Boquet, LeBlanc, Bascle, Aucoin, Rogers, Walker, Hotard, Blanchard, Arceneaux, Hebert, LeCompte, Martin, Matherne, and Whipple.

St. Anne became St. Ann when it became its own parish in 1908. It was formed as a mission of Sacred Heart Catholic Church in Montegut and its bell was purchased for $244.50 in 1872.

Farquard P. Guidry (1871-1965)

Farquard P. Guidry home on the corner of Park Avenue and Suthon Avenue in Houma 2016

Cattle brand of Ray Anthony Guidry May 26, 1939

Left, clockwise from left, St. Ann Catholic Church, Bourg c. 1935; St. Ann Church altar 1935, Monsignor Raphael C. Labit celebrant; rectory for St. Ann Church; Le Petit Presbytere, St. Ann Church, Bourg 1935

St. Ann Church 2016

The first pastors of St. Ann, in order, top left, the Rev. Adrien Van den Broek (1908-1926); top right, the Rev. Etienne V. Galtier (1926-1929); middle left, the Rev. Raphael C. Labit (1929-1938); middle right, the Rev. Clemens Schneider (1938-1939); bottom left, the Rev. Carl J. Schutten (1939-1941); bottom right, the Rev. Paul J. Gaudin (1941-1949)

The church bell was blessed on August 21, 1872 by Archbishop Napoléon-Joseph Perché of the Archdiocese of New Orleans, in the yard of Euphrosin Hotard. Land at what was then called Canal Belanger was purchased December 04, 1872 from Narcisse Labat of Bayou Lacache for $900, for the location of a small chapel. Father Jean Marie Joseph Dénécé celebrated the first Mass at *la petite Chapelle de Sainte Anne* on January 22, 1873, and the first sacraments of Confirmation and Holy Communion were offered on October 2 that year. The second pastor, Father Peter C. Paquet, had a *petit presbytere* built, by Ulysse Naquin, to serve as a parish hall.

Father Charles Richard oversaw the construction of a church measuring 45 feet by 90 feet, built of Louisiana virgin cypress, in 1900. Luke Bethancourt, his son Joseph and his son-in-law Joseph LeCompte constructed the building. The church was completed exclusive of interior ceiling at a cost of $550, and the ceiling was completed soon thereafter.

In December 1908 Father Adrian van den Broek, a native of Holland, was appointed the first pastor of the newly established St. Ann Parish. The hurricane of 1909 caused damage to the Bourg

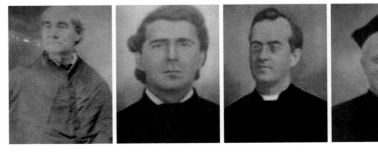

Sacred Heart Catholic Church priests who also served St. Ann, in order from left, the Rev. (Père) Charles M. Menard (1842-1864); the Rev. (Père) Jean-Marie J. Dénécé (1864-1890); the Rev. Pierre C. Paquet (1890-1895); the Rev. Charles Richard (1895-1911)

community, but parishioners continued to contribute their labor to the church. Farquard P. Guidry and others donated an organ to St. Ann. In September 1915 a hurricane demolished the mission chapel of Our Lady of Prompt Succor at Little Caillou, blew down St. Ann's steeple and caused considerable damage to the parish church at Bourg. The mission chapel was rebuilt, and St. Ann Church was repaired. Second pastor of St. Ann, Father Etienne V. Galtier, was responsible for establishing a station at Bayou Blue in which to celebrate Mass once a month, in a house offered by Farquard P. Guidry. The third pastor, Father Raphael C. Labit, oversaw construction of the Chapel of Saint Sophie in Bayou Blue in 1936. A succession of St. Ann pastors have continued the work of the Catholic Church in Bourg.

Among longtime business ventures closely tied to the land at Bourg was the Matherne Dairy, established in Bourg by Farmand J. Matherne and his brother Ernest in 1921. The dairy had previously been located in Bayou Cane, and when the business moved to Bourg, the cattle were herded through Houma down to a 112-acre tract between Bayou Blue and Bourg on Company Canal Road. The dairy herd of 60 cows consisted of Jerseys, Guernseys, and Holsteins. Raw milk was delivered in quarts, at a cost of ten cents each; the Mathernes also traded their 300 chickens' eggs for groceries. The former dairy, which closed in

St. Sophie Chapel lower Bayou Blue built 1927

Farmand J. Matherne, Sr. (1887-1980) c. 1920
Left, Matherne Dairy delivery truck; women from left, Ruby Bascle, Iola Duplantis, Della Matherne; children (from left) Kenneth Matherne, Gayle Duplantis, Richard "Dickie" Bascle 1941

Matherne Dairy in rear; horse named Patsy c. 1940

1943, was located at the current address of 203 Company Canal Road, Bourg, Louisiana.

Bourg is a quiet place where residences string themselves along the roads on both sides of Bayou Terrebonne that parallel the bayou's course. The area was noted for its expert waterfowl hunters for the New Orleans and other markets, and for those who served as guides for sports hunters.

A Houma newspaper article quoted (now deceased) retired Bourg librarian Elsie Whipple Lejeune's listing of prominent New Orleanians who visited Bourg for the waterfowl wintering in nearby Lake Long. Her father Tom Whipple and his brothers Walter, Mark, George and Willie "served as guides back then, taking the visiting sportsmen into the prime hunting areas south and east of Bourg."[6]

Miss Elsie said that the Whipple men "were known for their hunting/guiding expertise and for their decoy carving," pointing out Mark McCool Whipple as nationally known for his duck

Left, Farmand J. Matherne family, left to right: Floyd H. Matherne, Farmand J. "F.J." Matherne, Jr., Aaron W. Matherne, Bruce C. Matherne, Irvin P. Matherne, Stella R. Matherne Floyd c. 1985

The Charles Frederick Whipple family, from left: George Whipple, Charles "Willie" Whipple, Caroline Whipple (seated), nephew Walter, Jr. at her side. Standing in rear, Sarahlene Falgout Whipple; Mary Augustine Mars Whipple with grandson Eugene beside her; standing, Mark McCool Whipple; bearded Charles Frederick Whipple, Thomas Andre Whipple and Walter Mitch Whipple 1895

Roderick "Duck" Whipple (1912-1991) painting by Linda Sauls c. 1960

Decoy carved by Roy P. Whipple, Sr. (1903-1978)

decoy carvings.[4] Another family that carved excellent duck decoys were Lucien "Yak" Falgout and his son Loris Falgout.[5] (Still others were Curt Fabre and Carl David "Tabin" Ellender.)

Additionally well known Bourg hunters for market were Walton E. Champagne and Clement Lejeune (1896-1989), and scores of unknown others.

Men in pirogue from left, Richard Lauton of the Metro Life Insurance Company, Thomas A. Whipple, and in background, Teles Martin c. 1930; painting from above photo by Debbie Lirette

Bottom left to right, decoys carved by Mark McCool Whipple; decoy carved by Walton Champagne

BOURG | 305

Walter Mitch Whipple's (1867-1940) cabin boat Red Rover

Walton Ernest Champagne (1880-1963)

Roderick "Duck" Whipple left, and Mark McCool Whipple on the Red Rover *c. 1930*

Carl David "Tabin" Ellender, Sr. (1902-1986

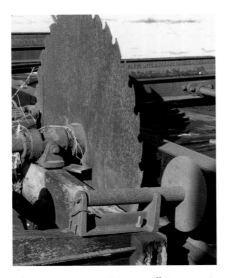

Lyes J. Bourg, Sr. purchased a sawmill in 1935 from Chatanooga, Tennessee, and moved it to Pointe-aux-Chênes. He cut lumber for six years and then moved the mill to Bourg in 1941. It remained in service from 1941 to 1961. It was closed due to the high cost of insurance.[7]

SOURCES

1. Battle of New Orleans Muster Lists, final copy, January 6, 2015
2. A Century of Lawmaking for a New Nation: U.S. Documents and Debates, 1774-1875, American State Papers, Senate
3. Thomas A. Becnel, *The Barrow Family and the Barataria and Lafourche Canal: The Transportation Revolution in Louisiana, 1829-1925*, Louisiana State University Press, Baton Rouge and London, 1989
4. *The Historical Encyclopedia of Louisiana*, Louisiana Historical Bureau, c. 1940
5. "Hunting was part of local Whipple family history," *The Courier* newspaper
6. Interview with Falgout descendant Dr. Jamie Hutchinson, 2016
7. Interview with Claude J. Bourg, 2016

Above Lyes J. Bourg, Sr.'s sawmill Bourg 2016
Below and opposite page photos 1955

1922 Bourg High School ring of Herman E. Walker, Sr.

Bourg Agricultural School opened 1913

Above, Bourg Elementary School 2016

Right, article on Bourg Agricultural School, The Rice Belt Journal January 20, 1911

Voted to Erect Agricultural School.
Houma.—The citizens of the Fifth Ward, Canal Belanger, being the Bourg School District, have voted a special 8-mill tax for fifteen years for the purpose of buying a location and erecting buildings and accessories for an agricultural school. Thirty-five taxpayers voted for the tax, with an aggregate assessment of $39,900, and four voted against the tax, with an aggregate assessment of $2,300. The state will furnish an experienced agriculturist in charge of the school, paying his salary.

BOURG SCHOOL DISTRICT
BOURG AGRICULTURAL HIGH SCHOOL

A very early mention of the first Bourg School Association appeared in the June 26, 1897 *Houma Courier* edition of the newspaper. The combination news item and advertisement named the "Committee of Arrangement" as T.P. Blanchard, chairman; C.P. Blanchard, Theophile Walker, J.L. Guidroz and C.A. Whipple.

The ad was for a Grand Fair at Canal Belanger on July 3 and 4, 1897 given by the association "for the benefit of their school."

The fair was to include a "Grand Ball Saturday night." On Sunday were to be bike and horse races, baseball, a boxing exhibition, dancing, refreshments "at moderate prices," a brass band, and "other amusements too numerous to mention."

Another *Houma Courier* newspaper article of April 13, 1901 announced another Grand Fair at Canal Belanger for April 27 and 28. The article succinctly stated, "Proceeds will be used in repairing Schoolhouse fence and other things, and for the continuation of the school.

"It has been four years since we have given a Fair for this purpose, and as the repairs are greatly needed we call on a generous public to assist us in this matter."

Assistance came in the form of a visit from the State Superintendent of Schools, T.H. Harris, in a 1910 visit with members of the Viguerie, Champagne, and Ellender families. Following his visit, the state created the Bourg School District.

An announcement in the *Rice Belt Journal* dated January 20, 1911 was titled "Voted to Erect Agricultural School." It read in part, "The citizens of the Fifth Ward, Canal Belanger, being the Bourg School District, have voted a special 8-mill tax for fifteen years for the purpose of buying a location and erecting buildings and accessories for an agricultural school. Thirty-five taxpayers voted for the tax, with an aggregate assessment of $39,900, and four voted against the tax, with an aggregate assessment of $2,300."

Architects Charles Allen Favrot & Louis A. Livaudais Ltd. designed the school building, for which the contractor was C.F. Garver. Cost was $7,993 for the school, which opened for the 1913-1914 session in District 5 of the Terrebonne Parish School Board.

Public records document that the Bourg Agricultural High School was a parish school also subsidized by the State of Louisiana in 1913, with A.J. Cormier as its first principal. It received accreditation as a high school on April 9, 1917. The school closed after the 1923-24 session when the facility became part of the Terrebonne Parish School System. The building, renovated a number of times, still serves the community as Bourg Elementary School.[1]

SOURCES
1. Terrebonne Parish School Board Minutes Books 1-3

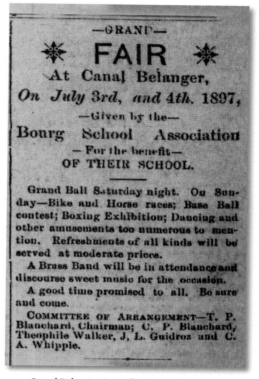

Canal Belanger Fair, The Houma Courier June 26, 1897

Architect Charles Allen Favrot (1866-1939)

Farmer's Market, 1982, Oil on canvas, 30 x 40 inches, George G. Rodrigue

FARMERS MARKET

BREAD 5¢

BEANS

LEMONS

TODAY'S FISH

LeCOMPTE PROPERTY-BILLIOT CLAIM

Etienne Billiot was an early Terrebonne Parish settler living on a Spanish land grant prior to 1800[1] in what is now Bourg. He registered his claim with the U.S. government after Louisiana became a state, on November 20, 1816.[2] His grant was confirmed on May 11 that year.[2] His land extended to both banks of Bayou Terrebonne. He was a planter, and built a home on the left bank that later became the east wing of the main house.[1]

Jean Baptiste Duplantis bought six arpents front of the Billiot claim in 1824, and the next year he and Joseph Monsan bought the rest of the Billiot claim at a Sheriff's Sale. The two new holders of the original claim sold an undivided half-interest to Agapi Hotard in 1838.

Planter Euphrosin Hotard, who had migrated to Terrebonne from St. John the Baptist Parish[1] in the early 1820s, became the sole owner of the entire original grant when he purchased an undivided half-interest from Jean Baptiste Duplantis, with all buildings, in 1837, and followed that with an 1841 purchase of the other undivided half-interest from his brother Agapi in 1841. Hotard's desire for land ownership was further gratified when he and John C. Watson purchased nearby swampland in 1845.

Hotard built the west wing onto the main house c. 1850.[1] The addition was a practicality prompted by the fact that he and his first wife, Marie Marcellite Folse, had 13 children together; after her death, Euphrosin married Marie Celine Mars, the widow of Hubert Madison Belanger, on April 29, 1850. She already had five children of her own, and she and Euphrosin added three more children to their family, for a combined total of 21 offspring in their household.[3]

Cattle brand of Euphrosin Hotard October 28, 1855

Jean Baptiste Duplantier (1770-1831) and Marie Francisca Brunet Hotard (1789-1838)

Below, Etienne Billiot Claim State Land Grant Map April 9, 1830

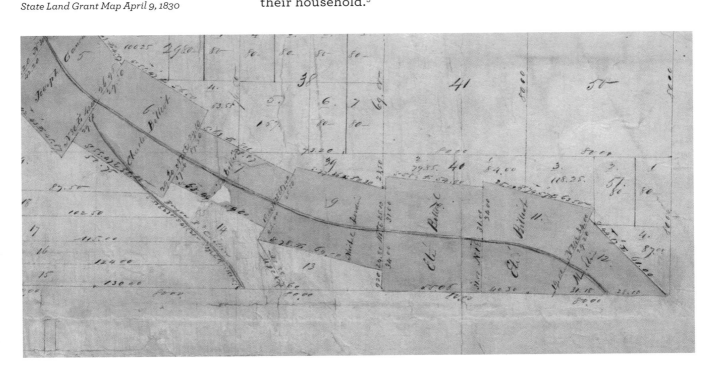

314 | HARD SCRABBLE TO HALLELUJAH

Home of Euphrosin Hotard with 65-foot long front porch c. 1925

Total acreage from the Hotard days was 1,500 acres.¹ The old house was torn down in 1940, and a new house was constructed at the same location.

Euphrosin's daughter Marguerite Olympe married Teles P. Blanchard³, who became an owner of the property after Hotard. On April 4, 1919 Joseph G. LeCompte (1874-1971) purchased part of the property of Teles Blanchard, and members of his family lived there long after his death in 1971. Joseph was appointed to the Terrebonne Parish School Board by Governor Oramel Hinckley Simpson in 1927 and was re-elected to the position for Ward 5 on the board into the late 1940s.⁴

Joseph and his wife Emily *nee* Bethancourt had ten children: Clifford, Bailey, Estelle (Mrs. Louis E. Jacobs), Emma, Athalie (Mrs. N.J. Oubre), Anna, Rudolph, Inez (Mrs. Dudley Patterson), Florence (secretary to Senator Allen J. Ellender, Sr.), Sydney (Mrs. R.K. Sheffield).⁴ Athalie, Anna, and Inez became longtime teachers in the Terrebonne Parish School System.

The current address of the LeCompte homesite is 4519 Highway 24, Bourg, Louisiana.

SOURCES
1. Helen Wurzlow, *I Dug Up Houma Terrebonne*, 1984
2. A Century of Lawmaking for a New Nation: U.S. Congressional Documents and Debates, 1774-1875, American State Papers, Senate
3. Ancestry.com
4. *Houma Daily Courier* article, Thursday, July 19, 1945

Euphrosin Hotard Jr. (b. January 1828) and Leonise Elodie Champagne Hotard (b. November 11, 1830)

Eugene and Inezida Hotard Carro and cattle brand of Eugene Carro of Canal Belanger May 18, 1907

Joseph LeCompte home site 2015

Joseph G. LeCompte and Emily F. B. LeCompte's 50th wedding anniversary photo: center front Joseph (1847-1971) and Emily (1877-1949); front from left, Athalia, Sydney, Florence, Inez; back from left, Bailey, Estelle, Emma, Clifford, Anna, Rodolph

Rural Retreat 1906

RURAL RETREAT
STORMY POINT

Marianne Marie Nerisse Iris, a triracial freewoman of color, received a Spanish land grant in 1788 for the property that became commonly known as Rural Retreat. Her son Etienne Billiot's claim was registered with the U.S. government on November 20, 1816, and he received confirmation on May 11, 1820.[1] His mother was Marianne and his father Jean Baptist Billiot of Plaquemines Parish, Louisiana.[2]

William Lackey Woods of Abbeville County, South Carolina in 1824 purchased land on both banks of Bayou Terrebonne ten miles south of Houma in what is now Bourg. His succession in 1839 listed his properties as 200 arpents front each side of the bayou, a dwelling house, sugar house, sugar mill, 26 slaves, and improvements, valued at $26,635.[3]

He bequeathed the plantation to his wife Phebe Olds Fuqua Woods. The widow Woods registered with the General Land Office of the United States for the purchase of 240 acres of public lands in Township 17 W, Range 18E in Section 51, and was granted confirmation of the purchase on June 13, 1844.[4] She married Dr. John Pierce in 1841. In 1856 the plantation was named Stormy Point after a strong hurricane wiped out the home there, and devastated the lower part of the parish, including Last Island. In 1861 her sons Rodney Shields Woods and Richard Covington Woods became co-owners with their mother of the 1,000 acres of the plantation.

Richard married Fannie Whitmell Pugh of Bellevue plantation in Assumption Parish. He was a Confederate Cavalry prisoner of war in Alabama during the American Civil War, but came home after the war in 1865 to Rural Retreat plantation.

Rodney, who married Fannie Pugh's sister Margaret Wood Pugh, was a Confederate Infantry prisoner of war at Vicksburg, Mississippi, and was exchanged in July 1864. He contracted typhoid fever and was sickly all his adult life. However, besides owning Rural Retreat, he

Phebe Olds Fuqua Woods Pierce (1802-1889)

Dr. John Pierce (d. 1845)

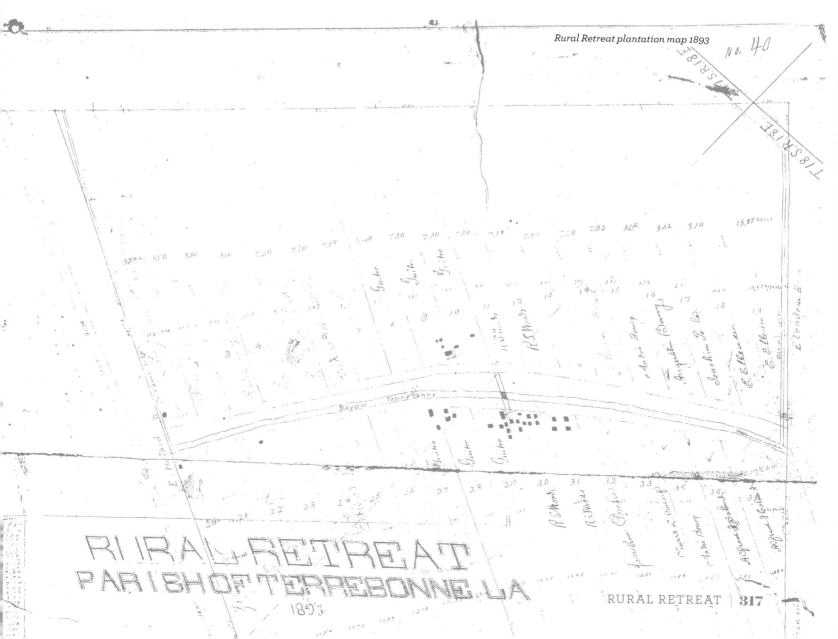

Rural Retreat plantation map 1893

Silvani Numa (1840-1924) and Elvina Bergeron Aucoin (1844-1924)

Richard Covington Woods (1835-1915) and daughters 1904

Left to right, Mildred, Virginia, Fannie, Maggie Woods at Ducros 1905

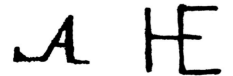

Left to right, cattle brand of Silvani Numa Aucoin March 17, 1894; cattle brand of Emile Hebert October 1, 1900

purchased Ducros plantation in 1872. His brother Richard bought half of Ducros from him in 1877. The two Woods families lived together at Ducros for decades.

Their mother, Phebe Woods Pierce, was sole owner of Rural Retreat beginning in 1877. The next year Dr. Andrew Van Woods and Rodney Shields Woods became co-owners of the Bourg property, until Mrs. Richard Covington Woods took ownership from 1890 until 1892. The New Orleans Bank assumed ownership in 1893.

Thereafter, lands on the right and left banks of Bayou Terrebonne were purchased by different planters. On the right bank, Rodney and Richard re-purchased the land. In 1897 Thomas C. Hall bought that property, and in 1898 Thomas R. Shaffer and Charles R. Morrison became the new owners.

Property on the left bank was sold to Silvani Numa Aucoin in 1893, followed by the purchases of Emile Hebert and Joseph Adam Bascle in 1897. Aucoin's cattle brand was registered as number 628 dated March 17, 1894, and Hebert's as number 646 dated October 1, 1900 in Terrebonne Parish records.

The first house at Rural Retreat was destroyed by the Last Island Storm in 1856. The grand second house on the bayou's right bank had a central section with two wings, consisting of seven bedrooms, a dining room, kitchen, butler's pantry, gallery, front porch, a second story with three dormers, eight French windows and a double French entrance door. One descendant described the house as "almost a duplicate of Madewood in Lafourche Parish".[5] The wings were removed prior to 1920, and the house burned and was dismantled in the 1940s.

The plantation sugar mill burned on December 13, 1854. A newspaper account of the fire commented, "The blow falls very heavily on Mrs. Pierce, as we learn that only five thousand Dollars of the actual loss was covered by insurance."[6]

A cemetery was located on the right descending bank of the plantation near the house. St. Agnes plantation and Klondyke plantation were carved out of the original Rural Retreat.

The Gabriel Montegut-Aubin Bourg agency divided and sold property on both banks in 1893, with Tristram Shandy Easton, Jr., of Eastonia plantation on Bayou Black, as surveyor. He was the father of Tris Easton, the future owner-editor of *The Houma Courier* newspaper.

Present address of where the plantation house stood is 4512 Country Drive, Bourg, Louisiana.

SOURCES

1. A Century of Lawmaking for a New Nation: U.S. Congressional Documents and Debates, 1774-1875, American State Papers, Senate
2. Audrey Westerman, Early Settlers Along Bayou Terrebonne Below Houma (map)
3. *Terrebonne Life Lines*, Volume 23 Number 2, Summer 2004
4. Certificate of approval, courtesy Mary Lou Eichhorn, Williams Research Center, The Historic New Orleans Collection
5. Helen Wurzlow, *I Dug Up Houma Terrebonne*, 1984
6. *Thibodaux Minerva* newspaper, December 16, 1854

Camille A. Aucoin (1887-1954) Eve Regina Robichaux Aucoin (1888-1982) c. 1930s

Rural Retreat Land Grant of Widow Phebe Olds Fuqua Woods June 13, 1844

St. Agnes aerial photo 1953

ST. AGNES
DelClaire

Stormy Point/Rural Retreat gave rise in part of its plantation lands to St. Agnes, on the right descending bank of Bayou Terrebonne. Charles J. Champagne of Thibodaux in Lafourche Parish purchased a section of Rural Retreat c. 1888.

Records from the 1880 U.S. Census record him and his household as living on Lorio plantation in his native parish. But by the time of the 1900 U.S. Census, their family was living in Ward 5 (Bourg), Terrebonne Parish.

Charles and his wife, Eve Cecile Guidry Champagne, were the parents of 16 children.[1] One of their sons, Albert Philip, married Marie Evelina Hotard in 1902 and in 1904 became owner of a section of St. Agnes. Albert became one of the largest independent sugar cane growers in the parish.[2]

Albert Philip's brother Charles Dalferes was ten years younger than Albert; Charles was the thirteenth of their parents' 16 children, and their first to be born in Montegut, Terrebonne Parish. He was born in 1889. Charles Dalferes also became owner of a section of St. Agnes. He died in 1943.[1]

Charles Dalferes' widow, Vaniola Duplantis Champagne, sold their part of the plantation to Paul M. Giovagnoli in March of 1949.[3] Giovagnoli was a native of Girard, Crawford County,

Charles J. Champagne (1850-1927)

Albert P. Champagne (1881-1958)

Kansas, who married Madeline Sanders.[1] No record of ownership transfer is recorded until 1960.

Preston Carl Roberts, Sr., originally of Texas, bought the Giovagnoli section of St. Agnes on June 14 that year.[3] He owned an oilfield service company, and used his land to raise livestock. He was married to schoolteacher Claire Fortmayer Roberts, a native of New Orleans. Roberts renamed their land DelClaire.[4]

After their children married, the parents donated parcels of the property to each of them: Carol Ann Roberts and Robbie E. Terrell; Preston Carl Roberts III and Kay Toups Roberts; Della Marie Roberts and Armand J. Domangue, Jr.; Courtney Claire Roberts and John Hollinshead, later Michael Flores.

The older three children built homes on their property. Courtney's portion was that on which the Roberts house was situated, but she and her husband never lived there. Eventually all of the Roberts children left Bourg to live outside Terrebonne Parish.

It is unknown if there was a sugar mill at St. Agnes and if there was, its demise is unknown. The site of any house built by the Champagnes or the Giovagnolis is also lost to history. The old house that was on site when the Robertses bought the property in the 1960s and in which they lived may have been the location of St. Agnes homes. The Roberts family renovated the house when they moved in. The home, currently occupied by Mrs. And Mrs. Morgan Brasher, is at 540 Country Drive, Bourg, Louisiana.

SOURCES
1. Ancestry.com
2. *The Historical Encyclopedia of Louisiana*, The Louisiana Historical Bureau, c.1940
3. Deed of sale from Giovagnoli to Preston C. Roberts dated June 14, 1960
4. Personal communication with Della Roberts Domangue, 2016

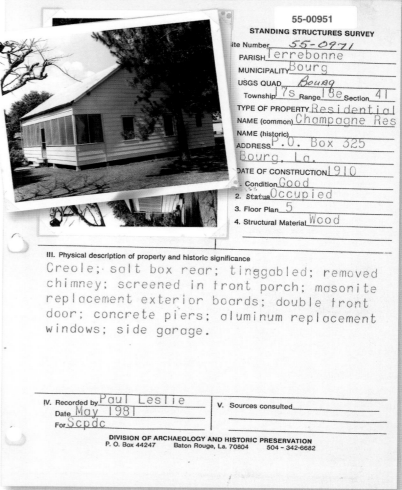

Champagne house built 1910
Standing Structures Survey 1981

St. Agnes Drive road sign, 2015

Klondyke aerial photo 1953

*Pierre Chaisson Claim
State Land Grant Map September 28, 1856*

> A correspondent informs THE COURIER that Mr. Chas. J. Champagne has purchased the tract of land at the rear of the Klondyke plantation, formerly belonging to Mr. Adam Boquet. This is a valuable addition to the Klondyke plantation and increases the size of the place considerably.

Notice of Charles Champagne purchasing Klondyke, The Houma Courier November 14, 1903

Albert P. Champagne (1881-1958)

KLONDYKE

Leon Cahn acting for The People's Bank was the first owner on record of the land that became Klondyke plantation.

The same Charles Joseph Champagne who was the owner of St. Agnes (formerly part of Rural Retreat on the right bank of Bayou Terrebonne) founded Klondyke on the left bank from the section of Rural Retreat located there. He did so in 1899 when he purchased from Leon Cahn 1,000 acres that formed the upper back section of Rural Retreat and then purchased the Adam Boquet tract in 1903. The plantation was about eight miles below Houma and a mile from the Bourg Post Office.

The name Klondyke may be derived from the 1896-1898 gold rush on the Klondyke River in the Yukon territory of Canada; Charles Champagne purchased the land a year after the gold rush.[1]

After Champagne's ownership, the Cambon Brothers, Sylvester, Ferdinand, and Maurice, were lessees until 1904, when Albert Philip Champagne took over from his father. Albert was married to Marie Evelina Hotard, a descendant of planter Euphrosin Hotard. Albert became one of the four largest independent sugar cane growers in the parish in the early part of the 20th century,[2] although Klondyke cane was sent elsewhere for grinding; no mill was constructed there. Klondyke consisted of 900 acres, with 800 acres in sugar cane. Fifty men worked

Klondyke Store 2016

the plantation year round, and between 175 and 200 people were necessary during the harvest season.[2]

Houma Cypress Co. purchased Klondyke swamp land in the early 1900s.[3]

Albert Philip served as a Terrebonne Parish Police Juror and a director of Terrebonne Lumber and Supply Company.[2] He was on the board of directors of Citizens Bank, and was also a member of the Board of Control established in 1934 to work with the Terrebonne Agricultural Extension Service for local implementation of the Jones-Costigan Act.[3] That act provided for the control of sugar cane production in continental America, with power vested in committees of local growers. Jones-Costigan was difficult to administer, and was in effect only two years. The Ellender Act of 1937 provided for payments on all sugar programs based on production and not yield. (Senator Allen J. Ellender of Bourg was author of the act.)[3]

The Klondyke plantation house on the left bank of Bayou Terrebonne burned on April 27, 1944.

Albert's son Bruce Philip Champagne, Sr. was manager of the Klondyke store; he became owner of the east half of Klondyke in July 1955. His sister Laura Champagne McComiskey received the west half of the property. A third child, Cloyd Champagne, died in 1939.

Albert died on May 11, 1958.

The wife of Bruce Philip Sr., Venus Vivian Eudy Champagne, and her sister-in-law Laura C. McComiskey leased both their halves of Klondyke to Wallace Joseph Ellender II beginning in 1958.

In 2005, Wallace J. Ellender III and Thomas Anthony Ellender purchased the land that had been Klondyke. It extends two miles on the Bourg-Larose Highway.

The present address of the Champagne house that burned is 4689 Highway 24, Bourg, Louisiana.

SOURCES
1. Leryes Usie, *Terrebonne Parish, Louisiana: Our Bayou Parish*, date unknown
2. *The Historical Encyclopedia of Louisiana*, The Louisiana Historical Bureau, c. 1940
3. Helen Wurzlow, *I Dug Up Houma Terrebonne*, 1984

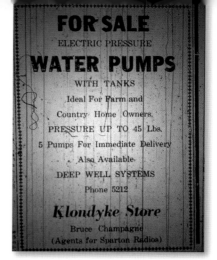

Klondyke Store ad, The Houma Courier *October 24, 1946*

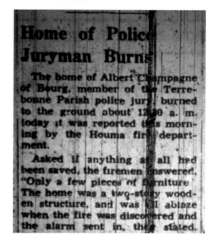

Albert Champagne home burns
The Houma Courier *April 27, 1944*

Klondyke map 1893

Below, left to right, second home of Albert P. Champagne 2016; original home site of Bruce P. Champagne on Klondyke Road 2016

HOPE FARM
HALF OF RICHLAND
BISLAND LANDING (1884)

Preceding all other claimants of land at what is still known as Hope Farm were Joseph Dianne, Etienne Billiot, and Henry Schuyler Thibodaux, all of whom registered their claims on November 20, 1816. They received confirmation on May 11, 1820.[1]

On October 13, 1828 Pierre Chaisson sold 240 acres of that property to William Bisland, Sr.. Pierre Cazeau[x] sold 250 acres to Bisland that same day.

Bisland was a native of Mount Repose plantation, Adams County, Mississippi. He was married to Mary Louisa Lavinia Witherspoon in 1820.[2] Bisland named his Terrebonne plantation Richland. Beginning 1832, Bisland was in partnership with his brother-in-law Thomas Rodney Shields, who had married the sister of Bisland's wife, Martha Jane Lenora Witherspoon.[2] The partners renamed the plantation Hope Farm.

An account written decades after an 1829 visit to Hope Farm by Seargent Smith Prentiss, who accompanied Thomas Shields there from Natchez, gives a picture of primitive living conditions. The account of the 1820s Hope Farm house written in 1883 describes the

Hope Farm Sugar House c. late 1800s

house when Prentiss visited there as "... exteriorly and interiorly rough; one room was for bed and board and parlor. There is a dim tradition that the boys had to be awakened so that the cook might get the table-cloth; this led to an investigation, and it was discovered that the sheet by night was the table-cloth by day!"

John Watson became Bisland's partner in December 1842. In 1844 Hope Farm produced 625 hogsheads of sugar.

William Bisland was sole owner again as of June 1845. He moved to Hope Farm from Mount Repose in Adams County, Mississippi in 1846 and built a home with the traditional four-columned facade on the left descending bank; at the time the plantation sprawled across both sides of the bayou. The house was constructed of long leaf pine and red cedar, shipped by sailboat from New York, the same material as that used for the Aragon house.[4] (The Hope Farm house was immediately across the bayou from the large Aragon home. The left bank house was torn down in the 1920s. It was located where the current Herman Walker, Sr. house is situated, at 280 Highway 55, Bourg.) Sugar production from Hope Farm in 1846 was 675 hogsheads.

Upon Bisland's succession dated October 19, 1847, the total acreage of Richland on both banks of Bayou Terrebonne was 7,613 valued at $115,436, and his 151 slaves were valued at $181,686.08. Moveable property valuation was $9,127.57.

William Bisland's son, William Alexander Bisland, bought a half of Richland on both banks at sheriff's sale on October 3, 1853 for $105,000, using a portion of his inheritance. He married Caroline Louise Baker at Fairfax plantation in St. Mary Parish in 1848. His Richland property was worked by a total of 92 slaves.[5]

Thomas Shields Bisland, brother of John Rucker and William Alexander, lived at Fairfax plantation in St. Mary Parish prior to and during the Civil War.[5] He purchased Fairfax from Judge Joshua Baker in 1857, including a steamboat, stock, and two flatboats.

Thomas had married Margaret Cyrilla Brownson, sister of Frances "Fannie" Brownson Bisland, wife of John Rucker Bisland of Aragon. Fairfax was the location of Fort Bisland, also called

Hope Farm, "The Big House" 1850s

Article on William Bisland's Sugar House storm damage Houma Ceres December 13, 1856

HOPE FARM | 325

Notice of Hope Farm available for sale, The Houma Courier, February 5, 1887

Thomas and Catherine Roddy Elinger's children: front row from left, Joseph, Thomas, Elizabeth, William, George; back row, from left, James, Ernest, Wallace, David

Cattle brand of Thomas Ellender (Elinger) Sr. May 17, 1841 and Thomas Ellender Jr. June 19, 1917

Hope Farm house in background, Elizie Trosclair on oyster lugger in Bayou Terrebonne c. 1920

Camp Bisland, during the War Between the States. A historical marker on U.S. Highway 90 near Calumet in St. Mary Parish reads in part, "Battle of Bisland, April 12-18, 1863, Gen. Nathaniel P. Banks' Union Army attacked Gen. Dick Taylor's Confederate forces entrenched at Fort Bisland."[5]

Thomas moved his family after the Civil War from Fairfax plantation to live in the second house at Hope Farm, on the lower right bank of Bayou Terrebonne.[4]

The Freedmen's Bureau returned Hope Farm to its owners in 1865.

James Bradford purchased 4,400 acres of Richland at a Sheriff's Sale in February 1887.

But Richland went the way of many large sugar estates in the decades after the Civil War. In 1897 Joshua Baker Bisland owned Richland on the right bank with wife Kathryn Nolan Cage of Woodlawn plantation in Grand Caillou. But by 1901 William Connell was dividing the land and selling it in parcels.

The Ellender brothers of Hard Scrabble purchased the left bank of Richland, then known as Hope Farm, on March 6, 1889 for $5,000 cash and a $10,000 note. Their land extended 15 arpents by depth, fronting on Bayou Terrebonne three miles above their father's Hard Scrabble property.

An 1897 Terrebonne Parish directory records the Ellenders' acreage at 2,500, with 300 cane acres producing 5,000 tons of cane. Acres in corn totaled 175.

The Ellenders purchased the property from James Bradford with John Bisland acting as intermediary.[6] That generation of Ellenders included non-participant Elizabeth, who had been married to Jules Homére Savoie; non-participant Henry, who died of measles during service to the Grivot Guards in the Civil War; Joseph, married to Telly Mae Deroche; Thomas Jr., married to Evelia Duplantis; bachelor George, who was killed by lightning and bequeathed his property to his remaining brothers; James, married to Elda Jarveaux; twin Ernest Fillmore, married to Mary Torr Lynch; twin Wallace, married to Victoria Jarveaux; non-participant William; and David, married to Emily Chauvin.

Twins Wallace and Ernest built identical new homes at Hope Farm and moved from Hard Scrabble in 1907. Ernest's house burned in the mid-1920s and was rebuilt slightly smaller and with a different façade. The houses still lie on each side of Hope Farm Road where it intersects with the highway that parallels Bayou Terrebonne.[7]

The eldest living brother, Joseph, died in 1923. His widow and her ten children then asked for a partition of the Ellender property. The partition was completed in 1924. According to Ellender scholar and historian Thomas Becnel, "Essentially each received a portion of the family lands, generally in the area where he was living. Each of the six received title to land and buildings, 11 or 12 mules, hay, corn, and a variety of farm equipment." James, Ernest and Wallace received Hope Farm and a northern part of Aragon. Joseph's family and Thomas, Jr., divided up the

Sawmill site, Hope Farm May 28, 1953

Hope Farm Bell, 2015

remainder of Aragon. David received all of Hard Scrabble.[6]

The Hope Farm Bell was cast in 1909 in St. Louis, Missouri by Stuckstide & Brothers Company that also cast Houma's St. Francis de Sales Catholic Church bell. The date of the Hope Farm mill's demise is unknown.

Livestock brand registrations of those who were also associated with Hope Farm were number 72 for Pierre Chaisson; 128, Bisland and Shields; 822, Joseph Ellender and Thomas Ellender Jr.; 821, Ernest and Wallace Ellender.

Hope Farm is 11 miles below Houma between Bourg and Montegut. It remains in Ellender ownership and is an active farm.

SOURCES

1. A Century of Lawmaking for a New Nation: U.S. Congressional Documents and Debates, 1774-1875, American State Papers, Senate
2. Ancestry.com
3. Bill Ellzey, "Story tells of journey to plantation home on Bayou Terrebonne," *The Courier* newspaper, March 20, 2016
4. Helen E. Wurzlow, *I Dug Up Houma Terrebonne*, 1984
5. Bisland Family Papers Collection, Louisiana State University Special Collections
6. "The Ellenders: Pioneer Terrebonne Parish Family, 1840-1924" by Thomas Becnel, Nicholls State University
7. Ellender family members accounts, 2016

Cattle brands, clockwise from left, brand of Pierre Chaisson April 30, 1828; brand of Bisland & Shields April 2, 1837; brands of twins Ernest and Wallace Ellender June 19, 1917; brand of Joseph Ellender June 29, 1917

Home of Wallace R. Ellender, Sr. and Victoria Jarveaux Ellender 2015

Home of Ernest Fillmore Ellender, Sr. and Mary Torr Lynch Ellender 2015

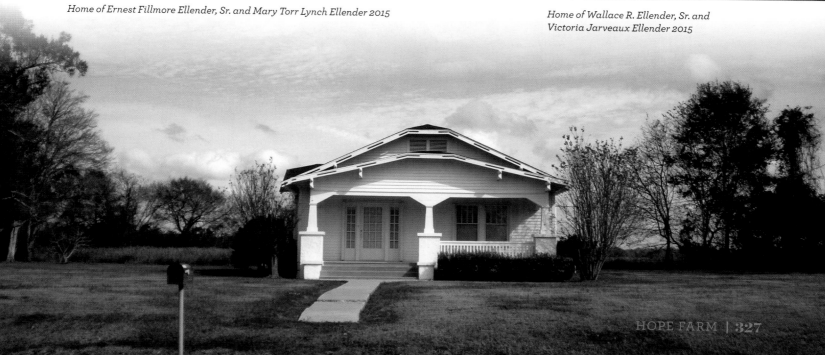

BAYOU POINTE-AUX-CHÊNES
EASTER LILIES

A break in the green of sugar cane fields of lower Terrebonne came from the white expanses of Easter lilies beginning in 1937. Farmers there took advantage of the Louisiana State University Agricultural Extension Service agent's guidance and expertise in growing Easter lilies for the bulb market. Milton J. Andrepont, county agent for Terrebonne, helped farmers to develop lilies' cultivation as a side crop.[1]

The state university's Louisiana Experiment Station developed a lilies breeding program, in part to help replace the influx of bulbs from former majority producers Japan and Bermuda. Terrebonne and Plaquemines parishes constituted the principal parts of the state for successful growth. Farmers were encouraged to grow lilies for bulbs because of a steady market. In the United States, 20,000,000 Easter lily bulbs were used each year. During the height of cultivation in Terrebonne, approximately 600,000 local bulbs went to market annually.[1]

Seedlings for planting had to undergo rigid inspection for various virus diseases, and bulbs ready for shipping underwent similar examination. Prices for bulbs were determined according to size of the bulbs, the largest category bringing nine cents apiece. One row 100 feet long normally produced 150 salable bulbs. Louisiana State University's experimental work on planting, spacing, and size of planting stock gave farmers important guidelines on producing their product. Harvesting of the bulbs occurred in August, to allow the shipped bulbs to bloom the next April.[1]

Claude J. Bourg of Montegut remembered his father's fields of Easter lilies from the 1930s and 1940s. Lyes J. Bourg, Sr. was a bulb producer on his land in Bourg and Pointe-aux-Chênes from

Junction of Highway 55 and Pointe-aux-Chênes Road 2015

Home of George Anthony and Amanda Ellender Hummel 2016

Easter lilies in Pointe-aux-Chênes field;
Milton J. Andrepont, county agent c. 1945

Inset, General Farmers Exchange ad
for Easter lily bulbs shipment boxes
The Houma Courier August 5, 1945

> If You are interested in Boxes for
> Shipping LILY BULBS, Contact Us
> We Have Them
> GENERAL FARMERS EXCHANGE, Inc.
> Church & School Sts.

Right, annual harvest of 600,000 bulbs, The Houma Courier *August 3, 1944*; far right, article on annual harvest of lily bulbs, Terrebonne Press *August 4, 1944*

Below, Easter lilies in Pointe-aux-Chênes c. 1945; from left, Henry Naquin, Lyes J. Bourg, Sr. and Evest Naquin far right (three children unknown)

FRIDAY, AUG. 4, 1944

Terrebonne Lily Bulb Growers Start Annual Harvest

Lily bulb growers of Terrebonne parish have begun the annual harvest of a South Louisiana product which gives promise of developing into one of the important phases of agriculture in this part of the state. Production of Easter lily bulbs brings growers of the parish something like $40,000 a year from approximately 600,000 bulbs marketed, despite the fact that the United States uses some 20,000,000 Easter bulbs each year. Formerly the bulk of these came from Japan and Bermuda. Since the war dealers have been turning to South Louisiana as the future source of supply. Terrebonne and Plaquemine parishes constitute the principal section where they can be grown successfully.

Unloading Easter lily bulbs at the train depot in Houma c. 1945; from left, a representative of the American Bulb Co., Charlie Pinel in cab of truck, Claude J. Bourg and unknown man in door of train car, "LouLou" Pinel on top of truck

Unloading Easter lily bulbs at the train depot in Houma c. 1945; from left, Dennis Ellender, representative of the American Bulb Co. in truck, Philip Pinel, "LouLou" Pinel, unknown man in door of train car, Claude J. Bourg on fender of truck, Charlie Pinel, Lloyd Pinel, Lloyd Bourg, Lyes J. Bourg, Sr. with hat leaning on truck far right

Packing bulbs for American Bulb Company in Montegut c. 1945; front row, a representative of American Bulb Company to the left and Henry Naquin to right. Others left to right, (child) Claude J. Bourg, man holding bulb, Lyes J. Bourg, Sr., right front, Edith Henry, Wilma Pinel behind her; Lloyd Bourg far back, man in center with metal hat, Lloyd Pinel and man far right with metal hat, Bruce Pinel behind Edith Henry, the rest unknown

Herman E. Walker, Sr. (1904-1980) left; Lyes J. Bourg, Sr. (1911-1981) right; middle unknown, Pointe-aux-Chênes, La. 1945

1935 until 1947, when mosaic disease became increasingly difficult to control. Another factor was that after the end of World War II, Japan re-entered the market with cheaper bulbs, which caused local farmers to cease growing lilies for market.[2]

SOURCES
1. *New Orleans Times-Picayune* newspaper article "State's Lily Harvest Nears End," August 13, 1944
2. Interview with Claude J. Bourg, Jr., 2016

Easter Lilies at Pointe-aux-Chênes, The Houma Courier April 27, 1944

Right to left, Ruby Boyne Gonsoulin, Onesia Avet (Mrs. Edward) Boyne, Mr. Deroche, Mrs. Deroche, Joyce Boyne Lasseigne holding Pat Gonsoulin, other children and girl unknown 1947

BAYOU POINTE-AUX-CHÊNES | **333**

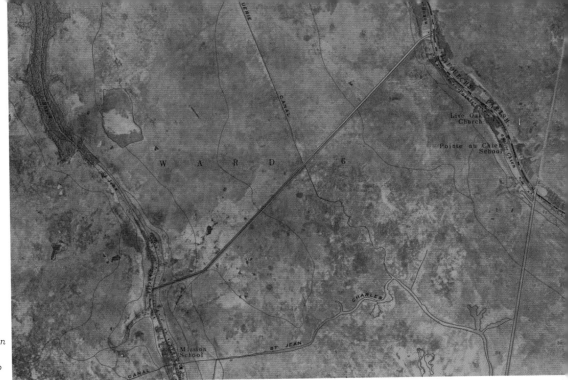

Soil Conservation Map, USDA, Ile à Jean Charles Road 1953 after completion of the first road from Pointe-aux-Chênes to Ile à Jean Charles

BAYOU POINTE-AUX-CHÊNES
ILE À JEAN CHARLES

Bayou Pointe-aux-Chênes originally met Bayou Terrebonne below Hope Farm plantation, its headwaters located at or near the current Hope Farm Road at its intersection with Louisiana Highway 55. After the bayou's siltation on its upper reaches, Hope Farm Road descended in an arc on the left bank of Bayou Terrebonne, roughly following the minor bayou's original upper course. It was the first road that reached the community at Pointe-aux-Chênes. The bayou reappears as a waterway at the intersection of Louisiana Highway 665 and Hope Farm Road, and farther south, Bayou Pointe-aux-Chênes becomes the boundary between Lafourche and Terrebonne parishes.

Jean Charles Naquin arrived in Louisiana from France aboard the *St. Remi* on September 9, 1785. After he married Magdeleine LeBoeuf in 1800, the couple settled at Ile à Jean Charles,[1] (also known as Ile de Jean Charles) according to local historian Laise Ledet. The "island"—actually a ridge of high land—is near the community of Pointe-aux-Chênes, French for Oak Point[2], the name acknowledging thick groves of oak trees in the area. (In documents and in signage, the bayou and place have in the past been identified inaccurately as *Point* or *Pointe-aux-Chiens*, which translates to "Point of the Dogs.")

Two natural waterways transect the ridge of high land called the "island." Bayou Jean Charles originates on Point Farm below Montegut Middle School, and flows from north to south to Lake Barré. Bayou St. Charles transects Bayou Jean Charles in the middle of the island, and runs from Pointe-aux-Chênes on the east to Montegut on the west.

Chief Victor Naquin's house on Ile à Jean Charles at the intersection of Bayou St. Charles and Bayou Jean Charles

Ile à Jean Charles Road May 2016

Right, Jean Naquin and Charles Naquin Claim
State Land Grant Map April 5, 1832

Ile à Jean Charles island trail; from left, Levest Naquin, Christopher Naquin, Henry Dardar, c 1940s.

Ile à Jean Charles Road approximately 50 years later, after the above photo on the same site

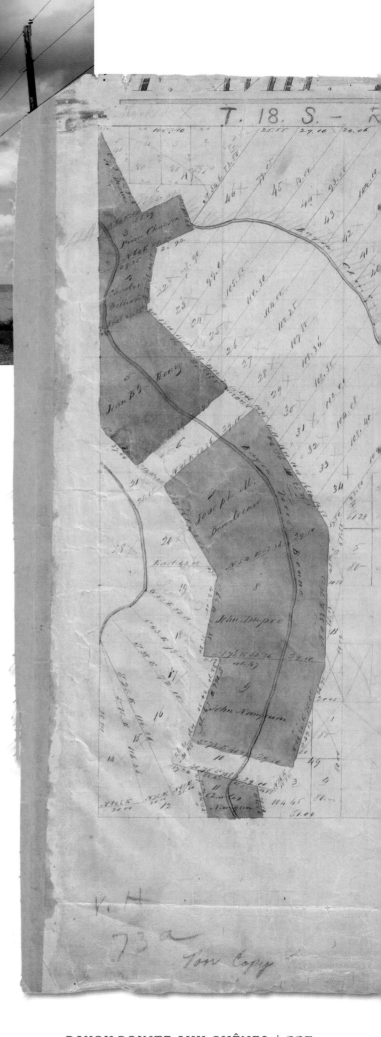

BAYOU POINTE-AUX-CHÊNES | **335**

Above, left to right, Loney Dardar with unknown child; Milton Billiot and Alcee Dupre (married to Noelia Naquin) in boat, c. 1965; Harry Dardar chickens, neighbor to Adam Naquin c. 1950s; Harry Dardar with Brahma bull raised at Ile à Jean Charles; Harry Dardar with unknown child c. 1950s

Basket hand woven from palmetto leaves 2000

Islanders at Timbalier Island c. 1950s, from left, Noelia Naquin Dupre, Alexander Billiot, Maryland Billiot Naquin, Pierre Alexander Naquin

Hand woven items by Denicia Marie (Naquin) Billiot and Maryland Billiot Naquin 2016

Denecia Marie (Naquin) Billiot and Maryland Billiot Naquin 2010

336 | HARD SCRABBLE TO HALLELUJAH

Chief Demé M. Naquin Sr. home, front by fence Noelia Naquin Dupre holding Frances, and Simon A. Naquin by back door left rear

Theodore and Marion Dardar c. 1950

Another account of the island's founding was given by a longtime local resident, Joseph Naquin, who told an interviewer in the 1970s that three men and their families settled the island. Walker Lorvin ("a big man with milk white skin and red hair" who spoke only English), Jean Charles [Naquin] and Jean Marie Naquin and his wife Pauline settled the island c. 1828 in local lore.[2] The Naquins were sons of the Frenchman Jean Charles Naquin and his wife Magdeleine LeBoeuf; the father died in 1820, and it may have been in his honor that the island was named, according to a Rootsweb.ancestry.com account. If not, the son of the same name was the place's namesake. Three of the children of Jean Marie Naquin and Pauline received land patents from the State of Louisiana in 1876, the first legal evidence of the place's settlement. The three siblings and one of their adult children are listed in the U.S. Census of 1880. By the 1900 census, the settlement was overwhelmingly made up of descendants of the three siblings.

Marie Eve (Naquin) Dardar (1894-1950) holding Rita Antoinette (Dardar) Falgout (b. 1936) 1939

Ile à Jean Charles school photo; teacher Laise Marie Ledet (1919-2002) far right, c. 1940s

Helen Verdin Dardar (1882-1957) holding Clarine Helen (Dardar) Lecompte (b.1953) 1953

BAYOU POINTE-AUX-CHÊNES | 337

Center, United Houma Nation tribal patch

Clockwise from left, 1970 Dixie Roto Magazine, Times Picayune: Craven Molinere working on a net; Louis Molinere paddling a homemade pirogue; at Hotard's General Store, Mrs. Emma Hotard at center, Louis Molinere at right, Levest Molinere (Louis' father) on step, and Noah Hotard in chair, Houma Indians

Antoine Naquin Grocery Store and Dance Hall also used to teach the people of Ile à Jean Charles; Chief Jean Victor Naquin (1868-1956) on left, Albert Dardar on right 1938

An instructor from France teaching French reading and writing at Antoine Naquin Grocery Store and Dance Hall Center 1938

By the turn of the 20th century, the island which had been connected to Pointe-aux-Chênes by a land bridge was accessible only by an elevated road. Road improvements and shelling in 1932 allowed Pointe-aux-Chênes children to attend classes at Montegut School, and in 1938 a bus began to take students to Terrebonne High School. The state built and completed what is locally termed the Island Road in 1953, giving Ile à Jean Charles residents access to better education and products from outside their own sphere. Prior to the road, islanders had to paddle pirogues through small bayous in the prairie to get to stores along Bayou Terrebonne for supplies. Conditions on Ile à Jean Charles were such that Mrs. Joseph Naquin in the 1970s recalled having lived in one of several "maisons de terre" (earthen or mud houses) which she described as warm in the winter and cool in the summer.[2]

Pointe-Aux-Chênes School c. 1920

Vital Naquin in pirogue, French instructor on bank in hat c. 1938

Boat in Bayou Jean Charles (from left) Augustine Dardar, Demé M. Naquin, Gabriel F. Naquin, Harry Dardar c. 1930s

Demé M. Naquin Sr.(1926-2009) holding oyster tongs c. 1960

At Chief Antoine M. Naquin's home, from left, Milton Billiot (b.1929), Chief Antoine M. Naquin (1896-1978), Wiley J. Dupre (1909-1982), Lillian (Naquin)Naquin (1935-1974), and Gabriel F. Naquin(1929-1993) c. 1950s

A2

Cattle brand of Alexandre Billiot April 16, 1860

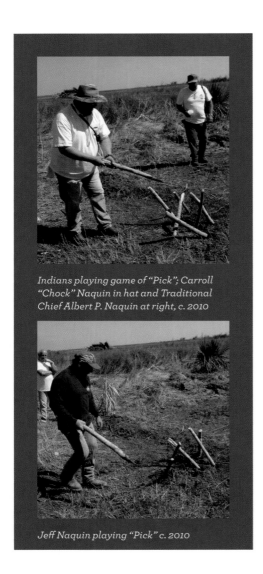

Indians playing game of "Pick"; Carroll "Chock" Naquin in hat and Traditional Chief Albert P. Naquin at right, c. 2010

Jeff Naquin playing "Pick" c. 2010

Pointe-aux-Chênes was settled by four brothers who reached it from the south. Alexandre, Temé [Barthelemy], Jean, and Celestin Billiot were considered the areas's first settlers.[2] Alexandre is first listed in official records, in 1850, as Chief of the Chitimachas.[1] The four were sons of Jacques Billiot and Rosalie Courteau. Alexandre Billiot died in 1908; two of his grandsons were interviewed in 1979 when they were ages 86 and 82. Joesph and Roger Billiot said that Alexandre and his family grew sugar cane, corn, and rice. He and Celestin built and ran a mill at Pointe-aux-Chênes, producing raw sugar and taking it to New Orleans for processing. As did many planters, Alexandre had large numbers of livestock. He registered his livestock brand number 469 on April 16, 1860 in Terrebonne Parish records.

Another source indicated that Alexandre and several of his brothers had come "down the bayou" to Pointe-aux-Chênes to use land that the federal government ostensibly gave to them. Joseph, another son of Rosalie Courteau and Jacques Billiot, followed his brothers to the area.[3]

Ile à Jean Charles became a subject of public scrutiny in 2016 when the U.S. Government's Department of Housing and Urban Development announced that nearly $50 million would be spent to relocate island residents whose land is being washed away by natural subsidence, wave action from the Gulf, and devastating effects of hurricanes. Houma's *The Courier* newspaper reported, "Once a thriving community with more than 100 families, hurricanes and constant flooding have dwindled that number to 25. An island estimated at 22,000 acres 60 years ago now stands at 320 acres. Experts suspect the island will be submerged within 50 years."

A controversy arose after the announcement when the island's Band of Biloxi-Chitimacha-Choctaw became designees of the federal funds, excluding United Houma Nation members from the grant.

SOURCES
1. Laise Ledet, *They Came, They Stayed*, Point-aux-Chênes timeline
2. Sherwin J. Guidry, "Accent on Pointe-aux-Chênes, Isle de Jean Charles" newspaper article in *The Houma Daily Courier and the Terrebonne Press*, April 18, 1971
3. Historical Report on United Houma Nation, Inc., the Bureau of Indian Affairs, 1994
4. *The Courier*, "State, HUD Near Agreement," May 12, 2016

Official patch and symbol of Isle de Jean Charles Band of Biloxi-Chitimacha-Choctaw

Albert P. Naquin, Traditional Chief of the Isle de Jean Charles Band of Biloxi-Chitimacha-Choctaw

Terrebonne Parish Consolidated Government aerial photo Ile à Jean Charles 2010

1" = 1000 ft

0 ft — 1000 ft — 2000 ft — 3000 ft

BAYOU POINTE-AUX-CHÊNES | 341

Verdin home FaLa Village 1983

Laurent M. Verdin Sr., (1908-1981) FaLa Village 1983

Opposite page, Fala-Lake Felicity Quadrangle USGS 1939

BAYOU POINTE-AUX-CHÊNES
FALA VILLAGE

A devastating hurricane in 1909 wreaked havoc on communities of lower Terrebonne Parish, including Point-aux-Chênes and Ile à Jean Charles. Many Native Americans who lived at Ile à Jean Charles and near Pointe-au-Barré left the wreckage of their former homes and moved south and east into Lafourche Parish, to a small community near Bayou Fala (Faleau).

Even before the 1909 hurricane, Alexandre Billiot and his family reportedly lived at Bayou Fala, described in one account as "on Bayou Blue behind Catfish Lake below the town of Golden Meadow, Louisiana."[1] Billiot family history recalled Alexandre finding many human bones when he plowed the earth; he gathered them into a mass grave. Older members of the community said they were the bodies of Native Americans killed when the "French made war on the Indians." Alexandre died in 1908.

One of the early families (in the late 1880s) who lived in Fala Village was that of Octave Verdin and his wife, Genevieve Armelise Billiot. Laurent Mitchell Verdin, Sr., the son of Octave Verdin, was born at Fala Village on August 10, 1908. The family homestead "had been in his family for three generations when his grandfather came from Terrebonne Parish to settle on the land then known as Fala Village."[1]

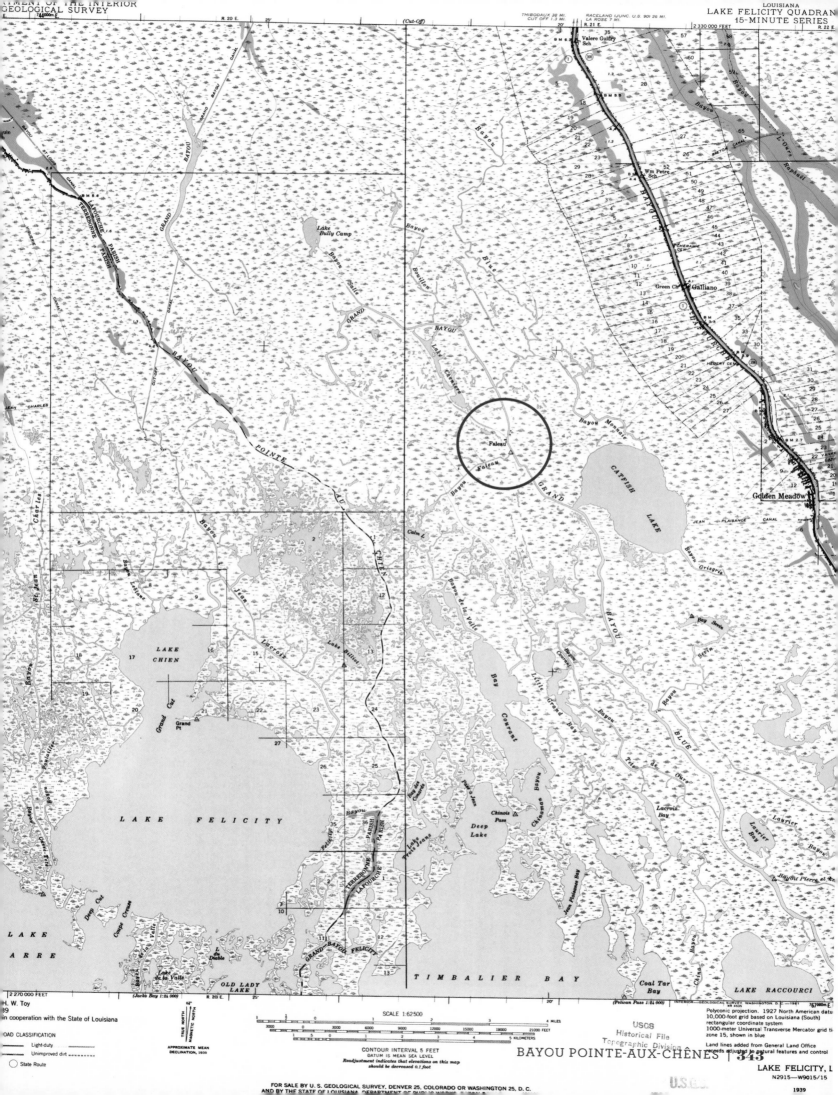

Verdin home Fala Village 1983

Verdin home Fala Village 1983

Laurent M. Verdin, Sr. and sons 1983

After the 1909 storm destroyed much of Fala, many people from the village sought new homes elsewhere, at the same time, ironically, that lower Terrebonneans moved to Fala.

Life in Fala was described by interviewee Laurent Verdin and his wife Felicie Verdin who were married in 1929 and raised a family of 11 children at Fala Village. As recently as the July 1983 interview[1], their home had "no electricity, natural gas, running water, indoor facilities, telephone or any conveniences of the modern world." The historian who transcribed the interview added, "They are very shy people and live a quiet life and are one of the last descendants of Fala Village."[1]

Laurent and Felicie's sons Laurent, Jr. (Lawrence), Forest, and Mayfield Verdin, remained at Fala after their father died in 1991. As their daughters married, they moved northward on Bayou Lafourche.

The Pointe-aux-Chênes Indian Tribe (PACIT) "community once incorporated several sub-communities including Fala Village and Les Quiens [L'Esquine[2]], but repeated hurricanes, coastal erosion, and saltwater intrusion have reconfigured the cultural landscape. The area that now makes up the PACIT community was previously the upper extent of a community that stretched into the marsh but is now for the most part open water."[3]

SOURCES
1. Interview of Laurent Mitchell Verdin by Lee Lafont, Royce Naquin, and Martin Cortez transcribed July 24, 1983 by historian Martin Cortez, Cajun Heritage, Inc.
2. Laise Ledet, *They Came, They Stayed* timeline
3. Dayna Bowker Lee, Folklife in Louisiana: Louisiana's Living Traditions, "Louisiana Indians in the 21st Century"

Verdin home Fala Village 1983

Aioli Dinner, 1971, Oil on canvas, 30 x 40 inches, George G. Rodrigue

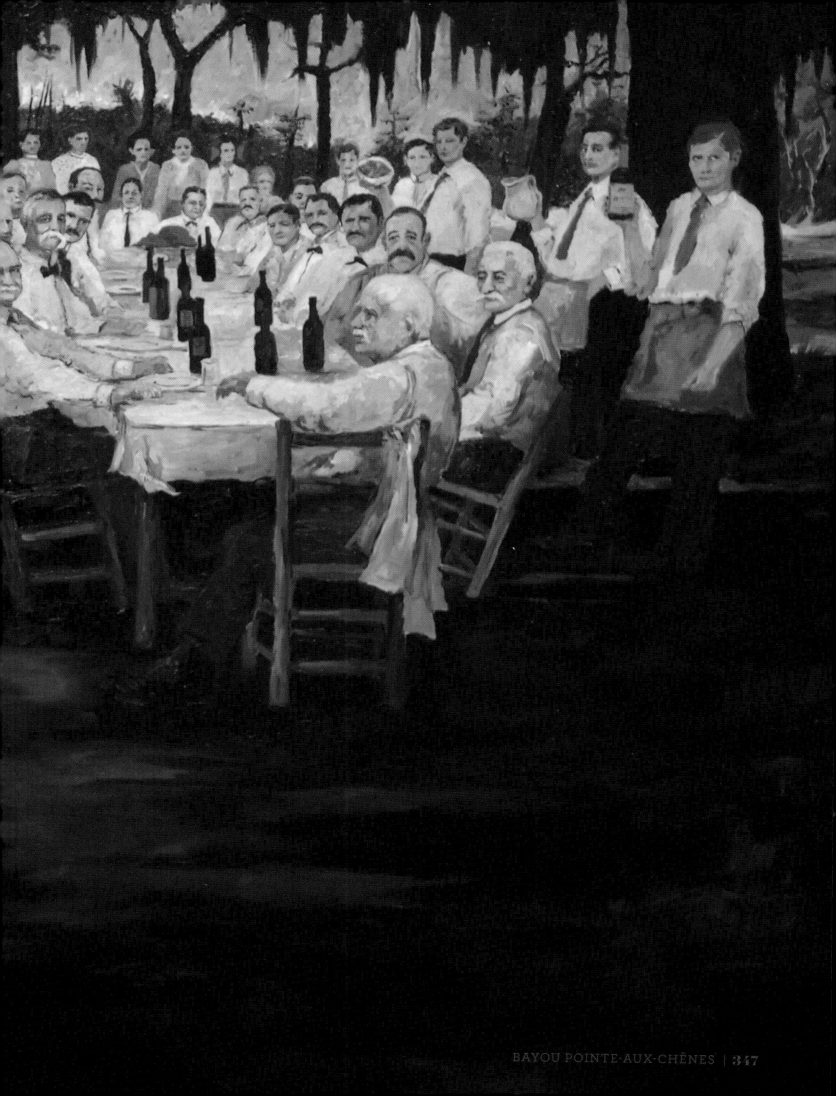

BAYOU POINTE-AUX-CHÊNES | 347

DEROCHE BROTHERS SYRUP MILL

Four Deroche Bros. Pure Cane Syrup was produced at a mill in Bourg, the product so named because four brothers of the Deroche family made the sweet product from their sugar cane crops at the current address 140 Lower Country Drive, Bourg, Louisiana.

Albert, Colbert, Walter and William Deroche began production in 1942 and closed the mill in the 1970s. The origin of their mill goes back to July 3, 1908 when Jean-Pierre Cenac, Sr. purchased the mill equipment from Filican "Tican" Duplantis for the price of $100.

From Cenac the mill was sold to the Porche Brothers on Bayou Dularge in 1914. The Porches sold the mill equipment to the Deroche Brothers, who used it for decades.

Local historian Sherwin Guidry, whose column "Xplorin' Terrebonne" appeared in *The Houma Daily Courier & The Terrebonne Press* newspaper for many years, wrote for the December 10, 1978 issue, "The open kettle method of syrup making is fast becoming a part of our historic past. 'Le Vin de Canne' [wine of the cane] passed through a series of cast iron kettles—'Le Grand , Le Prop, Le Flambeau, and Le Sirop.' The hot liquid was dipped from one kettle to another using large buckets swung on long poles set in rowlocks. The liquid finally ended in the smallest of the kettles called 'La Batterie.' From this kettle came the finished product plus sugary lumps called 'La Cuite.'"

David Jaubert, a descendant of the Deroches, wrote, "They would sell the syrup to the locals mostly instead of bigger

Deroche Brothers Syrup Mill c. 1955

Four Deroche Brothers Pure Cane Syrup jar never opened 2015

Robert Deroche getting into a truck c. 1955

Albert Deroche wearing hat c. 1955

Deroche Brothers Syrup Mill c. 1955

Waste roller and Clarence Truxillo c. 1955

Deroche Brothers Syrup Mill c. 1955

William Deroche on cane pile c. 1955

Deroche Brothers Syrup Mill 2015

companies because they liked to visit and talk." He said that the leftover cane was shipped to the sugar mill in Montegut, and that the Deroches kept the mill going until "they were too old to harvest the cane."[1]

Sherwin Guidry's account for the Deroches' closure was that regulatory requirements for their operation would have demanded a huge expenditure to meet regulations. One of the brothers told Guidry, "It wouldn't pay."

SOURCES
1. George J. Jaubert interview 2015

Deroche house built 1890, Standing Structures Survey 1981

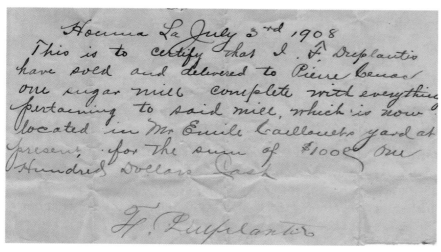
Jean-Pierre Cenac, Sr.'s bill of sale of sugar mill from Filican A. Duplantis July 3, 1908

DEROCHE BROTHERS SYRUP MILL | **351**

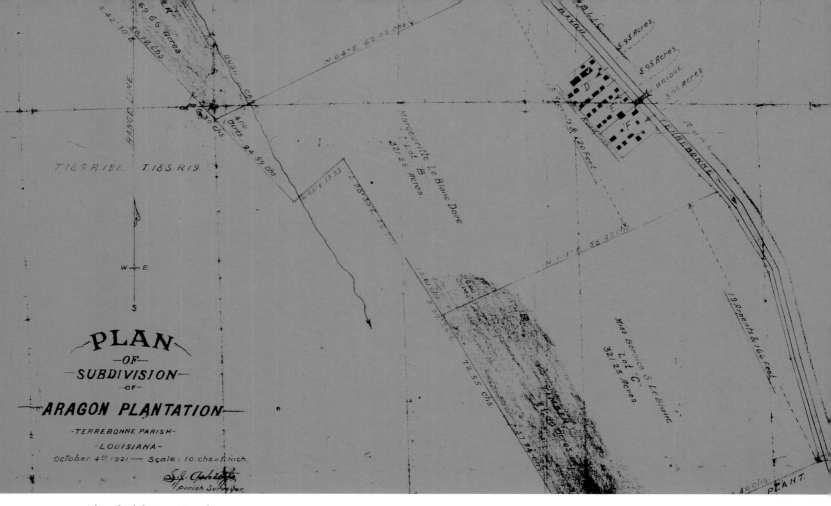

Plan of Subdivision Map of Aragon October 4, 1921

ARAGON
HALF OF ORIGINAL RICHLAND

William Bisland (1797-1847) and Mary Louisa Lavinia Witherspoon Bisland (1805-1873)

Pierre Chaisson, Charles Billiot, and Jean Baptiste Theodore Henry had claims that were the origin of Richland/Aragon on both banks of Bayou Terrebonne about 14 miles below Houma. Chaisson, Billiot, and Henry registered their claims with the U.S. government on November 20, 1816 and received confirmation of ownership on January 21, 1821.[1] Land baron Robert Ruffin Barrow, Sr. owned part of the property in the 1830s.

William Bisland of Mount Repose plantation in Adams County, Mississippi in 1836 paid $227,000 for the land that became Richland plantation. William (married to Mary Lavinia Witherspoon) had as his partner Thomas Rodney Shields (married to Martha Jane Lenora Witherspoon), his wife's brother-in-law.[2] In 1845 Bisland bought a Louisiana State land patent. Shields purchased 1,962 acres from H. Wilson. By 1842 William Bisland was sole owner.

After his father's death, William Alexander Bisland, acting as agent for John Rucker Bisland, purchased half of Richland on both banks of Bayou Terrebonne in 1853. John became owner in June 1857 and renamed that half Aragon plantation. The mortgage agreement in January 1858 with Dr. Stephen Duncan of Natchez, Mississippi was $29,436.00 in promissory notes. In January 1864 the mortgage was at eight percent for 3,071 acres and 104 slaves. John Rucker Bisland was married to Frances Ashton "Fannie"

Aragon c. 1850s

Brownson. His brother, Thomas Shields Bisland, married Fannie's sister, Margaret Cyrilla Brownson, and lived at Fairfax plantation in St. Mary Parish.[3]

John R. Bisland built a house on the right bank of Bayou Terrebonne of long leaf pine and red cedar from New York, shipped by sailboat. At the time of construction, 1857-1860, there was no local source for the desired wood materials. The façade of the grand home was built with six squared columns across the front, in the style of the time. It was similar to the four-columned house of his father at Hope Farm, immediately across Bayou Terrebonne.[2] The current address of the Aragon home, long since destroyed, is 276 Lower Country Drive in Bourg, now known as the site of the "Bisland Oaks."

John Rucker's family moved to Houma in 1865 to the former home of Emile A. Daigle after the plantation was confiscated by U.S. Government Union troops in 1862 during the Civil War.

Both the Bislands and the Brownsons had homes in New York, and after the confiscation of their property, Fannie used her family connections to have John P. VanBergen of New York become lessee in the spring of 1863 in an attempt to prevent destruction of the plantation. According to descendant Marianne Musgrave Brownson, "But by mid-May, the family connections must have been discovered because the lease was cancelled. VanBergen sued

Mount Repose, Adams County, Mississippi 1986

John Rucker "Jack" Bisland home, Barrow Street, Houma c. 1920

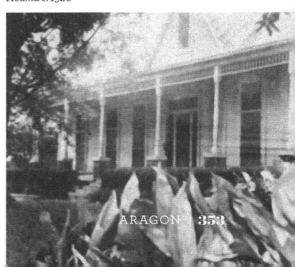

Bisland Oaks, 276 Lower Country Drive 2016

Fairfax plantation home of Thomas Shields Bisland and family, St. Mary Parish 1920

Cattle brand of Fannie A. Bisland March 3, 1864

John Rucker Bisland, Fannie Ashton Brownson Bisland, and unidentified grandchild c. 1902

Above left, J. Leonce LeBlanc (1853-1917) holding cane; below, trespass notice, The Houma Courier April 10, 1926

and got half the profits of the crops raised that year."[4]

Bisland and Shields registered their livestock brand as number 128 in Terrebonne Parish records, and Fannie registered brand number 483 dated March 3, 1864.

Thomas Shields Bisland and his family returned to live in one of the houses at Hope Farm in Terrebonne Parish, leaving Fairfax plantation in St. Mary Parish after the Civil War. Fairfax had been the site of Fort Bisland, or Camp Bisland, during the War Between the States. A historical marker on U.S. Highway 90 near Calumet in St. Mary Parish reads in part, "Battle of Bisland, April 12-18, 1863, Gen. Nathaniel P. Banks' Union Army attacked Gen. Dick Taylor's Confederate forces entrenched at Fort Bisland."[4] Fairfax was destroyed by Union forces, the house looted and the land laid bare.

John Rucker and his wife Fannie moved to Chicago before 1900 and are buried in the Brownson plot in Greenwood Cemetery in Brooklyn, New York. One of the Bislands' furnishings that has been preserved and currently located in Southdown Museum in Houma is the Gayoso bed which once was used in the home of Spanish Governor of Louisiana Manuel Gayoso de Lemos (1797-1799). It had come down to Fannie through Margaret Cyrilla Watts Gayoso, Fannie's grandmother.[2]

John Brownson Bisland and his sister Leonora Goode Bisland had temporary ownership of Aragon lands in 1881.

David R. Calder then took over Aragon. Calder was a native of Scotland, born in 1830, who fought as a Confederate soldier in the Civil War. He was a member of the Louisiana Sugar Planters' Association in 1877, took part in tariff rate negotiations in Washington, D.C. in 1878, and had a key role in establishing the Louisiana Sugar Exchange in 1883.[5]

Thomas Sefton was the next owner of Aragon, in 1882. On May 2, 1892, a tract of 3,070 acres was auctioned at a Sheriff's Sale. Another tract consisting of 302 acres—the Jean Baptiste Theodore Henry Tract—was also sold. Charles Benjamin Maginnis owned half-interest in the plantation, and purchased the other half-interest in 1893.

Aragon is described in an 1897 parish directory as consisting of 600 acres, with 457 acres planted in cane and 93 acres in corn.

In the late 1890s and early 1900s, ownership of the right bank land of Aragon underwent a number of changes as the 20th century progressed. In 1900 J. Leonce LeBlanc and Charles Benjamin Maginnis were partners in Aragon ownership. LeBlanc bought a third of the right bank Aragon below Bisland Cemetery.

The cemetery itself has an interesting history which goes back to 1768 when Spanish authorities consecrated the grounds to be used as a cemetery on the property of Jean Baptiste Theodore Henry, registered in 1803.[1] The first burial there is recorded as that of William Ross (1790-1853).[6]

Terrebonne Parish's Planning Department, under whose auspices the public cemetery lies, indicated in a 2016 interview that Bisland Cemetery was the site of more than 1,160 burials, some of them unpermitted. The wife of Edmond Guidry and daughter of Hubert Belanger and Sophie Comeau, Elmire Belanger Guidry, was buried at Bisland after her death at the age of 51 on April 17, 1861. Eleven soldiers who participated in the Civil War have CSA markers on their gravesites.

Elysée Alfred Trosclair and his wife Odelia Marie Deroche in 1905 bought "Le Grand Coeur" of Aragon—the main house residence, three arpents of land by depth, minus the stable and a small cabin. The sale was recorded on January 1, 1905, with his expenditure listed as $2,400. The Aragon plantation house was across from what is now the Herman Walker, Sr., home below Bourg at Hope Farm on the left descending bank of Bayou Terrebonne. In 1896 the Ellender Brothers had purchased from Edward LeBlanc land on the left bank of Aragon for $8,000 cash and a $12,000 note. That portion was added to existing Ellender Brothers property and farmed. It was partitioned in 1924 after the death of the oldest brother Joseph in 1923.

Changing economic times can be glimpsed by the fact that Odelia Deroche Trosclair, who lived in the large Aragon house, gave the handrail from the plantation house's second floor balcony to Dr. Steven Ernest Ellender, Sr. and Dr. Willard A. Ellender in exchange for health care.[2]

J. Leonce LeBlanc bought additional land in 1906, and held the right bank property until 1916. Edward A. LeBlanc-Doré is the next owner of record, in 1920. By 1935, the land passed from the Federal Land Bank to the South Coast Company during that year.

South Coast owned and worked what had been Aragon until 1968 when the Jim Walter Corporation purchased South Coast. The Walter Land Company is recorded as owner beginning in 1970, and in 1979, four different entities purchased the lands from each other in order: Mid-South Mortgage Co., Ag-Lands Investment Co., McCarthy Family Corp., and Cook Family Corp.

After that time, the land was subdivided and sold. The current address of where the Aragon plantation house stood is 276 Lower Country Drive, Bourg, Louisiana.

Lower Country Drive entrance Bisland Cemetery 2015

Odelia Deroche Trosclair (1863-1955) and daughter Cynthia Trosclair Hutchinson (1886-1972) c.1925

SOURCES
1. A Century of Lawmaking for a New Nation: U.S. Documents and Debates, 1774-1875, American State Papers, Senate
2. Helen E. Wurzlow, *I Dug Up Houma-Terrebonne*, 1984
3. 1971 addition to "A Southern Neo Colonial Home" by C. Mildred Smith, M.A. and G. Portre-Bobinski, Ph.D.
4. Bisland Family Papers Collection, Special Collections, Hill Memorial Library, Louisiana State University Libraries
5. *The Lafourche Country II: The Heritage and Its Keepers*, edited by Stephen S. Michot and John P. Doucet, Lafourche Heritage Society, 1996
6. Terrebonne Parish Planning Department Office registry book

Bisland mill, Aragon c. 1880

St. Peter's Baptist Church and 1938 meeting hall 2016

St. Peter's Church bell 2016

ST. PETER'S BAPTIST CHURCH

The same churchman who established the mother church in 1859 for the African American community of Terrebonne Parish, Little Zion Baptist Church (now New Zion Baptist Church), on Grand Caillou Road, also organized St. Peter's Baptist Church in lower Terrebonne in 1865. After serving the Houma community, Bishop Isaiah Lawson began missionary work, establishing Baptist churches throughout Terrebonne Parish, including St. Peter's at 669 Louisiana Highway 55, Montegut.

The first baptismal pool for the church was Bayou Terrebonne, and baptisms were well-attended, the baptized clad in white, the young ladies in long white gowns and the young men in white pants and shirts.

The first church of St. Peter's was constructed in 1865 of cypress harvested from the wooded areas of Hope Farm, owned at the time by the Bisland family. That building was moved, and rebuilt in 1938. It was moved again in 1977 when the current brick building was built. The wooden structure is used as a meeting hall for Masonic Lodge Number 5984.

The street that ran alongside the church was originally called St. Peter's, as was the neighborhood. St. Peter's Baptist Church celebrated its 150th anniversary in 2015. Current pastor of the church is the Reverend J.H. Thompson.

SOURCES
1. *The Houma Daily Courier and The Terrebonne Press* newspaper, October 8, 1972
2. Chronicling America, Historic American Newspapers, Library of Congress: *The Weekly Louisianian*, May 27, 1882

Aerial photo Point Farm 1953

Inset Jean Dupre Claim
State Land Grant Map September 25, 1856

Notice of Pointe Farm at Sheriff's Sale, in French and English, Houma Ceres *May 29, 1856*

POINTE FARM
POINT FARM

During Spanish control of Louisiana, Governor Esteban Rodríguez Miró y Sabater (1785-1791) laid claim to extreme southern Terrebonne lands 15 miles below Houma, part of which later became Pointe Farm. Miró became acting governor in 1782 and proprietary governor in 1785. This was at a time when residents of the new United States of America and its western boundary territories were vying for land with Spain in Louisiana and elsewhere.

Spain ceded Louisiana to France in 1800, and following the 1803 Louisiana Purchase, Jean Dupre had a claim on the land. He registered his claim on November 20, 1813 and it was confirmed by the U.S. government on May 11, 1820.[1]

Planter Robert Ruffin Barrow, Sr. purchased land on both banks of lower Bayou Terrebonne in 1828. In 1850 he took John McDonald in owner partnership. That was short-lived, because McDonald murdered a slave and fled the state, but not before Barrow sued him for the value of the slain slave.[2]

French native Jean Pierre Viguerie became R.R. Barrow's next partner in Pointe Farm beginning in 1855, a partnership that endured until 1872.

*Jules Martial Burguières
(1850-1899) c. 1897*

*Jean Pierre Viguerie
(1828-1888)*

*François Viguerie
(1829-1896)*

The geography of the plantation gave Point Farm its name. The land formed a point on the southeastern end, pointing in the direction of the island Ile à Jean Charles, while its western expanse fronted on Bayou Terrebonne across both banks. The land was originally called Pointe Viguerie, according to local historian Laise Marie Ledet (deceased).

Jules Martial Burguières became owner of record of the plantation in 1872. He was the son of Eugène Denis Burguières and half brother of Alfred Delaporte and of Marie Elvire Delaporte, Jean Pierre Viguerie's wife. Jules Martial Burguières, whose family was growing a literal plantation empire in St. Mary Parish in the Louisiana Sugar Belt, thus helped his brother-in-law Jean Pierre to retain his connection to the plantation he had owned with Barrow from 1855 until 1872.

Jules M. Burguières was first married to Marie Corinne Patout of the prominent Iberia Parish Patout family, and then to Ida Laperle Broussard of Breaux Bridge in St. Martin Parish. Ida was a descendent of Alexandre and Joseph Broussard *dit* Beausoleil who were among the first Acadians to settle in Louisiana after expulsion from Nova Scotia by the British.[3]

Jean Pierre Viguerie became sole owner of Pointe Farm on January 6, 1879. From that time on, the Viguerie family was closely identified with the plantation.

Jean Pierre and his brothers, François Camille and Alexis, were born in Hêches, Canton de Labarthe, Départment des Hautes Pyrénées, France. Their grandfather Guillaume Viguerie was a celebrated surgeon who had been involved in the Second Restoration, as a liberal, not as a Bonapartiste.[4] His son Jean Viguerie married Bertrande Bazerque Viguerie; they were the parents of Jean Pierre, François, and Alexis, and later of Guillaume, Jeanne, and Marie. As had many French families, the Vigueries suffered losses in the French Revolution of 1789 and subsequently in various Napoleonic campaigns. This fact no doubt figured in the Viguerie family patriarch's opting to send his three oldest sons to America.

Jean Pierre (born 1828) arrived in New Orleans on the vessel *Talma* from Bordeaux on January 17, 1846. François Camille (born 1829) arrived aboard the vessel *Victoria* in New Orleans on July 31, 1849.[5] François worked as a clerk in New Orleans for two years after first immigrating, according to the 1850 U.S. Census.

In the 1850 census, Jean Pierre is on record as a resident and merchant of Terrebonne Parish, residing with Isidore and Virginia Duthu. By 1855 Jean Pierre was a merchant in a building located at the corner of Main and Lafayette streets, which later was to be the first home of the Bank of Terrebonne and Trust Company.[6]

Together the Viguerie brothers first established themselves primarily as lumbermen, but also as merchants. In those early days of Terrebonne Parish, virgin cypress forests dominated the reaches of the parish. Jean Pierre operated his own sawmill at Ardoyne, and also served as Terrebonne Parish Tax Collector and later as Recorder.[6]

The third Viguerie brother, Alexis (born 1836), embarked upon the ship *Lemuel Dyer* in Bordeaux and arrived in New Orleans on October 28, 1854. Instead of joining his brothers, he struck out on his own. He traveled first to Cape Girardeau, Missouri, arriving there in 1876. He then moved to Trinidad, Colorado and purchased a land grant in 1888. Alexis, sometimes known as Alejandro, developed a cattle and horse ranch and had a general merchandise store. After 15 years in Colorado, he was gored by a bull and returned to New Orleans by private rail car with a doctor for medical treatment. On June 5, 1891 Alexis summoned a notary to the boarding house of Mrs. Joseph Gautreaux at 425 Bourbon Street where he lay in bed. He dictated his last will and testament witnessed by his brother François. The will was filed the next day, and Alexis died shortly thereafter. His older brother Jean Pierre having predeceased him, Alexis left his estate to François, his sisters Jeanne and Marie, Jean Pierre's widow Elvire, and his other brother, Guillaume.[7]

The year after Alexis arrived in the United States, Jean Pierre married Marie Elvire Delaporte of Houma on April 28, 1855. She was the daughter of Joseph Delaporte, a native of Locmine, Brittany, France, who had relocated to Terrebonne to be the tutor of Marianne Verret and her nine siblings. Delaporte and Marianne married. Marianne Verret Delaporte was the daughter of New Orleans native Jacques Verret, who was educated in Europe, served as second lieutenant under General Don Bernardo de Galvez during the Revolutionary War, was an engineer and owned vast real estate holdings in Lafourche and Terrebonne.[6]

Jean Pierre and François Viguerie became sugar planters and land owners together and separately in the 1860s. Viguerie plantations were Pointe Farm, Orange Grove, and Belle Farm (formerly Porche Hermitage plantation) on Bayou Black, Boutelou in Gibson, all in Terrebonne; Louisa in St. Mary Parish, and Evergreen in Lafourche Parish.[6]

The 1860 U.S. Census placed François living with Jean Pierre and his family at Pointe Farm and working as a merchant. Jean Pierre, the census recorded, owned real estate worth $12,000 and personal possessions valued at $12,000. Soon after that census, the Civil War broke out. Both brothers served the Confederate States of America during the conflict, but Pointe Farm remained in operation.[8]

Jean Pierre became a naturalized citizen of the United States on April 27, 1852.[9] His witnesses were Henry Larne and Martial Verret, the brother of Jean Pierre's future wife. François became a naturalized citizen on June 24, 1854. His witnesses were Samuel Wolf and Casimir Tremoulet.[9] The brothers received certificates of amnesty from the French government for avoiding conscription by emigrating, in 1870. Jean Pierre went back to his homeland in 1873. There he met his siblings born after he and his brothers had emigrated, and with his son and nephew made a grand tour

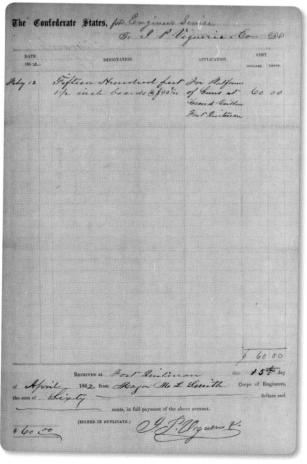

Confederate States of America's $60 purchase of lumber to build a gun platform at Fort Quitman on Grand Caillou, Terrebonne Parish April 15, 1862

The Grand Isle, *an excursion boat to the coast c. 1920s*

Ad for Caillou Island resort The New Orleans Bulletin, *July 4, 1875*

of the continent.[6] His sister Jeanne, her husband Jean Lay, and son Jean Baptiste Lay later visited from France and remained a number of years before returning to France. Sibling Marie (Madame Joseph Consalle) never visited America.[10]

On March 31, 1875, the U.S. Postal Service opened the Point Farm Post Office run by postmaster Peter P. Flynn. Service was discontinued on November 13, 1876.

As sole owners of Pointe (the French spelling of Point) Farm, Jean Pierre and his wife Marie Elvire owned 2,000 acres and had 500 acres in cultivation. The 1880 census recorded that the two Viguerie brothers and their families were living together on the left bank of Bayou Terrebonne in the first big plantation house. In 1880 Viguerie sold to Thomas Ellender (Elinger) a tract of his land called Caillou Field (which became Hard Scrabble) for $2,500. It fronted Bayou Terrebonne five arpents on both banks.

In 1882 François Camille Viguerie became half-owner of Point (the French spelling having been dropped) Farm. He had married Maryland native Georgianna Marie Metcalf (1847-1887), who was of English ancestry, in St. Mary Parish on June 13, 1868. In adulthood their daughter Clara became a nun in the order Sisters of Charity and took the name of Sister Marie Corinne.[6]

On February 26, 1883, François Camille Viguerie became a founder and president of the Lower Terrebonne Refinery, an enterprise that would undergo many alterations in name and ownership, but which proved important to the sugar industry for decades. James Monroe Sanders was vice-president and board members were John R. Bisland, Jean Marie Dupont, Victor Buron, Paul Faisans, and M. Neuville Fields.

The refinery's site on the right bank of Bayou Terrebonne was purchased from Jean Pierre and François Viguerie on March 28, 1883.

Lower Terrebonne Refinery was liquidated on July 11, 1891 and on the same day the Lower Terrebonne Refinery & Manufacturing Co. became incorporated. Henry Chotard Minor was president; board members were Emile A. Daigle, Charles B. Maginnis, William P. Tucker, and Jean Norbert Caillouet.

Besides their plantation lands, the Viguerie brothers acquired expanses of land in Terrebonne, St. Mary, Calcasieu, and Vermilion parishes, much of it for their stands of sought-after cypress, according to one historian.[6] They also owned property on Timbalier Island and Brush Island, and nearly 400 acres on Caillou Island,[6] which at the time was quite large. For land grant/patent purposes, an 1832 survey was conducted, measuring out the island's 1,290.51 acres in sections. Cattle could be driven from the mainland at Sea Breeze to the islands at low tide. The Vigueries had a large hotel and rental cottages on Caillou Island, as well.[11] They dug the Viguerie Canal in Montegut for timber transport in 1870, and a canal on the bay side of Timbalier Island to accommodate landing and loading cattle. By 1909 storms had so damaged the coast that traveling by horse and buggy at low tide to Caillou Island and Timbalier Island was no longer possible.

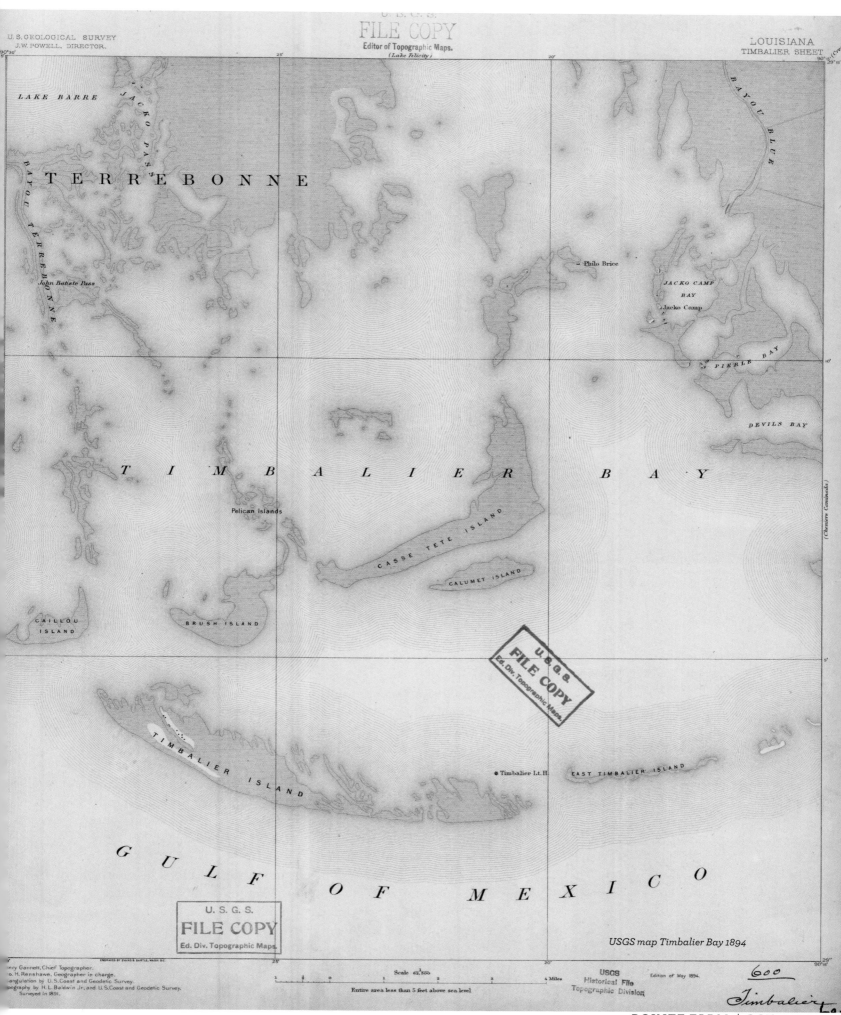

USGS map Timbalier Bay 1894

POINTE FARM | 365

Opera House (1859-1919) by Adrien Persac (1859-1873) corner of Bourbon and Toulouse streets

Death notice of François Viguerie,
The Houma Courier *May 30, 1896*

> **DIED.**
> VIGUERIE—In Kansas City, on Monday, May 25, 1896 FRANCOIS VIGUERIE, aged 97 years, a native of France and a resident of Terrebonne Parish for many years.
>
> she passed must have been as dreadful as the imagination can picture.
> Soon after her father received the stroke, Miss Viguerie telegraphed to her relatives here but owing to some delay in the telegrams they were not received following morning.
> Mr. Viguerie the eldest deceased accompanied es D. Wilson, and Miss ierie left Tuesday for Kansas City, but on reaching New Orleans they learned that the remains had been sent to New Orleans.
> The remains of the deceased were brought to Houma and interred in the Catholic cemetery last Thursday at 1 o'clock p. m.
> Mr. Francois Viguerie was born in France 67 years ago. He came to this Parish many years ago, and soon became a prominent figure in the social and business circles of Terrebonne. At the time of his demise he was the owner of a large sugar plantation on bayou Terrebonne. He was a brother of the late John Pierre Viguerie and leaves a large family to mourn his loss.
>
> For Sale.

The Vigueries' lifestyle was that of the educated and cultured. The wealthy brothers and their families attended the French Opera in New Orleans often, where they acquired librettoes for the brothers to sing at home. "They had beautiful singing voices," according to an account in Helen Wurzlow's *I Dug Up Houma Terrebonne*. They lived a gentrified existence and engaged in altruistic pursuits.

Jean Pierre was a founder of the Houma Academy which the Marianites of Holy Cross order later staffed as a "convent school" that became St. Francis de Sales School.[6] François Camille Viguerie donated land for the building of Sacred Heart Church October 1, 1893, and the family sold land for the Montegut School September 16, 1907 for $400.00. A.R. Viguerie donated land for the rectory December 17, 1897.

The eldest Viguerie brother, Jean Pierre, had a fatal heart attack on December 30, 1888 in Montegut. After his death, the original Point Farm house on the left descending bank of Bayou Terrebonne was dismantled. In 1893 Jean Pierre's family lost most of their assets following weather disasters and an economic depression.

The year before that occurred, on May 21, 1892 an ad in the

Houma Courier newspaper gave details of a Sheriff's Sale at Point Farm Store, "formerly owned by D.R. Calder and J.D. Wilson." Calder owned Aragon plantation at the time. The sale was to be Saturday, May 28, at 11 a.m., and the store's stock of merchandise to be sold consisted of "Dry Goods, Clothing, Notions, Groceries, etc... Also 2 mules, 1 peddler wagon, 1 covered wagon, 1 buggy, 1 sulky, 2 sets of harness, 2 iron safes and 1 lot of store fixtures. Terms—cash."

François' sons Albert Robert and Arthur Camille ran Point Farm beginning in 1896. That year, François was aboard a train en route to Colorado with his daughter, Emma, in hopes the weather there would improve her health. They disembarked in Kansas City, Kansas, where a doctor diagnosed the father's Bright's disease. He died in Kansas City on May 15.[12] Emma died of consumption at the age of 21 in El Paso, Texas, in December of the same year.[13]

Arthur Camille became manager of the plantation store, which survived many more years on the left bank of Bayou Terrebonne near Sacred Heart Church (1113 Highway 55) in Montegut. An 1897 parish directory recorded that Point Farm consisted of 450 acres with 300 in cane and 100 in corn. Manager was Arthur Camille Viguerie and overseer, Trasimond Duplantis.

A.R. (Albert Robert) Viguerie of Point Farm and Evergreen in Schriever is recorded as having 800 acres, 450 in cultivation, at Schriever in 1900. The average annual output of the plantation was 900,000 pounds.[14]

A natural gas well was drilled near Point Farm in 1901, by H.G.

Point Farm Store 2015

Sheriff's Sale merchandise at Point Farm Store, The Houma Courier *May 21, 1892*

Tax sale, State of Louisiana vs. David E. Calder, The Houma Courier *February 20, 1892*

Articles, top to bottom, ad for high blood pressure remedy, The Houma Courier *October 16, 1897; Hoof Oil ad* The Houma Courier *July 17, 1897; ad for Louisiana oranges* The Houma Courier *November 29, 1913*

Albert Robert Viguerie (1871-1954) and Arthur Camille Viguerie (1877-1949)

Article about priest's ordering a collection for Emma Viguerie's funeral Mass to be sung, The Houma Courier *January 23, 1897*

POINTE FARM | 367

Point Farm, third house of A.R. Viguerie c. 1900

House of A. R. Viguerie dismantled after 1928 by South Coast Corporation

Bush near the refinery owned by Bush and Maginnis at the time.

Point Farm Planting Co., Ltd. was formed on February 27, 1905 and continued operations until 1920. President Albert Robert owned 1,137 shares; vice president Miss Clara Viguerie (Sister Marie Corinne Viguerie) owned 599, with Gabriel Montegut acting as proxy; secretary-treasurer Arthur C. Viguerie owned 1,147 shares. Shares were valued at $10 each.

Continuing the public service spirit of the previous generation, Albert Robert Viguerie was president of the Terrebonne Parish School Board for more than 30 years. He was married to Irene Bascle, daughter of Felix A. and Ada Lester Bascle. Albert had the second Point Farm house built; it was dismantled by the South Coast Company in the late 1930s. The house was located at the intersection of Point Farm Road and Louisiana Highway 55 paralleling Bayou Terrebonne. Albert's family then moved to the "Harry G. Bush House." It is located at 94 Mill Site Road, on the right bank of Bayou Terrebonne. Albert was recorded as belonging to numerous trade organizations as well as the fraternal organizations the Maccabees and the Elks in Houma.[15]

By the 1920s the Viguerie family was no longer associated with Point Farm.

In December 1927, Whitney-Central Trust & Savings Bank foreclosed on the property. Carl F. Dahlberg & Co. took ownership on January 19, 1928, and seven years later South Coast Company became owner, holding title until the 1960s.

Point Farm Road sign 2015

In rather rapid succession, Point Farm lands were transferred to the Jim Walter Corporation when it bought South Coast in 1968, Walter Land Co. in 1970, Mid-South Mortgage Co. in 1979, Ag-Lands Investment Co. in 1979, McCarthy Family Corp. in 1979, and Cook Family Corp. in 1979.

The Harry G. Bush House that Albert R. and his family resided in at 94 Mill Site Road is now occupied by Tim and Cheryl Cenac Cronan.

Arthur Camille Viguerie's home is near Sacred Heart Church at 1105 Highway 55, on the left bank of the bayou in Montegut. It was formerly the plantation overseer's house, which was moved to the front of the property in 1903. Arthur Camille was married in 1883 to Elodie Chauvin, who died in 1918 leaving the widower with eight children under the age of 15. That house is currently occupied by Anthony R. Viguerie and his wife Joyce.

SOURCES

1. A Century of Lawmaking for a New Nation: U.S. Documents and Debates, 1774-1875, American State Papers, Senate
2. Thomas Becnel, *The Barrow Family and the Barataria and Lafourche Canal: The Transportation Revolution in Louisiana, 1829-1926*
3. Donna McGee Onebane, *The House that Sugarcane Built: The Louisiana Burguières*, University Press of Mississippi, 2014
4. Robert Alexander, *Re-writing the French Revolution*, 2003
5. Ships' passenger manifests
6. Helen Wurzlow, *I Dug Up Houma Terrebonne*, 1984
7. From research by Laura Browning, 2016
8. Confederate States of America documents
9. Terrebonne Parish Naturalization Records
10. Viguerie family recollections
11. Letter to the editor of *The New Orleans Bulletin*, dated June 29, 1875
12. Obituary in *The Houma Courier*, May 30, 1896
13. Obituary in the *El Paso Daily Herald*, December 27, 1896
14. Henry Rightor, *Standard History of New Orleans,* The Lewis Publishing Co., Chicago, 1900
15. Henry E. Chambers, *A History of Louisiana, Volume 27*, The American Historical Society, Chicago and New York, 1925

Arthur Camille Viguerie home 2015

Point Farm tenant houses, Standing Structures Survey June 1981

Top, Lower Terrebonne Refinery with sugar cane barges in bayou c. 1900

LOWER TERREBONNE REFINERY
CELOTEX CORPORATION
SOUTH COAST CORPORATION

The refinery at Montegut had its genesis in Jean Pierre and François Viguerie's lands at Pointe Farm, and grew to be one of the largest sugar refineries in the state and nation. The location of the refinery was on the land claim of Jean Charles Naquin of 1795, registered on November 20, 1816 with the U.S. government and confirmed on May 11, 1820.[1]

The Lower Terrebonne Refinery was founded on February 26, 1883. Jean Pierre and François sold a segment of their plantation lands for the refinery site on the right descending bank of Bayou Terrebonne in the sale transaction dated April 26, 1883.[2] It was described as "a certain tract of land on the right descending bank of Bayou Terrebonne about 16 miles below Houma, having a front on the bayou of 310 feet by 350 feet in depth."[2] In time the refinery complex expanded to the left bank, where offices, a store, and quarters were located.

The (open kettle) mill benefitted not only the Viguerie brothers after its founding in 1883, but also numerous other plantation and small sugar cane farm owners along lower Bayou Terrebonne—all part of the widespread Sugar Belt of Terrebonne Parish and south central Louisiana.

Lower Terrebonne Refinery employees 1927

   Production by lower Bayou Terrebonne area planters recorded in the 1884-85 annual sugar report for the parish[3] indicate the need for a major refinery operation in the area, although many of the plantations had their own individual mills at the time. The totals for area plantations produced that year (in hogheads—barrels holding approximately 1,000 pounds of sugar each) were Rural Retreat, 500; Hope Farm, 193; Aragon, 150; Pointe-Aux-Chênes, 33; Hard Scrabble, 65; Point Farm, 240; Magenta, 165; Angela, 108; and Live Oak, 185. Planters with smaller farms in the area produced two, five, seven, and 24 hogsheads that year.

   François Viguerie was the refinery's first president, and his neighbor James Monroe Sanders at Live Oak was designated vice-president. Secretary was "Capt. T.S. Bisland, Secretary; Messrs. Eugene Fields, Neuville Fields, John R. Bisland, Paul Fazende, Victor Buron, and David Calder, directors. Capt. T.S. Bisland is general manager at the refinery," according to a Thibodaux newspaper article of the time.[4]

"Negro Quarters"
Lower Terrebonne Refinery

LOWER TERREBONNE REFINERY | 371

Left to right, Jean Marie Dupont (1835-1904); Emile A. Daigle (1842-1913); Albert M. Dupont (1861-1943); Henry Chotard Minor (1841-1898);

Godchaux Sugar sack 100 lbs. New Orleans, La. c. 1920s

The mill, which had been built for $85,000 and had undergone an $11,000 addition, burned on November 18, 1890. A Houma newspaper reported about the mill's loss, "Among the planters who lost by the unfortunate disaster are Ernest Picou, Viguerie and Marmande, J.N. Robichaux, S. Woods, the est. [estate] of Sanders, est. of Fields, F. Viguerie, Leo Lirette, and Eugene Fields."[5] The article continued that the mill "was developing the Lower Terrebonne and Little Caillou Country, as the planters in that section were having their sugar cane manufactured there."[5]

The Lower Terrebonne Refinery was liquidated on July 11, 1891, and Lower Terrebonne Refinery & Manufacturing Company was incorporated on the same day. The liquidators sold the plant to Ferdinand Kane and Albert M. Dupont, who sold to Jean Marie Dupont and Emile Daigle. Jean Marie Dupont sold his interest to Daigle, who sold to the new corporation.[2]

Henry Chotard Minor was president of the new company, and board members were Emile A. Daigle, Charles B. Maginnis, William P. Tucker, and Jean Norbert Caillouet.[2] In 1893, the company bought 26.43 arpents from François Viguerie.

In 1891, when the Lower Terrebonne Refinery & Manufacturing Company built the post-fire mill, "other sugar mills in Louisiana were grinding from 450 to 800 tons of cane each twenty-four hours. Terrebonne, with its 2,000 ton grinding capacity," had much greater daily productivity.[6]

W.E. Butler's book *Down Among the Sugar Cane: The Story of Louisiana Sugar Plantations and Their Railroad* described the transportation system used to and from the mill:

"Barges traveling over the bayou delivered cane to the mill, supplemented by cane cars on one of the most extensive plantations in Louisiana. The waterway system used twenty-three 40-ton barges, pulled by three towboats...."

Left, loading sugar cane in rail cars from barges on Bayou Terrebonne using oxen-driven cane hoist c. 1910

"The 36-inch gauge plantation tramroad was constructed in 1891 at the time the new mill was built. It extended thirty-five miles into the cane fields. The main line followed Bayou Terrebonne, operating five miles down the bayou from the mill and terminating ten miles up the bayou at a point along a branch line of the Southern Pacific above Ashland and Woodlawn plantations. The transfer point from the small-gauge tramroad cars to standard gauge railroad cars was at Colley Switch. At that time, the little narrow gauge line operated year round due to the more or less isolated location of the mill and the town of Montegut. There were no roads, and the bayou was the only other means of transportation to the outside world. It was at Colley Switch that supplies for the plantation commissary and equipment for the mill were received from the Southern Pacific Line and transported over the narrow gauge railroad to the factory. The sugar and molasses manufactured at the mill also moved in 36-inch gauge home built boxcars to the switch for further shipment."[6]

By 1897 the Lower Terrebonne Refinery and Manufacturing Company had new officers and board members. Charles B. Maginnis was president, Harry Garland Bush was vice president, and Reuben Gresham Bush was secretary and treasurer.

E.C. Wurzlow's parish directory published that year described the mill's scope, which underlined its importance to Terrebonne Parish. The summary described the operation as "Refiners and manufacturers of sugar. Product 1896, 10,100,000 pounds. Capacity of factory 1,000 tons of cane per day; sixteen miles

Above, loading sugar cane on barge using oxen-driven cane hoist c. 1910

Left and below, contract for cane hauling, A.R. Viguerie and Oliva Pelegrin, August 24, 1922; Lower Terrebonne Refinery Train Engine #1 c. 1900

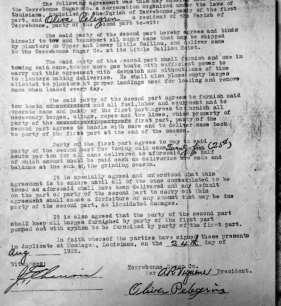

Lower Terrebonne Refinery Montegut, La. 1953; Inset, Lower Terrebonne Refinery 1946

Louisiana Celotex Co. mill Marrero, La. March 15, 1927

LOWER TERREBONNE REFINERY | 377

Top left, Lower Terrebonne Refinery saddle back (tank) locomotive c. 1900

J. Farquhard Chauvin (1878-1959) bookkeeper at LTR

Notice of "Dr. McBride" discussion of oil excitement in Terrebonne Times Picayune June 1, 1901

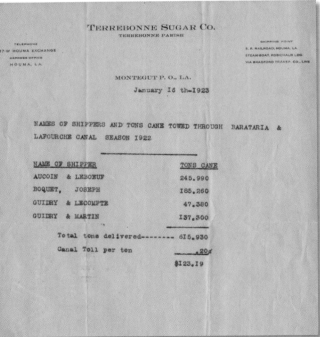

Terrebonne Sugar Co. shipping statement January 16, 1923

Left to right, Albert R. Viguerie (1871-1954); Charles J. Champagne (1850-1927); Allen A. Sanders (1868-1945); Farquard P. Guidry (1871-1965); Ernest F. Ellender (1857-1925); J. Leonce LeBlanc (1853-1917) seated with cane

of narrow gauge [rail]roads; three locomotives, tug boats and barges used in bringing crop to factory. Gives employment to 200 mechanics and laborers. Crop of 1893 was 25,000 tons; crop of 1897, 60,000 tons. The production of cane has more than doubled itself since the establishment of this refinery. Number of acres (1897), 3,300."[7]

Locomotive engineers were Emile Doré and J.C. Dupre.[6] The dummy locomotives (their horns and whistles "dumbed" so as not to frighten working animals in the fields) were named Clara and Laura (a saddleback locomotive), the names of François Camille Viguerie's daughters. The mill railroad continued until closure in 1972.[7]

At the mill, necessary personnel were a general manager, head sugar boiler, superintendent, blacksmith, carpenters, bookkeeper, coopers, a chief engineer, a chief chemist, a mechanic, locomotive engineers, a painter, and a steam threader.

The Southern Manufacturer publication of 1901 described the Lower Terrebonne Refinery and Manufacturing Company as "one of the largest and most complete sugar factories in the world. There were two immense mills, each equipped with a Krajewski crusher and a six-roller mill. They had the combined capacity to grind over 2,000 tons of sugar daily."

The refinery ran its own light plant, and in 1901 expected to refine over 17 million tons of sugar. In that year the company had contracted with 150 sugar planters to purchase their cane.[8]

Also in 1901, Houma Oil & Mining Company drilled the Bush & Maginnis Well to 1,000 feet with a diamond bit on mill property, an enterprise which created local excitement about the future of oil exploration in Terrebonne Parish. The storm of 1909 heavily damaged the mill and surrounding plantations, residences, and businesses.

Terrebonne Sugar Company bought the refinery for $376,175.76 on March 23, 1912. It was purchased from Frank B. Williams, who had a mortgage claim of $186,532.[9] Officers were Albert Robert Viguerie, president; Charles J. Champagne, vice president; and Allen A. Sanders, secretary. Board members were Farquard P. Guidry, Ernest Filmore Ellender, Hugh S. Suthon, and J.L. LeBlanc.

The refinery encountered hardships in 1927 and the Whitney Central Trust and Savings Bank foreclosed. On January 19, 1928 the bank sold the land and plant to Carl F. Dahlberg and Company. Dahlberg and Company sold it to the South Coast Company on September 27, 1935.

The Carl F. Dahlberg Company name was repeated numerous times in the ownership rolls of many sugar cane plantations along lower Bayou Terrebonne and in other areas of Terrebonne Parish. The reason for the Dahlberg concern's buying up such extensive acreage is that Bror G. Dahlberg, Carl's brother, discovered that the bagasse byproduct from sugar cane grinding contains "cellulose in its toughest, strongest form."[10] Bror had previously been in the lumber business in Minnesota when he recognized the practicality of manufacturing a synthetic board out of cellulose "taken from a plant that can be cut and regrown each year as a crop." His experiments led him to sugar cane bagasse[10]

Swedish natives Bror and Carl established Celotex in 1921, and began to acquire sugar cane lands. Subsidiaries of Dahlberg Sugar Cane Industries in 1929 included the Celotex Company and South Coast Company, the latter a name familiar in Terrebonne Parish. Celotex produced fiberboard for insulation and other products using bagasse; the original Celotex manufacturing plant, built in 1920, was in Marrero, Louisiana, at the site's current 7500 Fourth Street address.

Carl F. Dahlberg, First Vice President, Celotex Co. and Bror G. Dahlberg, President, Celotex Co. c. 1929

Dahlberg Sugar Cane Industries ad in Saturday Evening Post *November 30, 1929 and Celotex ad from* American Builder Magazine *June 1929*

Tokens for 5 and 25 cents, Celotex Corporation, Marrero, La.

Below Celotex Corporation New Orleans 1959

Louisiana Celotex Co. mill Marrero, La. March 15, 1927

LOWER TERREBONNE REFINERY | 381

Tokens of South Coast Corporation

In 1927, many plantations that had produced millions in profits for their owners were being sold at sheriff's sales for taxes, or had landed in the hands of creditors. The Celotex Company purchased four Louisiana plantations: Georgia at Matthews, Oaklawn at Franklin, Ashland at Houma, and Terrebonne at Montegut. Besides its holdings in Louisiana, Celotex sugar cane growing and processing operations expanded to 150,000 acres of fertile lands near Lake Okeechobee, Florida. South Coast and Celotex were closely intertwined, with Carl Dahlberg performing as president of the South Coast Corporation, one of the largest cane sugar producing companies in the country, for the final ten years of his life, according to one of his obituaries in 1942.[11] When the South Coast Company was formed, the magnitude of its properties amounted to 39,046 acres at Ashland, Georgia, Oaklawn, and Terrebonne plantations.

South Coast ran the Lower Terrebonne facility for 44 years, until it closed its operation in 1972. Records show that Jim Walter Corporation bought the property in 1968, and the Walter Land Company purchased it in 1970. Refinery lands in southern Terrebonne passed from the Mid-South Mortgage Company to the Ag-land Investment Company, the McCarthy Family Corporation and the Cook Family Corporation.

The mill in its entirety was deconstructed in 1972 and reconstructed in Guatemala.

What is left at the site on the left bank of the bayou are three chemists' homes occupied during grinding seasons—at 1135, 1137, and 1139 Highway 55, Montegut, Louisiana. The store site is at 1125 Highway 55, and the overseer's homes at 101, 159, and 199 Recreation Drive.

On the right bank remains the Harry G. Bush house at 94 Mill Site Road that became the Albert Robert Viguerie home and then the Earl Gravois residence. A manager's house at 92 Mill Site Road was the Jim Redmond residence. The former office is at 90 Mill Site Road. A mill manager home was the residence of

Information about Terrebonne Factory, The South Coast Corporation, in Gilmore's Louisiana Sugar Manual *1942-43*

Guatemala Sugar Refinery reconstructed from dismantled Lower Terrebonne Refinery December 30, 2013

J.J. Munson Mechanical Sugar Cane Harvester and J.J. Munson, General Manager, South Coast Corporation 1930

Lower Terrebonne Refinery chemists' houses 2015

LOWER TERREBONNE REFINERY | **383**

Left to right, Point Farm Store 2015; home of James Redmond, former manager of Lower Terrebonne Refinery, occupied by Murphy J. Savoie 2015; Lower Terrebonne Refinery office 2015

Montegut resident and singer J.P. Richardson "The Big Bopper" died with Buddy Holly and Ritchie Valens in a February 3, 1959 plane crash in Iowa.

Lower Terrebonne Refinery ruins 2016

Richard Fryou at 752 Sarah Road. (His daughter Adrianne Fryou was married to J.P. Richardson, known as the Big Bopper and singer of "Chantilly Lace," who was killed in the plane crash in Iowa on February 2, 1959, that also killed Buddy Holly—"That'll Be the Day," "Peggy Sue"—and Richie Valens—"La Bamba," "Donna." Adrianne was pregnant with a son at the time of the crash.)

Remnants of the mill operation are the weigh station on Crochetville Road, quarters houses on Crochetville Road, and two quarters homes that were moved in the 1960s to their current location. One is currently occupied by John and Celeste Gravois Tieken. The other at 4139 Country Drive, Bourg, had been the home of Delbert L. (since deceased) and Anita Gravois Leggett before they moved to Youngsville, Louisiana. Murphy J. Savoie is the current occupant of the Jim Redmond home at 92 Mill Site Road.

The Mission of St. John's Church had its genesis in the parlor of the H.G. Bush home when members of the Protestant Episcopal Church of the area had no formal place of worship. The small congregation then used a cottage on loan by Allen A. Sanders. After Episcopalians of that area used temporary facilities for 18 years, Lower Terrebonne Refinery & Manufacturing Company donated a building, which congregants carried across the bayou to its new site near the Montegut School. The building

was remodeled, and stood on land which was a gift of Allen A. Sanders. The August 1910 "St. Matthew's Record" publication of Houma claimed the church as its first daughter church, which had been completed in July and first used for service on Sunday, July 24. An altar cross and vases were memorial gifts from Mr. and Mrs. John D. Shaffer of Terrebonne Parish, the organ from the Charles Janvier family of New Orleans, and the lectern Bible a memorial gift from the family of Charles Williams. It is unclear as to when the Mission of St. John's in Montegut closed its doors.

SOURCES

1. A Century of Lawmaking for a New Nation: U.S. Congressional Documents and Debates, 1774-1875, American State Papers, Senate
2. Helen Wurzlow, *I Dug Up Houma* Terrebonne, 1984
3. Alcée Bouchereau, *The Louisiana Sugar Report*, 1884-1885
4. *Thibodaux Sentinel* newspaper, December 22, 1883
5. *Houma Courier* newspaper report of the fire
6. W.E. Butler, *Down Among the Sugar Cane: The Story of Louisiana Sugar Plantations and Their Railroads*, Moran Publishing Corporation, Baton Rouge, Louisiana, 1980
7. E.C. Wurzlow, *1897 Directory of the Parish of Terrebonne*
8. *The Southern Manufacturer*, 1901
9. *New Orleans Times-Picayune* newspaper articles of February 4, 1912 and March 23, 1912
10. *The Saturday Evening Post* magazine ad, November 30, 1929
11. *Chemical Engineering News* magazine of the American Chemical Society, August 25, 1942

Terrebonne Refinery fire bucket used as planter 2016

Below, St. John's Chapel, Montegut, c. 1910

MONTEGUT
LE TERREBONNE
SANDERSVILLE
CROCHETVILLE

The area along the southernmost reaches of Bayou Terrebonne developed from original Spanish land grants, some of which go back as far as 1788. Sugar cane plantations developed along the bayou at *le Terrebonne* (The Terrebonne) just as they had along the bayou's course from its headwaters.

However, the area known as *le Terrebonne* and then as Montegut, later developed, to a great extent, around the central-factory model sugar mill established as Lower Terrebonne Refinery in 1883, and also known by its later name incarnations. As is the case with such large concerns, a whole community grew up around it and Montegut had a concentrated population before many of the other bayou communities did.[1]

Lower Terrebonne Refinery & Manufacturing Co. c. early 1900s

Left, Ernest Robichaux (1914-1957), fourth Montegut Postmaster, who served 42 years and right, Henry J. Klingman (1889-1914), third Montegut Postmaster

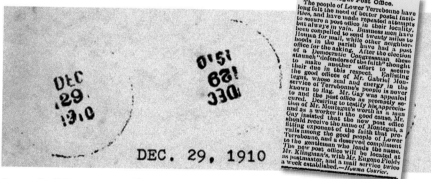

Postmark of Montegut Post Office December 29, 1910 and article in the New Orleans Times Picayune on post office being named in honor of Gabriel Montegut, August 3, 1885

The community received its name from the post office established in 1885, for prominent Houma resident Gabriel Montegut. The post office was named for Montegut at the request of U.S. Congressman Edward James Gay. Eugene Fields was the first postmaster; he served from July 22, 1885 until October 24, 1889, followed by James D. Wilson (October 24, 1889-April 14, 1899), Henry Klingman (April 14, 1889-June 8, 1914), and Ernest L. Robichaux (June 8, 1914- March 8, 1957).

Henry J. Klingman General Merchandise Store invoice January 1, 1911 and Henry J. Klingman complimentary plate 1910

Henry J. Klingman home at right in photo; former store location at left, photo 2015

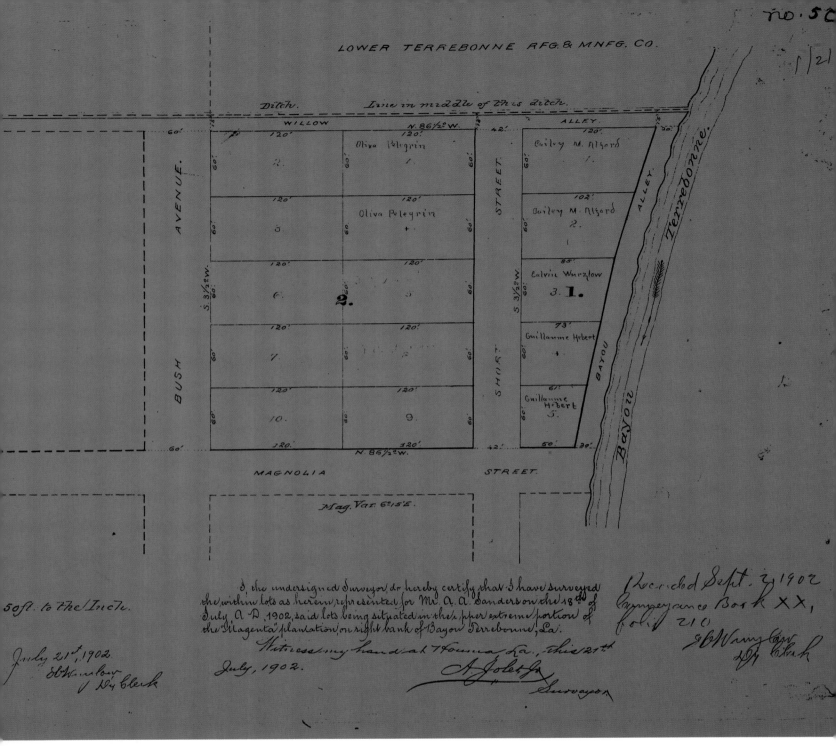

Sandersville map July 18, 1902

Notice of Sheriff's Sale of Sandersville lots
The Houma Courier *June 11, 1910*

Saturday the 11th day of June, A. D. 1910, between the hours of 11 o'clock A. M. and 4 o'clock P. M. the following described property, to-wit:—

"Two certain lots of ground, situated in the Parish of Terrebonne, about 16 miles below the town of Houma, on the right descending bank of the Bayou Terrebonne, measuring each, sixty feet front on Short Street by a depth of 120 feet; designated on a plan of Sandersville made by A. Jolet, Jr., Surveyor, on file and of record in Clerk's Office as Lots Nos. 6 and 8 in Block No. 2, with the improvements thereon."

Terms: Cash in United States Currency.

A. W. CONNELY,
Sheriff.

In 1901 a natural gas well was drilled on mill property, which led to further exploration by oil and gas companies in Terrebonne. By 1915 Montegut was the location of a consolidated school, the central mill, churches, several stores, and two subdivisions carved from plantation properties of the Fields and Sanders families.

Sandersville was on the right descending bank of Bayou Terrebonne just below the mill. Streets in Sandersville were christened Bush Avenue, Magnolia Street, Bayou Alley, Willow Alley, and Short Street.[2] Crochetville developed immediately below Sandersville on the right descending bank of Bayou Terrebonne. Purportedly named for property owner Walter "Bud" Crochet, it was not identified on maps until 1965.[3]

Aerial photo of Montegut and refinery 1953

The second Sacred Heart Catholic Church in Montegut, completed 1870

Sacred Heart Catholic Church (built 1909) and rectory c. 1917

Catholic missionary priests visited the bayou areas of Terrebonne in earlier years, and in 1864 Sacred Heart Parish was established in lower Terrebonne. Father Jean Marie Joseph Dénécé was appointed as pastor November 9, 1864. A small chapel was constructed on Little Caillou June 2, 1866.

Montegut native Arthur Robert Viguerie served as Terrebonne Parish School Board President for 25 years, and was instrumental in establishing one of the first consolidated schools in the parish, which earned a spot in the National Register of Historic Places in October 1993. Before its formation, the Fields School, Robichaux School, and other farm and plantation schools provided education for children of lower Bayou Terrebonne.[1]

Montegut Elementary School opened 1913-14 school year

Montegut Elementary School 2016

National Register of Historic Places plaque in front of Montegut Elementary School October 1993

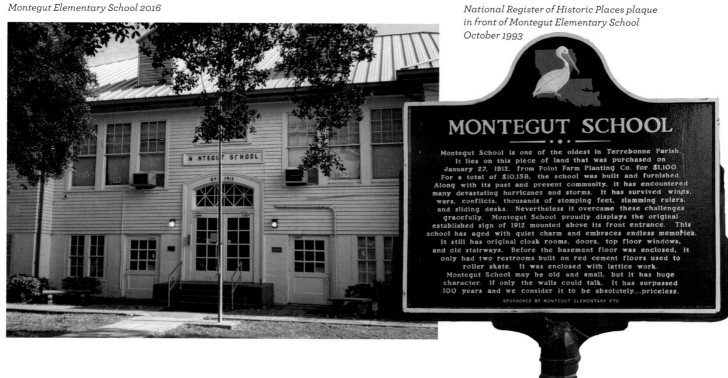

MONTEGUT | **391**

Oak/Gem theatre Montegut 2015; closed in 1967

Oak/Gem Theatre ads, left 1941 and right 1936

Band concert at Montegut Elementary School, Band Director Mr. Grisman, 1931

Walton James Quick (1881- 1951), contractor of Montegut School, home and family in Lake Charles, La. c.1920

John (1874-1956) and Ludovia Lirette (1880-1919) Guidry home 1163 Highway 55, 2015

Sherwin J. "Chabbie" Guidry (1920-2014) with his first car in front of John Guidry General Merchandise Store in Montegut, 1943

Paul Michel Cenac (1878-1947) and Marie Clothilde Klingman Cenac (1878-1969) c. 1930

A small Montegut school existed in 1879. Fundraising for a larger school began in 1907, and on January 27, 1912, the Terrebonne Parish School Board purchased property on Point Farm for $1,100. The school board passed a millage for District 7 on April 2, 1912. Architects Charles Allen Favrot & Louis A. Livaudais Ltd. designed the structure, and the contractor was Walton James Quick. Construction cost totaled $10,148. The school opened for the 1913-14 school session, and is still in use as the Montegut Elementary School at 1137 Highway 55, Montegut.[4]

Prominent family names in Montegut include Klingman, Pellegrin, Guidry, Robichaux, Stoufflet, Crochet, and Redmond.

Although sugar cane production began to be curtailed in the 1920s, the operation of the sugar refinery continued in Montegut until the mill around which the community had grown was dismantled and shipped to Guatemala in 1972.

With the waning of the sugar industry, Montegut's major economic sustainer was the result of Texaco's creation of its largest land base of operations there, in lower Montegut, in the 1950s.

SOURCES
1. Sherwin Guidry oral history conducted by Glen Pitre for the Terrebonne Parish Consolidated Government; Guidry's publication *Le Terrebonne*
2. Sandersville map of July 18, 1902 by A. Jolet, Jr.
3. Toby Henry, Montegut Fire Chief, 2016 interview
4. Terrebonne Parish School Board Minutes Books 1-3

Left to right, Ernest Stoufflet (1874-1939) 1904; Alidore Stoufflet (1848-1930) (wife Melice née Walker (1844-1924)) and nephew Wallace Stoufflet (1886-1912); Odelia Stoufflet Robichaux, Easton Robichaux, and Annette Duplantis Stoufflet, c. 1904; Enos Redmond (1878-1942) c. 1920s; Angele Laurentia Bourg Redmond (1882-1941) c. 1900

Eugene Pierre Dalmace Fields, Sr. (1829-1913) and Marie Augustine Tasset Dardeau Fields (1835-1894) c. 1880

ANGELA

Four years after the end of the Civil War, Eugene Fields and Jean Pierre Viguerie founded Angela plantation in 1869 on both banks of Bayou Terrebonne 19 miles below Houma. The subsequent main house was situated on the left bank. The land upon which Angela rose was an original claim by Jean Batiste Dugas, confirmed by the U.S. government in 1833.[1]

Eugene Pierre Dalmace Fields was the twin of Neuville Demetrius Fields, and son of native Dubliner William Fields who married Theotiste Ludivine Dugas of Plattenville in Assumption Parish, Louisiana.[2] Eugene partnered with Jean Marie Dupont in developing Magenta, and he did the same with Dupont in 1872 in partnership at Angela plantation. Eugene's second wife Marie Augustine Tasset (born in Brittany, France) was the sister of Dupont's wife, Lydie Marie (Astugue) Tasset.[2] Marie Augustine's first husband, Alexandre André Dardeau, had been killed in a right of passage dispute by Jean Pierre Viguerie in April 1862.

The plantation was in the hands of Neuville Demetrius Henry Fields from 1872 until 1878. Neuville's first wife was Felonise Celeste Bourg and his second wife was Elmire Chauvin.

Jean Marie Dupont and Eugene Fields again took ownership of Angela in 1886, followed by Eugene Fields alone in 1890. Allen Andrew Sanders of Magenta was also the owner of Angela from 1900 until 1920.

Seven years later, along with many other plantations along lower Bayou Terrebonne, Angela succumbed to Whitney Central Trust & Savings Bank's foreclosure in 1927. Carl F. Dahlberg & Company purchased Angela in 1928, followed by South Coast Co.'s cultivation of its land until 1968, when Jim Walter Corporation purchased the property. Two years later Walter Land Co. was the owner of record, followed by 1979's four owners in succession: Mid-South Mortgage Co., Ag-Lands Investment Co., McCarthy Family Corp., and Cook Family Corp.

Currently the address of what was Angela is on the left bank of Bayou Terrebonne at 1213 Highway 55, Montegut, Louisiana.

SOURCES
1. A Century of Lawmaking for a New Nation: U.S. Congressional Documents and Debates, 1774-1875, American State Papers, Senate
2. Ancestry.com

Article of Eugene Fields visiting Houma The Houma Courier August 7, 1897

Eugene Pierre Dalmace Fields, Sr., seated; standing center, Rev. Elie Fields, S.J.; others are unknown, possibly Augustave, Clovis or Arthur Fields c. 1900

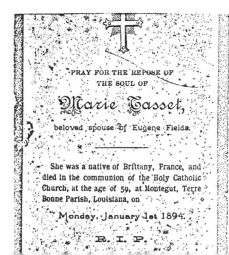

Marie Augustine Tasset Dardeau Fields death notice January 1, 1894 and Eugene Pierre Dalmace Fields death notice April 15, 1913

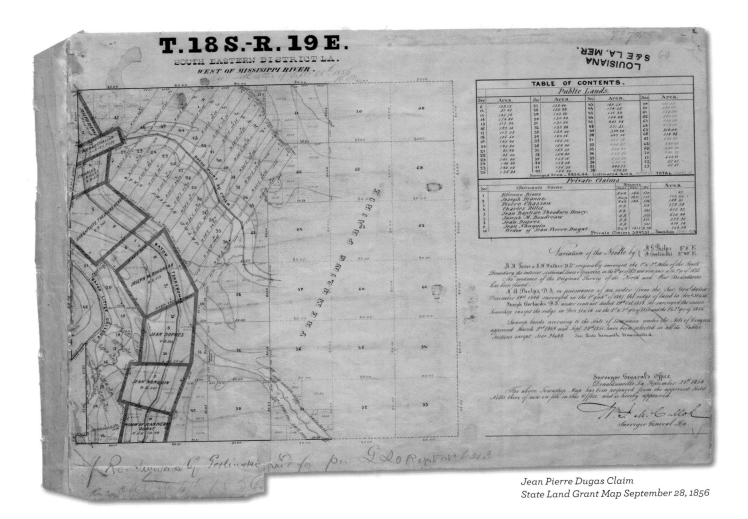

Jean Pierre Dugas Claim
State Land Grant Map September 28, 1856

MAGENTA

Seventeen miles below Houma along Bayou Terrebonne Jean Charles Naquin in 1795 and Jean Pierre Dugas (Dugat) in 1792 had original Spanish land grants which they registered on November 20, 1816 and which were confirmed by the U.S. government on May 11, 1820. Jean Baptiste Theodore Henry also had a claim that he registered and for which he received confirmation on the same dates.[1] That land became the foundation of Magenta plantation, named for a wildflower native to the West Indies.

William Fields, a native of Dublin, Ireland, married Theodiste Ludivine Dugas on January 19, 1829 in Thibodaux, Louisiana. He was born in 1796 and she on March 12, 1802. The property they owned in lower Terrebonne Parish came by way of Theotiste's mother, Reynalda Naquin Dugas. It had been purchased from Charles Naquin by Jean Pierre Dugas on February 3, 1795 and confirmed by the U.S. government in 1833.[1]

Thirty-eight-year-old William built the original house on his estate in 1834. He died at the age of 42 while on a visit to his mother in Ireland in 1837. Lake Fields in Lafourche Parish is named for the Irishman. His widow and family farmed the land until Jean Marie Dupont purchased the property in 1858. Dupont (1835-1904) was a native of Benac, France who on February 2, 1859 married Lydie Marie (Astugue) Tasset (1829-1898), also a French native, born in Forgeres, Brittany.[2]

Jean Pierre Viguerie of Point Farm entered the roll of that land's owners in 1873.

But Jean Marie Dupont and Eugene Pierre Dalmace Fields began to assemble lands from different tracts over the years dating back to 1869.[2] They were related by marrying sisters. Eugene was first married to Celeste Fidelise Pitre, and after her 1862 death, in 1864 he married Marie Augustine Tasset, sister of Jean Marie Dupont's wife Lydie Marie.[2] Marie Augustine and Lydie Marie's brother François was a priest who was living in Montegut when he died in 1891, after spending decades south of the U.S. border. A footnote about the Tasset sisters is that Marie Augustine, who married Eugene Fields, had first married Alexandre André Dardeau who had been born in Noyers, France. He was killed on a bank of Bayou Terrebonne by Jean Pierre Viguerie in April 1862 in a dispute about right of passage.

Dupont and Fields bought their first tract of land together in 1869, from Jean Pierre and François Viguerie. A large amount of land that was to be Dupont's and Fields' plantation was purchased from the Dardennes and the Naquins. One tract from the Hippolite Naquin succession measured two and a half arpents front; another tract purchased in 1876 was from Rosalie Naquin, widow of Hippolite Pitre.[2] Dupont and Fields bought two tracts from the Dardenne family.[2]

Dupont and Fields held Magenta lands until 1892. An 1897 parish directory describes the plantation as consisting of 700 acres, with 123 acres in sugar cane.[3]

In 1900 Magenta became the property of Allen Andrew Sanders, who was owner of the plantation until 1930. Sanders was married to Ella J. Trahan, and was the son of James Monroe

Lydie Marie Tasset (1829-1898)
Jean Marie Dupont (1835-1904)

Trespass notice, Magenta plantation The Houma Courier *October 17, 1917*

Below, Allen A. Sanders (1868-1945) in sugar cane field c. 1920

Randolph August Bazet, Sr.(1899-1982), Terrebonne Parish Clerk of Court 1924-1964 and Inza Mae Sanders Bazet (1899-1983)

Sanders who migrated to Terrebonne from Mississippi. The father bought land below Montegut that became Live Oak plantation, and at one time owned Brush Island which is now under water.[2]

Local historian Sherwin Guidry said that Mr. Sanders of Magenta cut a slip (1870)—the Sea Breeze Cut—in the natural land bridge preventing Bayou Terrebonne from flowing into Lake Barré (*Lac Barré*, meaning locked lake). He did this to open the lake to let steamboats through, to take his sugar to Bayou Lafourche and New Orleans. Before the lake was opened, he said, planters used portage from Bayou Terrebonne to the lake, using horses and wagons and manpower.[4]

May Sanders Bazet was the daughter of Allen Andrew Sanders, who in 1932 arranged for two oaks from Magenta to be planted aside Main Street, Houma, in honor of her father. She was the wife of Randolph A. Bazet, Sr., Terrebonne Parish Clerk of Court from 1924 until 1964. The two trees were named George and Martha Washington; Martha was removed for construction of a new post office in 1935.[2] Her father, A.A. Sanders, had also made a land donation on the upper line of Magenta on the left descending bank of Bayou Terrebonne to St. Matthew's Episcopal Church of Houma in January 1899.[2]

The original Magenta plantation house was constructed on the left descending bank of Bayou Terrebonne. The second home

Magenta 2015

built in the 1850s had undergone multiple revisions and even partially burned. Two rows of live oak trees are on the grounds of the homesite. Mrs. Bazet recalled the life on the plantation as one of comfort and hospitality, with guests staying weeks instead of days, and tables generously laden. She recalled that the plantation had its own cemetery and a bell that rang to summon workers to the fields.[2]

After the Sanders family no longer owned Magenta, the plantation went through several ownerships in rapid succession beginning with Whitney-Central Trust & Savings Bank's foreclosure on December 5, 1927, and on January 19, 1928 Carl F. Dahlberg & Company's purchase of the plantation. South Coast Company purchased it on September 27, 1935, and farmed the cane until 1968.

That year the Jim Walter Corporation purchased it on June 19, and two years later Walter Land Company became owners on December 31, 1970. Mid-South Mortgage Company took over, followed by Ag-Lands Investment Company, McCarthy Family Corporation, and Cook Family Corporation, all in 1979.

Early owner Eugene Fields had his livestock brand registered in Terrebonne Parish records, number 495 dated December 22, 1864.

A natural gas well was drilled at Magenta in 1901.

The major motion picture *Beasts of the Southern Wild* was filmed on Magenta in 2010; Terrebonne native Quvenzhané Wallis was nominated for an Academy Award as best actress for her role in that movie in 2011.

Sylvia Masters was owner of Magenta around the turn of the 21st century. Mike Keene and Eva Dover Cenac Keene owned the property from the year 2000 until before hurricane Katrina hit in 2005. William A. Hart purchased the former plantation after 2005.

Hart renovated the house which he now uses as a private fishing camp. The current address of the Magenta homesite is 1233 Highway 55, Montegut, Louisiana.

SOURCES
1. A Century of Lawmaking for a New Nation: U.S. Congressional Documents and Debates 1774-1875, American State Papers, Senate
2. Helen Wurzlow, *I Dug Up Houma Terrebonne*, 1984
3. E.C. Wurzlow, *1897 Directory of the Parish of Terrebonne*
4. Sherwin Guidry, "Xplorin' Terrebonne" column, 1972

Lottie Winnifred Sanders Gazzo (1895-1972) wedding photo June 8, 1920

Notice of H. G. Bush Petroleum Company drilling on Sanders place The Houma Courier *April 11, 1903*

Cattle brand of Eugene Fields December 22, 1864

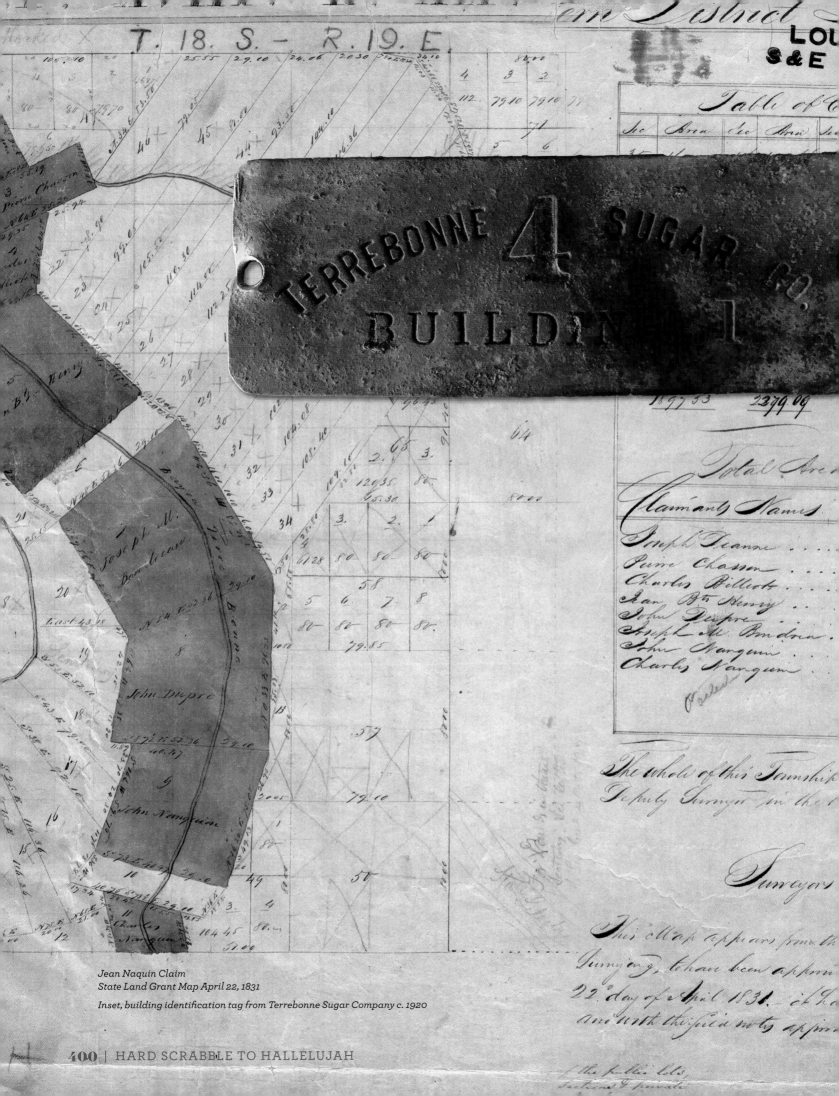

Jean Naquin Claim
State Land Grant Map April 22, 1831

Inset, building identification tag from Terrebonne Sugar Company c. 1920

ELIZA

Jean Naquin's land claim registered in 1816 and confirmed by the U.S. government in 1820[1] was the land upon which Eliza plantation developed, as had Magenta and Sunbeam.

William Franklin Price was the founder of Eliza plantation below Montegut in 1869, which he cultivated in cane until he died in 1892. Price was born in Thibodaux, Lafourche Parish, in 1835, and married Mary Louise Ross (born c. 1842) in 1856.[2]

Census records of 1870 for Terrebonne Parish describe William, age 35, a farmer, in Ward 8; his wife Mary, 27 years old; and their sons Ernest, age 8; Lee, age 4; and Falden, age 2. The 1880 census has the household living in Ward 6, "may be Montegut area; dwelling #279." The children of the couple were listed as William Ernest Price, Forest Jackson Price, Lee Frederic Price, Fielding George Price (probably the "Falden" of the 1870 listings), Marie Eliza Cordilia Price, John Wallace Price, and Franklin "Frank" Joseph Price.

A partnership described as Sanders & Price was owner of Eliza from 1904 until 1916, the year that Terrebonne Sugar Co. purchased the plantation. Terrebonne Sugar Co. cultivated the land in cane until 1927.

The same succession of ownership that transpired with other lower Bayou Terrebonne plantations then occurred with Eliza: Whitney-Central Trust & Savings bank foreclosure in 1927, Carl F. Dahlberg & Co. purchase 1928, South Coast Company purchase 1935 and its subsequent ownership and cultivation of the lands until 1968, followed by Walter Land Company purchase in 1970, and 1979 transfers of the property from Mid-South Mortgage Co., Ag-Lands Investment Co., McCarthy Family Corp., and the Cook Family Corp., all in 1979.

It is not known if a typical large planter's house was built at Eliza. Current address of the house on the plantation's site is on the left bank of Bayou Terrebonne at 1305 Highway 55, Montegut, Louisiana.

SOURCES
1. A Century of Lawmaking for a New Nation: U.S. Congressional Documents and Debates, 1774-1875, American State Papers, Senate
2. Ancestry.com

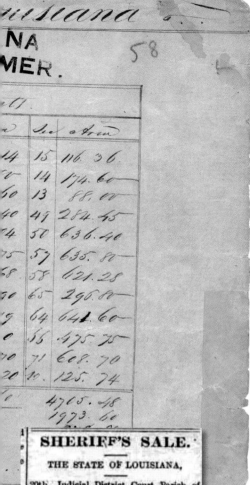

Ad for the Sheriff's Sale, A. Adler & Co. vs. George W. Price The Houma Courier *March 4, 1903*

PARISH OF TERREBONNE.

Jno. T. Moore Pltg Co. Ltd	Schriever		Julia	Bayou Terrebonne	*Schriever	55			
R. S. Wood & Bro	"		Ducros	"	"	55			
Jno. T. Moore Pltg Co. Ltd	"		Waubun	"	"	55	D E V and C	B W and M	3,750,000
Jno. T. Moore Pltg Co. Ltd	"		St. George	"	"	55			
Henry Gauthreaux	"		"	"	"	55			
David Levy	"		Isle de Cuba	"	"	55	D E V and C	B W & C	1,126,440
J. J. Ayo	Gray		Ayo	"	"	55			
O. E. Pelletier	"		Evergreen	"	"	55			
J. P. Landry									
Foret Landry	"		Half Way	"	"	55			
Eno Landry									
Town of Houma	Houma		Parish Seat	"	" †Houma	71			
H. W. Connelly	"		Homestead	"	"	71			
Barrow & Duplantis	"		Myrtle Grove	"	"	71	T E V and C	W and M	2,501,640
§Gueno Bros	"		Presqu'ile	"	"	71	D E V and C	W and M	1,798,200
Heirs Alfred Boudreaux	"		Front Lawn	"	"	71	S T O P		
Cambon & Champagne	Montegut		Klondyke	"	"	71			
Chas. J. Champagne	"		St. Agnes	"	"	71			
Ellender Bros	"		Hope Farm	"	"	71			
J. L. LeBlanc	"		Aragon	"	"	71	S and K	B W Sh & Sl	
Ellender Bros	"		Hard Scrabble	"	"	71			
Point Farm Pltg. Co. Ltd	"		Point Farm	"	"	71	S and K	Wood	
†L. Terrebonne P. & M Co	"		Central Factory	"	"	71	D E V and C	W M C & Sl	6,989,391
A. A. Saunders	"		Magenta	"	"	71	S T O P	Wood	
Eug. Fields	"		Angella	"	"	71	S and K	"	
Mrs. M. J. Saunders	"		Live Oak	"	"	71			
A. A. Saunders	"		Sunbeam	"	"	71			
Saunders & Price	"		Eliza	"	"	71			
Jos. A. Robichaux & Bro	"		Red Star	"	"	71	S and K	Wood	
Robichaux & Carlos	"		Argene	"	"	71			
Leo Lirette	"		Orange Grove	"	"	71	S and K	Wood	
Alidor M. Guidry	"		Eloise	"	"	71			
Emile Daigle	"		Pecan Tree	"	"	71	S and K	Wood	
Ernest Guidry	Bourg			"	"	71			
Ashland P. & M. Co. Ltd.	Houma		Ranch	Bayou Petit Caillou	"	71		B W & Sh	

* Wells, Fargo & Co Express.   † Southern Pacific Route, Houma Branch, change cars at Schriever — Wells, Fargo & Co Express.
§ Now operated by the Lower Terrebonne Planting and Manufacturing Company.   ‡ Lower Terrebonne Planting and Manufacturing Company.

Louisiana Sugar Report *1906-07*
listing Sunbeam

SUNBEAM

A claim by Jean Naquin which he registered with the U.S. government on November 20, 1816 and for which he received confirmation of ownership on May 11, 1820 was the root upon which Sunbeam plantation grew.[1]

James Monroe Sanders bought the property in 1870. Allen A. Sanders of Magenta, his son, also farmed Sunbeam plantation on the left descending bank of Bayou Terrebonne. He cultivated Sunbeam in cane for 30 years, from 1890 until 1920.

After his ownership ended, the same progression of landholders occurred on the same dates as for Magenta and other plantations along that stretch of Bayou Terrebonne: from Whitney-Central Trust & Savings Bank's foreclosure on December 5, 1927, to South Coast Corporation from 1935 until 1968, to Jim Walter Corporation in 1968, to Walter Land Company in 1970, to Mid-South Mortgage Company, Ag-Lands Investment Company, McCarthy Family Corporation, and Cook Family Corporation, all four in the year 1979.

The present address of the house on the property is 1325 Highway 55, Montegut, Louisiana.

SOURCES
1. A Century of Lawmaking for a New Nation: U.S. Congressional Documents and Debates, 1774-1875, American State Papers, Senate

Jean Naquin Claim
State Land Grant Map April 22, 1831

LIVE OAK

Jean Pierre Dugas had a land claim confirmed by the U.S. government in 1833.[1] That land was the basis for both Live Oak and Angela plantations.

James Monroe Sanders of Madison County, Mississippi, migrated to Terrebonne Parish in the 1850s and married Mary Jane May of Thibodaux at Rural Retreat plantation house in Terrebonne Parish on February 12, 1855. Sanders bought land along lower Bayou Terrebonne, below the site where his son Allen A. Sanders later developed Magenta.

The 1860 census records J.M. Sanders as age 30, a planter with $10,000 real estate property and $17,700 personal property. His wife Mary was 21, and in the household were their children Elizabeth Armogene, born 1856; Marthe Alice, born 1857; (daughter Marie Lucretia, born 1859 possibly died before the census, because she is not listed); and Mrs. Elizabeth May, Mary's mother.

The couple went on to have a total of 14 children, one source listing them as Annie E., Elizabeth "Annie" Armogene, Marthe Alice, Marie Lucretia, Yancey John Lee, Ada Ashley, Laura Jane, Allen Andrew, Robert Hunley, Charlotte, Hattie Altia, Fannie Mary Eudora, Eleonora Benton, and Warren James Goode Sanders, all born in Montegut.[2]

Former Live Oak home site 2015

Allen A. Sanders (1868-1945)

James Monroe Sanders fought in the Civil War, and in 1883 served as the original vice-president of the first cooperative sugar mill in lower Terrebonne Parish. He dug the Sea Breeze Cut (1870) for passage from Bayou Terrebonne to Lake Barré. He also owned Brush Island for a time.

Sanders' plantation was known for its hospitality, his granddaughter Mae Sanders Bazet recalled in later years. She said that guests visited for as long as six weeks or more, and that at Live Oak, meal service was for 24 instead of the usual 12. She also said that each child of James Monroe and Mary had his or her own nurse maid.[3]

James died in 1888, and his heirs then ran the plantation. They leased lands to James Martin & Breaux beginning 1900. A natural gas well was drilled on Live Oak plantation in 1901.

The 1900 census records the household of Mary J. Sanders, widow age 60, as including sons, daughters, and grandchildren: Alice Sanders, 40, single; Laura Sanders, 34, single; Robert Sanders, 30, single; Hattie Sanders, 26, single; Warren Sanders, 19, single; Yancey Sanders, 38, widower; Jasper Concannon, 8 or 9, grandchild; and Jim Sanders, 5 or 6, grandchild.

Live Oak property went the way of other area plantations when Whitney-Central Trust & Savings Bank foreclosed in December 1927. Carl F. Dahlberg & Co. bought the land in 1928, and South Coast Co. in 1935, which they farmed in cane until 1968. Walter Land Co. purchased it in 1970, and in 1979 the property passed from Mid-South Morgage Co. to Ag-Lands Investment Co. to McCarthy Family Corp. to Cook Family Corp.

Live Oak was 18 miles below Houma; the present address of the old homesite is 1337 Highway 55, Montegut, Louisiana on the left bank of Bayou Terrebonne.

SOURCES
1. A Century of Lawmaking for a New Nation: U.S. Congressional Documents and Debates, 1774-1875, American State Papers, Senate
2. Roots Web, ancestry.com
3. Helen Wurzlow, *I Dug Up Houma Terrebonne*, 1984

Dugas Cemetery Arch 2015

DUGAS CEMETERY

The land that became the site of Dugas Cemetery was settled by a native Frenchman named Jean Pierre Dugas (born c. 1776) from Chattelerault, France. He immigrated to Louisiana in 1785 with his parents and siblings aboard the *St. Remi*.[1]

His future wife, Renée/Reynalda Naquin was aboard the same ship with her widowed father and siblings. Jean Pierre and Renée married on February 18, 1800 in Plattenville at Assumption Church. Records in Albert J. Robichaux, Jr.'s *Colonial Settlers Along Bayou Lafourche, Louisiana Census Records: 1770-1798* show that in 1798 still-single Jean Pierre lived with his family of origin on Bayou Lafourche.[1]

At some point after his marriage in 1800 he settled the land described as in Township 18 South, Range 19 East, Section 10.[1]

The first mission chapel was built January 1, 1844. The Dugas family donated the church site November 27, 1859, 15 years later. Father Jean Marie Charles Menard as associate pastor of St. Joseph Church in Thibodaux dedicated the chapel to St. John the Baptist in 1859. The first Mass was celebrated there January 26, 1868.

The memoirs of Catholic priest Jean Marie Joseph Dénécé record that he built a chapel there in 1867, about where the cross currently stands in the Dugas Cemetery. The chapel was destroyed by the hurricane of September 20, 1909.[1]

Lead grave marker in French script for J.N. Robichaux (1820-1898) and Marie Robichaux (1822-1897)

Jean Pierre and Renée's son Jean-Baptiste Placide Dugas lives on in legend connected to the land his parents donated for the church and cemetery. The stories that arose about him called him the Hermit of Terrebonne. Jean-Baptiste dropped out of society, supposedly because he had been rejected by his fiancée for another man. He built a cabin deep in woods and briars on the rear of Dugas property and rejected all contact with other human beings. Although it was known that he had other clothes, he wore a gray blanket closed by a thorn in all seasons. He ate what he could hunt, trap, forage, and grow; if given other fare, he shared it with wild creatures that crawled into his cabin.

People caught sight of him occasionally, but he shunned contact. His roamings were restricted to the boundaries of Dugas lands. The only public appearance he made in decades was after a dance was held during a fair on St. John the Baptist Church grounds to raise funds for its further construction. At an 1866 trial in Houma, he was accused of attempting to tear the church down with his bare hands. Jean-Baptiste was released after he explained that a dance profaned the purpose for which his family had donated the land. He died of pellagra on November 18, 1891 and was buried in an unmarked grave in Dugas Cemetery.[2]

SOURCES
1. *Terrebonne Life Lines, Volume 17 Number 2*, Summer 1998
2. Jean-Baptiste Placide Dugas obituary, *Daily Picayune* newspaper, New Orleans, December 13, 1891

Below, the Hermit of Terrebonne Jean-Baptiste Placide Dugas. He lived in a rustic cabin in the woods for 40 years, and was 76 when the photos were taken on November 15, 1891. He died three days later, on November 18, 1891, and was buried in an unmarked grave in Dugas Cemetery.

Top to bottom, grave markers of Dominique J. Carrere (1834-1893); Francois Gallay (d. 1836); and Leo N. Robichaux (1870-1931)

Hard Scrabble c. 1950

Land Patent of Henry and Thomas Roddy #225, June 13, 1844

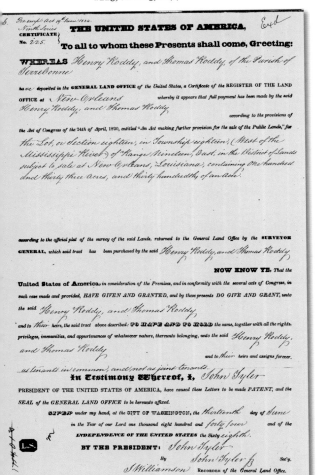

HARD SCRABBLE
CAILLOU FIELD

Spanish Governor of Louisiana Estevan Miro first laid claim prior to 1803 for Spain to the land along lower Bayou Terrebonne that evolved into Hard Scrabble plantation. Jean Pierre Dugas then claimed the land, which he registered with the U.S. government, which confirmed his claim in 1833.[1]

Robert Ruffin Barrow, Sr. became owner of the property in the 1840s. Frenchman Jean Pierre Viguerie held it in his possession after Barrow.

Ohio native Thomas Ellender (Elinger), the next owner of Hard Scrabble, migrated to Louisiana, first settled in Donaldsonville and later moved to Terrebonne Parish. He was born in 1818, the son of Joseph Elinger and Sarah Compting of Lebanon, Warren County, Ohio.[2] The woman who was to become Thomas' wife, Catherine Roddy, was the daughter of William Roddy and Marie Catherine Thomas Roddy.[2] A conveyance record of December 29, 1823 shows that William Roddy purchased four arpents fronting on both sides of Bayou Terrebonne from Hubert Bellenger (Belanger) for $5,000. Other records indicate that William bought a number of other tracts of land in Terrebonne Parish in subsequent years.

William's 19-year-old daughter Catherine received a land patent on December 5, 1839 for land between Bayou Terrebonne and Bayou Petit Caillou below Collins (Sarah) plantation. Catherine married Thomas Elinger on February 17, 1840.

According to the 1850 census, Thomas and his young family were living at Bayou Petit Caillou, where he farmed and owned 650 acres above Lacache plantation. Thomas was 33, and could not read or write, but his wife Catherine, 30, could. Their four children at the time were listed as Elizabeth, age 9, Henry, age 7, Joseph, age 6, and Thomas, age 1. Catherine's mother, also Catherine, was 80 years old and living in the same household.

The 1860 census recorded that Thomas Ellender (Elinger) owned property valued at $6,000 and personal assets worth $1,000. By 1863 his family had expanded to ten children, all born at Little Caillou. George was born in 1852, James in 1854, twins Ernest and Wallace in 1857, William in 1860, and David in 1863.

In 1874 Ellender bought two and a half arpents by depth on the left bank of Bayou Terrebonne for $828. He then purchased Hard Scrabble from Jean Pierre Viguerie, the "Caillou Field" part of Point Farm five arpents by depth on both banks of the bayou in 1880 for $2,500.

Thomas Ellender built the Hard Scrabble house, constructed by Samuel Hornsby, on the left bank of the bayou. As was the case for all local carpenter-artisans of the day, Hornsby used tools that today would seem rudimentary, and yet their clever construction skills built sturdy homes that endured. One example of Hornsby's expertise was described by local historian Sherwin Guidry in one of his newspaper columns "Xplorin' Terrebonne" that ran for years in the *Houma Daily Courier*. In one article, he described a "lightning" joint in Thomas Ellender's home that was" almost directly beneath the door leading from the hallway to the dining room, a main line of traffic." The 8 x 10 joist with a square peg driven in the center of the joint to tighten it needed no support beneath, the joinery was so sturdy.

Jean Pierre Dugas Claim State Land Grant Map September 28, 1856

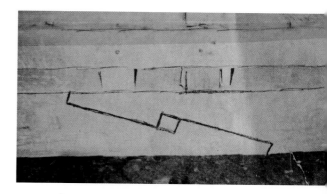

"Lightning joint" joinery in Hard Scrabble plantation house

Below left, Thomas Ellender (Elinger) family, front row from left, Joseph, Thomas Jr., Elizabeth, William, George; back row from left, James, Ernest, Wallace, David

Cattle brand of Thomas Ellender (Elinger) Sr. May 17, 1841

HARD SCRABBLE | 409

Thomas died in 1884, but his heirs expanded his holdings when they paid $2,000 to J.M. Dupre in 1886 for two and a half arpents by depth on both banks below Hard Scrabble.

Seven of the eight surviving sons of Thomas Ellender and Catherine Roddy Ellender formed Ellender Brothers in 1886. They were Joseph, Thomas, Jr., James, twins Ernest and Wallace, David, and bachelor George who was killed by a lightning strike riding on his horse and willed his property to his remaining brothers.[2] The Ellender brother non-partner was William. Their oldest brother Henry was a Confederate soldier in Company H, 26th Louisiana Infantry, "Grivot Guards" who died of measles in Mississippi Springs during the Civil War in 1862.

In 1900 Ellender Brothers purchased from the State of Louisiana 881 acres of swampland for $110.13. The brothers sold the cypress lumber on Hope Farm, also Ellender property, to the St. Louis Cypress Co. of Plaquemine, Louisiana, in 1903 for $35,000. The company dug St. Louis Canal across Ellender land to Bayou Terrebonne to float logs upstream to mills.

The brothers purchased sugar mill equipment from Ralph Bisland for $725 in 1891 and Albert Viguerie's sugar mill and land three years later at a cost of $13,000. The right bank of Hard Scrabble became the property of the Estate of Thomas Ellender as of May 5, 1892; and the left bank was owned jointly by the Thomas Ellender Estate and Jean Pierre Viguerie as of that date, as well. A narrow gauge railroad connected Hard Scrabble to Hope Farm farther up the bayou in Bourg.[2] Louis Bouchereau's annual *Louisiana Sugar Report* for 1894-1895 records that "On the night of December 28th [1894], the sugarhouse of the Ellender Bros., parish of Terrebonne, valued at $17,000, and insured for $10,000, burned to the ground. The fire started near a juicer pump."

The Ellenders did not live as did the plantocracy at other estates, according to Allen Ellender biographer Thomas A. Becnel.

St. Louis Cypress Company logging canal c. 1920s

Newel post from Hard Scrabble plantation now in the home of Dr. Craig M. Walker and Tina Hebert Walker

Henry Ellender (1843-1862) *Thomas (Elinger) Ellender (1818-1884)* *Elizabeth Ellender (1841-1894)*

Wallace Richard Ellender, Sr.(1857-1925) and Victoria Jarveaux Ellender (1869-1946) family 1910

They lived frugally and without luxuries, much as the neighboring Champagnes, Robichaux, and Vigueries did. When they worked with their father, the sons did not receive pay, relieving the estate of the expense of laborers' salaries. The sons and their families lived close to one another and the extended family members were close.[2]

Becnel pointed out that the Ellender brothers "used plantation funds to educate their children....Ernest and Wallace took the lead in establishing a school" for which they engaged Thomas, Jr.'s son Henry Jefferson Ellender as teacher. Henry later became principal of Lapeyrouse Public School in 1898. Henry also ran the Hard Scrabble store.

Longtime U.S. Senator Allen Ellender (1937-1972) was born to Wallace and his wife Victoria Jarveaux Ellender on September 25, 1891 in a small Acadian cottage upstream from Hard Scrabble. Victoria was born in Terrebonne Parish on June 28, 1869, to French native Auguste Jarveaux, who settled in Bourg in 1854 and married Elodie Guidry. Victoria was the seventh of 13 children. She married Wallace Richard Ellender (born October 31, 1857) on December 13, 1889, and together they reared four sons and a daughter. Their children besides Allen were Claude J. Ellender, Walterine Ellender Caillouette (Mrs. Charles Caillouette),

Henry Jefferson (1874-1951) and Jessie Chauvin (1885-1970) Ellender

Henry Thomas Ellender Sr., DDS (1909-1994)

Allen J. Ellender, Sr., brothers and sister c. 1960, left to right, Claude J., Dr. Willard A., Walterine E. Caillouette, Allen J., Sr., Wallace R., Jr.

HARD SCRABBLE | 411

Allen J. Ellender, Sr., report card March 31, 1907

U.S. Senator Allen J. Ellender, Sr., (1890-1972)

Wallace R. Ellender, and Dr. Willard A. Ellender.[2]

Allen was known to locals as "Sous-Sous." He was named for Allen A. Sanders, a neighbor who owned Sunbeam, Eliza, Magenta, and Angela plantations. His father Wallace and his twin Ernest moved their families from Hard Scrabble land and built identical houses next door to each other on Hope Farm land in 1907. Ernest's house burned in the mid-1920s and was rebuilt slightly smaller and with a different façade. The houses still lie on each side of Hope Farm Road where it intersects with the highway that parallels Bayou Terrebonne. Descendants of each twin still reside in those houses.[3]

The Ellender family registered many cattle brands in Terrebonne Parish records beginning with Thomas Elinger, number 280 dated May 17, 1841. Others were Hyman Ellender, number 703 dated June 30, 1906; Thomas Ellender, Jr., number 820 dated June 19, 1917; Ernest and Wallace Ellender (twins), number 821 dated June 29, 1917; Joseph Ellender, number 822 dated June 29, 1917; Wilson E. Ellender, number 830 dated April 6, 1918; Fay Chauvin (husband of Josephine Ellender), number 857 dated September 27, 1919; Wallace R. Ellender, number 951 dated September 9, 1929; Paul Ellender, number 959 dated June 13, 1930; and Carl Ellender, number 982 dated March 30, 1935.

A successful natural gas well was drilled on Hard Scrabble grounds on September 23, 1901.

David Ellender received Hard Scrabble in partition of properties in 1924. The house had been badly damaged in the 1909 hurricane, and was eventually torn down by Lloyd Ellender c. 1960.[3] The present location of the one-time Hard Scrabble home is 1531 Highway 55, Montegut, Louisiana.

Whoever named Hard Scrabble could have chosen it because of the word *hardscrabble* that means involving hard work and struggle. Allen Ellender biographer Thomas A. Becnel said that "where the name came from remains a mystery," but family members recall hearing that the Ellender patriarch "had come by it the hard way."

SOURCES
1. A Century of Lawmaking for a New Nation: U.S. Congressional Documents and Debates, 1774-1875, American State Papers, Senate
2. Thomas A. Becnel, *Senator Allen Ellender of Louisiana: A Biography*, Louisiana State University Press, Baton Rouge and London, 1996
3. Interviews with Ellender descendants, 2016

Vice President Spiro T. Agnew attends funeral of Senator Allen J. Ellender, Sr., who died July 27, 1972 at Bethesda Naval Hospital, Bethesda, Maryland

U.S. President and Mrs. Richard M. Nixon attend the Houma funeral of Senator Allen J. Ellender, Sr., who was buried July 31, 1972

Dignitaries gathered in Houma to honor Terrebonne native Allen J. Ellender, Sr. at his funeral July 31, 1972

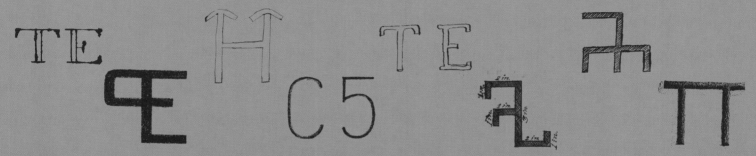

Left to right, cattle brands of Thomas Elinger May 17, 1841; Hyman Ellender June 30, 1906; Thomas Ellender, Jr. June 29, 1917; Wilson E. Ellender April 6, 1918; (second row) Paul Ellender June 13, 1930; Carl Ellender March 30, 1935; Fay Chauvin (husband of Josephine Ellender) September 27, 1919; Wallace R. Ellender September 9, 1929

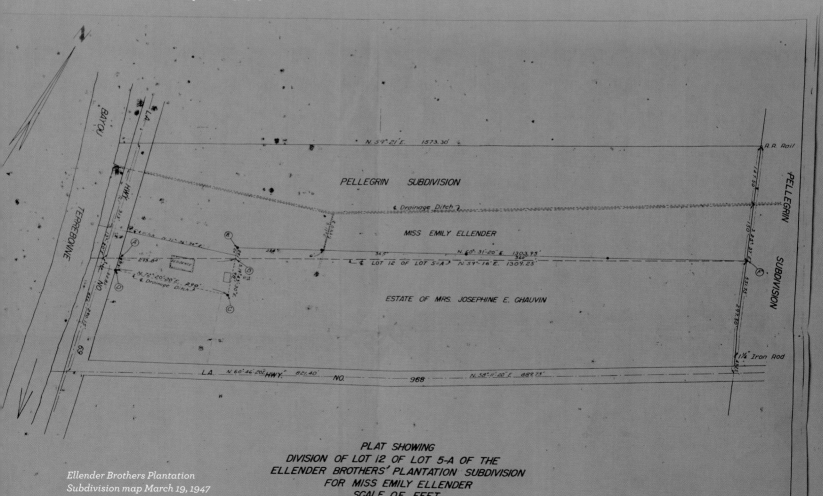

Ellender Brothers Plantation Subdivision map March 19, 1947

HARD SCRABBLE | 413

Humble Canal looking east from Bayou Terrebonne 2015

HUMBLE CANAL

Humble Canal was dug across Hard Scrabble land in the late 1940s after oil and gas exploration had proliferated in Terrebonne Parish. Brown & Root, Inc. had it dug to access the Lirette Field and its pipeline exchange area. The canal connected Bayou Terrebonne to Wonder Lake to Bayou Barré. It is located below Canal Belanger and above Lapeyrouse Canal.

Hard Scrabble land looking west from Bayou Terrebonne 2015

ARGENE

Pierre Bourg laid claim to the lands that became Argene and inhabited it prior to 1810. He registered the claim with U.S. government on November 20, 1816, and received confirmation of ownership on May 11, 1820.[1]

Clement Carlos, born c. 1824 in France, founded Argene plantation below Montegut in 1878. He had married Marie Louise Molinere, but by the 1860 census, Marie is not listed in the household of Clement. He is described as 36, a farmer, possessing $3,000 in property, and $500 in personal belongings.

Members of his household in 1860 were his children Louisa Carlos, age 8; Charles, age 3, and Virginia, age 1; Lavinia Devin, mulatto, age 26, probably a housekeeper; and Jean Bertin, age 33, a day laborer.[2]

Alcide Ferjus Carlos is listed as the next owner of Argene, from 1890 until 1892. He was married to Aurelia Letitia Gautreaux, and they named their daughter Argene with the plantation as namesake. She spanned historical eras by living from 1886 until 1980. Alcide Ferjus was the son of Clement Charles Carlos and Marguerite Anaise Landry. Alcide and Aurelia had five other children besides Argene—Ambrose, Marius, Margueret Naise, Wycliffe, and Joseph Maurice Carlos.

From 1904 until 1912 owners were Robichaux & Carlos (Alcide). His name may be found in Terrebonne Parish records as registering livestock brand number 663 dated July 23, 1903.

After 1912 there is no record of Argene's existence as a plantation.

There is no house remaining at Argene to mark the plantation's homesite on Bayou Terrebonne's left bank, but the Carlos Oak stands aside Bayou Terrebonne at 1603 Highway 55, Montegut, Louisiana.

SOURCES
1. A Century of Lawmaking for a New Nation: U.S. Congressional Documents and Debates, 1774-1875, American State Papers, Senate
2. 1860 United States Federal Census

Headstone of Argene Carlos (1886-1980) Dugas Cemetery 2015

Cattle brand of Alcide Carlos July 3, 1903

Carlos Oak at Argene home site 2015

Opposite Page: Pierre Bourg claim Tobin map 1938

The South Coast Corp
Coast Corp
Henry J. Ellender
85
Pirogue Canal
Lee P. Lottinger
Olivo J. Pellegrin
Wid. Luby
Pellegrin
Lottinger et al
Guidry
20
21
CHAS. HANQUIN
Henry J. Ellender
South Coast Corp.
W. W. Price
H. J. Klingman
CHAS. BILLOT
Henry J. Ellender
Mrs Francois Caillet Jr.
PIERRE BOURGUE
Delta Securities Co.
23
22
J. N. Thatcher
United Gas Public S. Co.
Jess R. Miller
Mrs. L. M. Albright
Mrs R. Sanders
Alcid
Carlos
LOUIS SANVAGE
Humble Ellender
A. Lirette
Mrs F. Belonger
Olivo J. Pellegrin
JEAN BILLOT
H. J. Ellender
E. A. Robichaux
J. H. Thatcher et al.
Marcel Billiot
L. J. Lapeyrouse
Henry J. Ellender
160 Ac.
Mrs. Emile Lirette
Est. of E. P.
John J. Guidry
V. Lirette
Arthur Lirette
Mrs E. Lirette
Alfred Lirette
Alec. Lirette
Mrs W. Duplantis
C. Wurzlow
Robichaux
J. J. Lapeyrouse
Frank A. Guidry
Terrebonne
32
33
La-Terre Co. Inc.
La-Terre Co.
J. C. Dupont
Pickard & Haywood
Lirette
Gulf
R. Verret
C. Wurzlow
Dr. R. W. Collins
Harris Gagne
J. P. Miller
J. N. Thatcher
La Terre Co. Inc.

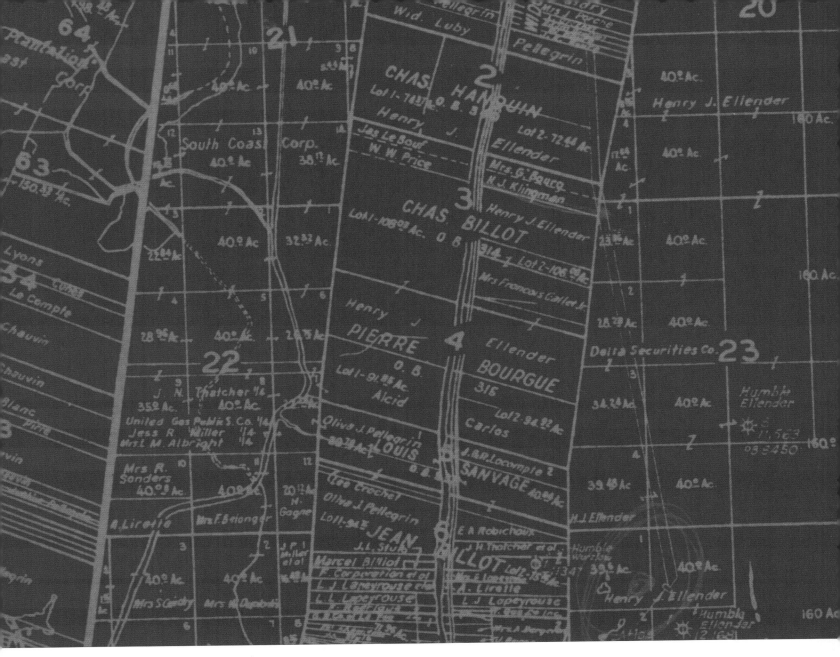

*Pierre Bourg and Louis Sauvage claims
Tobin map 1938*

RED STAR

Pierre Bourg and Louis Sauvage both held claims to the land that became Red Star plantation, inhabited prior to 1810 and ownership confirmed by the U.S. government on April 22, 1831.[1] Sauvage was the brother of Marianne Marie Nerisse Iris, holder of one of the earliest Spanish land grants in the Terrebonne area.

Twenty miles below Houma on both banks of Bayou Terrebonne, Red Star was owned exclusively from beginning to end by the Robichaux family whose long ago roots were in France, then Nova Scotia, before Louisiana.

Patriarch Jean Baptiste Marin Robichaux was born in Donaldsonville in 1794 and married Marie Madeleine Breaux (b. 1794) at Plattenville, Assumption Parish, in 1818. They lived in Lafourche Parish for a time, and in 1841 moved to the virtual wilderness of lower Bayou Terrebonne. Houma at the time consisted of only several houses, and "down the bayou" there were woods teeming with deer, rabbits, and all kinds of wildlife to hunt.[2] The profusion of waters nearby provided fish, shrimp, oysters, and crabs to feed his family.

The *Biographical and Historical Memoirs of Louisiana* in 1892 recorded, "He then commenced to make a home for his family, and by great industry and perseverance he became a wealthy planter. . . . They reared a family of thirteen children, and their eldest son, Honoré is still living, as is also the youngest child, Frozine…."[2] Jean Baptiste died at age 58 in 1852, and Marie Madeleine died in 1861 at age 67.[2]

The memoir went on to detail the efforts of their child Joseph Narcisse Robichaux (born 1820), who married Ursula Robichaux (no relation to Joseph Narcisse) of Lafourche Parish in 1840. In 1841 they moved to Terrebonne to "his present plantation. This was a dense canebrake and forest, but with the determination and perseverance for which he is remarkable, he cleared up one of the finest plantations in the parish, and is to-day classed among the substantial and honored citizens of the same. In 1852 he put up a mill with oak rollers and operated by two horses and the first year made seven hosgheads. Last season he made 261 hogheads, or 283,000 pounds, and is gradually increasing the capacity of his mill."[2]

Because his plantation sprawled on both sides of Bayou Terrebonne, Narcisse built a bridge to allow his wagons to reach the fields on both banks. A wench operated the bridge, which opened by sliding, "like a box of matches in its case," according to local historian Sherwin Guidry.[3]

Joseph Narcisse's son Etienne Paul was born at Red Star in 1841 and died in 1896 on the plantation after spending his entire life there. He married Marguerite Camilla Hebert, born in 1845. Marguerite's father, Louis Hebert, was a farmer in Terrebonne

Robichaux men at Red Star c. 1920

Robichaux wedding, from left, Joseph Luby Robichaux, Allie Pellegrin Robichaux, Bruce Pellegrin, and Eula Lebouef Fabre February 27, 1924

Article on death of Gustave Hebert, first oil field related casualty in Terrebonne Parish Houma Courier September 14, 1901

Robichaux tomb at Dugas Cemetery, (from left) Udalize Badeaux Robichaux d. 1894, Honorine Robichaux d. 1861 and Honore Robichaux d. 1907 (2015)

Honore Robichaux d. 1861, tomb at Dugas Cemetery 2015

Short newspaper social items, Narcisse Robichaux to New Orleans The Houma Courier *April 3, 1892; Joseph A. Robichaux to Houma,* The Houma Courier *April 30, 1892; Alfred Robichaux's return from trip around the world* The Thibodaux Sentinel *August 28, 1910*

> Mr. Narcisse Robichaux, a prominent planter of Lower Terrebonne, visited New Orleans this week.

> Mr. Jos. A. Robichaux, of Lower Terrebonne, was in town this week.

> Mr. and Mrs. E. G. Robichaux and son Alfred, returned Monday from a trip around the world.

Parish who served in the Seminole War as a hunter of game for the officers. Etienne and Marguerite were the parents of five; their unmarried daughter Ada Marie in 1925 still lived at the old homestead.[4]

Etienne's son Joseph Alfred Robichaux purchased assets of the succession of Joseph Narcisse at Red Star from the siblings, and became the sole owner of the plantation in 1899. He was vice president and assistant cashier of The People's Bank & Trust Company of Houma and secretary-treasurer of The People's Sugar Company as well as being a planter.[4]

Joseph Alfred attended Terrebonne Parish public schools and completed the junior year at St. Charles College at Grand Coteau, Louisiana. In 1885 he graduated from the L.C. Smith Commercial College in Lexington, Kentucky.[4]

He married Ada Ashley Sanders, daughter of James Monroe Sanders of Live Oak, in 1888. The couple had no children. Joseph Alfred's second marriage was to Frances N. Lewis in Birmingham, Alabama in 1913. She was a graduate of Peabody College of Nashville, Tennessee.[4]

The first oil field-related death in Terrebonne Parish occurred at Red Star in 1901 when Gustave Hebert, age 24, was oiling machinery of the Houma Oil Company which was drilling a well on Red Star land. He slipped and fell 24 feet to the ground, struck his head, and died about an hour later, according to a coroner's inquest.[5]

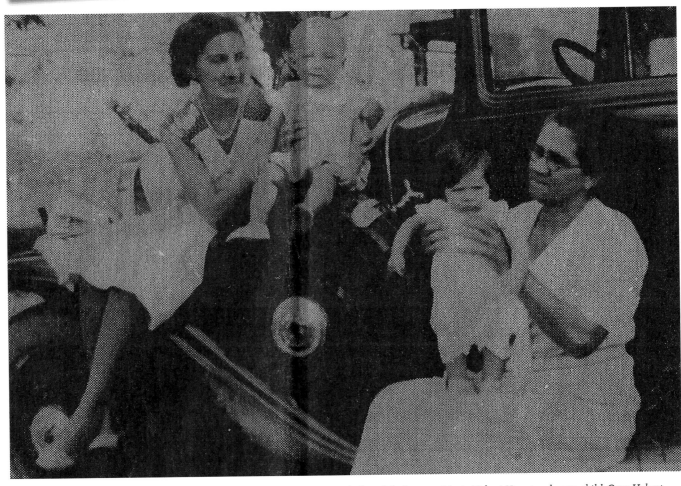

Family of Gustave Hebert, first oil field related casualty in Terrebonne Parish; from left, Augusta Marie Hebert Verret, unknown child, Cora Hebert, widow of Gustave Hebert holding Rena Verrett (Labit) 1932

Tombstone of Mr. and Mrs. Etienne Robichaux in Dugas Cemetery 2015

James A. Robichaux & Co. became owners of Red Star in 1912.

One of Jean Baptiste Marin Robichaux's sons, Jean Baptiste Honoré, is recorded in Terrebonne Parish official documents by registering his cattle brand as number 274 dated November 11, 1840. Joseph Narcisse, his brother, had his brand recorded as number 423 dated May 29, 1855.

A small house now stands on or near the site of the original home at 1635 Highway 55, Montegut, Louisiana.

SOURCES

1. *A Century of Lawmaking for a New Nation: U.S. Congressional Documents and Debates 1774-1865*, American State Papers, Senate
2. *Biographical and Historical Memoirs of Louisiana, Volume II*, Goodspeed Publishing Company, Chicago, 1892
3. Sherwin Guidry newspaper column "Xplorin' Terrebonne" in *The Houma Courier & The Terrebonne Press* newspaper, July 2, 1978
4. Henry E. Chambers, *A History of Louisiana, Volume 2*, The American Historical Society, Inc., Chicago and New York, 1925
5. *Terrebonne Life Lines*, Volume 10 Number 2, Summer 1991, Terrebonne Genealogical Society, Houma, Louisiana

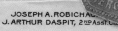

Cattle brands of (from left), Joseph Narcisse Robichaux May 29, 1855, and Jean Baptiste Robichaux May 20, 1887; tokens from Peoples Sugar Co. Stores, 50 cents; letterhead of The People's Bank

Oil Property Changes Hands.

The Houma Oil Co., Limited, this week, turned mineral options on many acres of Terrebonne land over to capitalists represented by Prof. Caracristi, the noted geologist who, during the past month or so has been examining the mineral resources of this parish. Prof. Caracristi has been accompanied in his investigations by his assistant, Dr. James D. Bouroughs, who has had charge of the oil, gas, phosphate and marl deposits of Southern Alabama, and also by Mr. S. P. McKenzie, a leading attorney of Birmingham who is considered an expert in mineral leases.

Prof. Caracristi will not talk oil. He will discuss other subjects, but retains a Sphynix-like silence when the Terrebonne oil field, with its possibilities and impossibilities is broached. However, the transfer of large interests to the company he represents would indicate that he is satisfied with the examinations thus far made. That would be the logical conclusion. The entrance of outside capitalists into the Terrebonne field marks a new era in the development of the mineral resources of this parish, and we hope that it will prove an era of great discoveries and of incalculable benefit to the parish.

The principal conditions of the transfer of the Houma Oil Co's. holding are as follows:

The Houma Oil Co., transfers all of its options except that on the two acres front of the Robichaux place jointly owned by the Houma and Peoples Companies.

The machinery barring the piping is loaned to Prof. Caracristi's company to be used exclusively on the lands transferred by the Houma Oil & Mining Co., and has to be returned in good order.

The Houma Oil Co., agrees to assist Caracristi's company when requested so to do to secure other options, which will be under the control of the Caracristi company and subject to the same conditions as the options transferred.

Prof. Caracristi's company agrees to pay the royalty of 12½ per cent of the product of the wells, mines, etc., to the landlord and a royalty of 12½ per cent additional to the Houma Oil & Mining Co.

Prof. Caracristi's company agrees to begin the drilling of an oil well within 60 days from date of contract. In case of unavoidable circumstances an extension in the time limit not to exceed

1903 article on the Houma Oil Co. Ltd. turning over mineral rights to Prof. Caracristi except for Robichaux place owned by Houma Oil & Mining Company
The Houma Courier *March 4, 1903*

*Death certificate of Leo Lirette
August 31, 1917*

ORANGE GROVE

*Alexander Lirette
(1812-1883)*

*Leo Lirette (1839-1917)
The Daily Picayune
June 9, 1901*

Jean Billiot held a land claim (later Orange Grove) which he inhabited prior to 1810. He received confirmation of ownership from the U.S. government on March 22, 1831.[1]

Jean Pierre and François Viguerie of Point Farm also owned Orange Grove located several plantations below Point Farm, during the 1860s. However, the plantation is most known for ownership by the Lirette family.

Oranges were a lucrative commodity in Terrebonne of the late 1800s, and a widely grown crop besides sugar cane. Orange Grove may have logically received its name from this fact. It was one of three plantations of the same name, another on Bayou Black and yet another on upper Bayou Terrebonne in Gray, that plantation later taking the name of Pilié. Freezes around the turn of the 20th century decimated orange orchards for many years.

Alexander Lirette & Sons purchased Orange Grove in 1872. Alexander was born in 1812 and married Marie Angela Angeline Gautreaux, eight years his junior.[2] Alexander served as sheriff of Terrebonne Parish from 1838 until 1844.[3]

His son Leo Lirette, born 1840, was a sugar planter most of his life. He first married Adele Guidry in 1859 and after her death in 1868, he married Marie Celestine Louviere in 1870.[2]

Leo and Marie Celestine were the parents of Alex, Alfred, Arthur, Alvin, Emile, Leo and Bagley Lirette, and daughters Mrs. Robert H. Sanders, Mrs. Sidney Guidry, Mrs. Wallace Duplantis and Miss Ella Lirette. Oscar Lirette was his only child by his first wife Adele.[4] They were reared in Orange Grove's main house on the left bank of Bayou Terrebonne.

Left; Leo Lirette Well and unexpected blowout The Daily Picayune *June 9, 1901*

Le Danois gas well The Daily Picayune *June 9, 1901*

THE LE DANOIS GAS WELL.

Joseph Bernard Songe discovered gas in Montegut The Daily Picayune *May 26, 1918*

Lirette spent the final 20 or more years of his life as a pioneer in exploring oil field possibilities, and he sank considerable amounts of his fortune in test work to that end.[4] The first natural gas well in Terrebonne Parish was drilled on Lirette's land in 1899 by prospector P. LeDanois, with L.E. Marion. Multiple wells then were drilled, in 1900, 1901, and 1902.

Leo Lirette's obituary in a local newspaper said that he "was considered a political leader in the affairs of the parish and was a member of the jury commission at the time of his death."[4] He died on August 31, 1917.

His estate took over ownership of what had been Orange Grove in 1920. Current address of the house that marks the property is 1639 Highway 55, Montegut, Louisiana.

SOURCES
1. A Century of Lawmaking for a New Nation: U.S. Congressional Documents and Debates, 1774-1865, American State Papers, Senate
2. Ancestry.com
3. Terrebonne Parish public records
4. Leo Lirette obituary, dateline Houma, Louisiana September 2, 1917

Orange Grove home site, 2015

WAY BACK IN '80'S BEGINS HOUMA'S ROMANCE

FAITH WINS GAS FROM WILDS

"Old Man Songe' Defies Superstitions and Denizens of the Swamp, Holds Hat Over "Devil Water" Spring, Waits, Strikes Match—Pfoof! and Whiskers, the Growth and Pride of a Lifetime, Singed Off.

One summery day in the early 80's an aged man, known to his friends as "Old Man Songe," crept mysteriously out into the marshes about eighteen miles south of Houma, La. Apparently unmindful of the billions of mosquitoes that tore at him, and heeding the danger of alligators, snakes and other swamp-infesting reptiles no more than he feared the evil spirits popularly supposed to haunt the place, the old man pressed steadily forward until, far back from the winding ruts that formed the roadway, he reached a spot where a small pool of water, stirred by some unknown force, bubbled and boiled—devil water it was known as in those days.

Stooping over the pool of devil water the old man removed his massive straw hat and grasping it by the crown, held it down close over the bubbling water. For several minutes he stood thus. Then, striking a match he shoved the lighted stick beneath the hat. Puff! A cloud of flame rose upward around the hat. The old man jumped back—but too late. The fiery cloud enveloped his head and, in the flash of an eye, completely burned off a set of whiskers he had been a life time growing.

Thus did "Old Man Songe" lose his whiskers; and thus did the world first learn that there lays, just south of Houma, what appears to be one of the largest gas fields in the world. Mr. Songe's discovery came as quite a shock to the natives who for years had clung tenaciously to the belief that the pools of water in the marshes were being made to bubble by evil spirits. It was even a hard blow to the wise men of the district, for their theory, long advocated, and supported by convincing argument, was that the pools bubbled because some sort of volcanic action was taking place beneath the surface of the earth; and they cocked their eyes knowingly as they opined that it would not surprise them to see the "whole blamed district go up in smoke" some day.

LOSS WAS CONVINCING

But the loss of "Old Man Songe's" whiskers convinced all that it was some sort of gas, seeping upward that caused the bubbling. Mr. Songe dreamed of a fortune in oil—just as hundreds since his time have dreamed. He never made a cent from the "oil fields," however; nor has anyone else so far. Yet public confidence in the Houma field—now generally referred to as the Terrebonne fields—never wavered from Mr. Songe's time to the present. Of late years they

Methodist Mission Chapel, 2015

METHODIST MISSION CHAPEL

Roman Catholicism was the faith of most of Terrebonne Parish's majority French population from the time of the parish's settlement and beyond its establishment as a civil entity in 1822. Not until substantial numbers of Anglo-Saxons and other European ethnicities arrived in the parish did Protestant religious denominations begin to establish places of worship. One exception was the Methodist church in Gibson built in 1825, the oldest recorded church congregation in Terrebonne's boundaries, even before St. Francis Church was built in Houma in 1848.

The Louisiana Conference of the Methodist Church sent missionaries to Terrebonne Parish beginning in the 1840s. Ministers of that faith began to preach in homes and whatever larger venues they could engage, in order to serve whatever small numbers of Methodists who lived in the area, and especially to convert people who were steeped in what they termed "Romanism."

Among the Terrebonne communities the circuit preachers reached were Montegut and Pointe-aux-Chênes, and later farther south in Pointe-au-Barré. The ministers, as well as deaconesses sent to the area, often needed translators when their entire prospective congregations on any given day or evening spoke nothing but French.

Outside of Houma, one of the early Methodist churches in Terrebonne was at Bayou Blue. The first southern Terrebonne community to become the site of a Methodist church was Dulac. The denomination had made sufficient inroads in the 1930s to build small Methodist mission chapels in several bayou areas, including one established at Pointe-au-Barré. The Methodist mission building closed in the 1970s, in part because locals had better transportation opportunities to reach larger churches "up the bayou." The current address is 1659 Highway 55, Montegut, Louisiana.

SOURCES
1. *Methodism Along the Bayou* Chapter 4: "Methodism 'Returns' to Terrebonne Parish," Louisiana Conference of the United Methodist Church, www.la-umc.org/chapter4

The First Cajuns, 1986, Oil on canvas, 36 x 36 inches, George G. Rodrigue

Stoufflet Store at Pointe-au-Barré; Horace "Petch" Lebouef, Emery Stoufflet, Ferris Courteaux c. 1935

Pointe-au-Barré Road junction with Highway 55 (2015)

426 | HARD SCRABBLE TO HALLELUJAH

POINTE-AU-BARRÉ

Pointe-au-Barré is French for "closed off point," a phrase that accurately described Lac (Lake) Barré in southeastern Terrebonne Parish. It was a freshwater lake closed off by land from brackish coastal waters, forcing portages of boats and goods over the land bridges until men cut through the natural levees to open the lake to transportation.

Local sugar planter James Monroe Sanders of Live Oak had his crew cut through the land from Bayou Terrebonne to Lake Barré c. 1870 to connect lower Bayou Terrebonne to Barataria Bay via Lake Barré for transporting sugar cane and other commodities to New Orleans. Unfortunately, the cut allowed for saltwater intrusion and expedited unfortunate coastal erosion of the area over the years.

Smithsonian anthropologist John Swanton recognized, in his field notes, Pointe-au-Barré in 1907 as the seat of the Houmas Indians.[1] The descendants of the founding families there spread not only into Ile à Jean Charles and Bayou Pointe-aux-Chênes, but also to neighboring bayous.[2]

Tombstone of Rosalie Courteau, Dugas Cemetery 2015

Rosalie Courteau (c.1787-1883), daughter of Houma Courteau, was a key ancestor for the Houmas Indian tribe, some oral history accounts mentioning her as a leader. She resided at Pointe-au-Barré consistently after 1815, according to her own testimony. In 1859 she bought land in her own name, "back of" Bayou Terrebonne on Bayou Barré. Her father had purchased land on Bayou Little Caillou earlier, in 1836. By 1849 two of Rosalie's sons were small-scale sugar planters on Bayou Pointe-aux-Chênes. Her grave is in Dugas Cemetery in Montegut.

As more and more European and Acadian settlers entered the parish, the Native Americans were forced farther south. John Swanton counted almost 900 in several settlements, among them 160 at Pointe-aux-Chênes and 65 at Pointe-au-Barré.[3] Four families considered ancestral families of the Houma Nation held Spanish land grants: Jean Billiot, Tacalobe Courteau, Louis Sauvage (Rosalie's brother) and Alexander Verdin.[4]

According to John Swanton's journal of 1908, 65 families of Native Americans lived at Pointe-au-Barré. In the 1940s that number had dwindled to about two dozen families, a small graveyard and store. As successive storms destroyed homes and property, many of the residents moved away.

SOURCES
1. John Swanton, *Bureau of American Ethnology, Indian Tribes*
2. Summary under the Criteria and Evidence for Proposed Finding Against Federal Acknowledgment of the United Houma Nation, Inc., 1994
3. Swanton, Bulletin 43, p. 291
4. Mark Edwin Miller, *Forgotten Tribes: unrecognized Indians and the Federal Acknowledgment Process*, 2004. Board of Regents, University of Nebraska

ST. PETER CATHOLIC MISSION CHAPEL

St. Peter Mission Chapel was built in 1950 as a mission of Sacred Heart Catholic Church of Montegut to serve Catholics in lower Montegut and Pointe-au-Barré.

Pastor of Sacred Heart, Father Gerard Pelletier described the need for the mission in a letter to Archbishop Joseph Francis Rummel, S.T.D., dated August 1948. Fr. Pelletier felt the population below Sacred Heart was being neglected because of their lack of transportation to the church. He mentioned that the whites, as well as the Indian population there, could not afford the bus fares for their large families to reach the church. He proposed boat transportation for people above and below the church, which a parishioner would provide free of charge, except for fuel and driver. For "the Indians, living mostly all in the same neighborhood, around the little School below Montegut, I would give them a Mass in the Schoolhouse," Father Pelletier proposed.

He later petitioned the Archdiocese of New Orleans to provide a mission chapel to be built on the Lirette property next to the Lower Montegut Indian School. He celebrated Mass for the first time in the school after November 7, 1949. On that date he wrote to Archbishop Rummel, "If we ever build a Chapel there, it would be understood there would be no segregation, that all residents of that section would be invited regardless of their origin."

Before the chapel was built, the priest offered Masses in the Pointe-aux-Chênes school. In a September 1950 letter, he acknowledged to the archbishop the receipt of a check for $6,000 for the construction of the mission chapel, to be named in honor of St. Peter, at Pointe-au-Barré near the Indian school. The Sacred Heart Extension Society was one of the donors for the chapel.

Because transportation improved and attendance declined at the chapel, in 1968 then-Sacred Heart pastor Ivern Bordelon asked that the chapel be closed. It was decommissioned in 1969, the property surrounding it returned to lessors Arthur Lirette and his wife Elise Lecompte. The building was sold, rolled across the highway, loaded onto a barge, and taken to Chauvin where it now serves as part of a nondenominational church. The former address was 1745 Highway 55, Montegut, Louisiana.

St. Peter Catholic Mission Chapel exterior c. 1950

The Rev. Gerard Pelletier 1948-1952

Archbishop Joseph Francis Rummel, S.T.D.

The Rev. Ivern M. Bordelon 1966-1976

SOURCES
1. Houma-Thibodaux Diocese Archives Collection

St. Peter Catholic Mission Chapel interior c. 1950

Altar of St. Peter Catholic Mission Chapel, first Mass celebrated 1951

Lower Montegut Indian School 2015

LOWER MONTEGUT INDIAN SCHOOL

The earliest limited education to Native Americans living along the extreme southern reaches of Bayou Terrebonne was at the Stoufflet School from about 1900 to 1926. The reason for the school's closing is unknown.

Marie Molinere, a Houma Indian woman, recalled the original 1904 school in a 2005 interview as she approached her 100th birthday as a tiny, one-room building that included both Native Americans and whites. She said she was not discriminated against and that she was allowed to speak French in class.[1]

One or more schools were available to Native Americans residing below Montegut on Bayou Terrebonne, according to the U.S. Census of 1900.

The Terrebonne Parish School Board approved a recommendation on August 8, 1939 by Superintendent H.L. Bourgeois to establish four public schools for Native Americans, "so-called Indians," at Ile à Jean Charles, on Bayou Terrebonne below Montegut, on Bayou Grand Caillou, and on Bayou Dularge.

The schools, including that at Lower Montegut, were open for eight-month sessions, with one teacher per school to be paid $50 per month. In 1939, a building that had served as the first Little Caillou School below Chauvin was moved by barge to Bayou Terrebonne to become the Lower Montegut Indian School. On October 3, 1940, the School Board authorized that a certified teacher be hired at the school. A school board report of 1941 stated that 26 students were registered at the school.[2]

The estate of Arthur Lirette negotiated with the School Board a lease on a tract measuring 185.87 feet by 200 feet deep on Highway 69. Dated June 1, 1956, the lease was for 20 years at $60 a year, the $1,200 payable in advance. Four years later, a lease for an addition to the school site was negotiated by the Board with the Lirette estate, to endure from December 1, 1960 to December 1, 1980, for $2,400 payable in advance. After 1960 a frame building was moved to the site from Little Caillou, and in 1963 the semi-circular driveway to the buildings was blacktopped.[3]

In 1964 Terrebonne Parish ceased operation of the tri-racial system of schools, and schools for minorities were closed.[4] The School Board leased the school buildings to the Houma Tribe, Inc. to be used for educational purposes in return for the tribe's furnishing all necessary insurance coverage and maintenance of the building, on August 8, 1975. On January 20, 1981, the School Board declared the school as surplus, and cancelled the lease with the Houma Tribe, Inc.

The United Houma Nation in late 1981 bought from the Lirette heirs the land and buildings of the former school, with $36,000 in federal funds obtained through the Terrebonne Parish Consolidated Government. The land and facilities are still being utilized; the tribe puts it to use as a community gathering center.[5] The current address is 1739 Louisiana Highway 55, Montegut.

Pointe-au-Barré oak ridge 2016

SOURCES

1. Interview by Kimberly Krupa, Grand Bois, March 30, 2005, cited in *Louisiana History: The Journal of Louisiana Historical Association, Vol. 51, No. 2, Spring 2010,* "So-Called Indians" Stand Up and Fight: How a Jim Crow Suit Thrust a Louisiana School System into the Civil Rights Movement" by Kimberly Krupa
2. Terrebonne Parish School Board September 25, 1941 Record Book, Vol. 4, and November 4, 1941 Record Book Vol. 4
3. Terrebonne Parish School Board records
4. Procedures to be filed in Complying with August 28, 1963 Order: "Plan for Desegregation of Indian and White Schools," August 13, 1964, Naquin v. Terrebonne School Board, in the Journal of the Louisiana Historical Association
5. Interim chairman of the United Houma Nation Kirby A. Verret, 2016 interview

Funerals in lower Terrebonne (these photos 1922-1936) often required that coffin and funeral attendees be transported by boat to the cemetery on the oak ridge. Pictured here is the Rev. Joseph M. Coulombe, pastor of Sacred Heart Church in Montegut. c. 1930

Eloise 2015

ELOISE

Jean Billiot had an early land claim on what developed later as Eloise plantation. Billiot had his ownership confirmed by the U.S. government on April 22, 1831.[1]

Eloise was founded by Isadore Dupre and Edmond Guidry two years before the start of the Civil War, in 1858. Eloise was at the extreme southern reaches of Bayou Terrebonne, five miles below Montegut and below Pointe-au-Barré Road.

Guidry (1813-1894) was born in St. James Parish, and in 1835 married Elmire Irma Azel Belanger (1819-1861), daughter of eventual Terrebonne planter Hubert Belanger and his wife Sophie Marie Comeau. Elmire was the sister of Hubert Madison Belanger, recognized as co-founder of Houma, with Richard H. Grinage. Edmond and Elmire became the parents of 16 children.[2]

Guidry worked as a carpenter on the very first buildings in Houma, among them the Rockwood-Grinage Store and Post Office.[3] He built St. John the Baptist Catholic Church (dedicated February 22, 1874) near Dugas Cemetery in Montegut, and St. Ann Catholic Church (dedicated January 22, 1873) at Company Canal in Bourg. Guidry also built his family's home, which predates the Civil War. The current address of the house site is 1851 Highway 55, Montegut, Louisiana.

Left, John (1874-1956) and Ludovia Lirette (1880-1969) Guidry family: front, Ludovia and John; second row, left to right, Wenzel "Manan," Iris, Claudia "Toot," Louella "Nunae," Velma, Robley "P.K."; back row, from left, Horace "Jack," Gillis, and Sherwin "Chabbie" Guidry c. 1950

Dupre and Guidry are owners of record until 1878, when title shifted to Charles Guidry, who held the land for more than two decades.

Alidore Noel Guidry (1850-1935) then had title to the land at Eloise from 1904 until 1912. Alidore was one of the 16 children of Edmond and Elmire Guidry. He operated the A.N. Guidry & Sons General Store adjacent to the family home on the left descending side of the road, and at bayouside ran a combination bait shop, bar, sandwich shop, as well as selling gasoline. He was married in 1871 to Anazile Helena Robichaux (b. 1854).[2]

Alidore Guidry (1850-1935)

John Joseph Guidry (1874-1956) was the son of Alidore and Anazile. He and his wife Ludovia Valerie Lirette (1880-1969) were the parents of 11 children, two of whom died in infancy. John opened John Guidry General Merchandise Store in 1931 where he operated a meat market, ice cream parlor and gas station managed by his son, Wenzel A. "Manan." John's son Robley ("P. K.") also helped with the business. Another son, Sherwin J. "Chabbie" became an unofficial historian of Montegut, much of Terrebonne Parish, and the wider southern portion of Louisiana through his *Houma Courier* and *Bayou Catholic* columns Xplorin' Terrebonne, Xplorin' Acadiana, LeTerrebonne, and Land of the Houmas, later compiled into books.[2]

Two of John Joseph's sons, Horace "Jack" Guidry and Robley "P.K." Guidry, married two daughters of Farquard Guidry from Bourg. Jack married Loretta and P.K. married Edith. Their father was a planter on his large land holding in Bourg, who also owned a sawmill on lower Bayou Terrebonne, and for some years owned a substantial interest in the Lower Terrebonne Refining Company. He later became president of the Bank of Terrebonne and Trust Company.[4]

John (1874-1956) and Ludovia Lirette Guidry (1880-1969), c. 1950

From left, John J. Guidry invoice to Leo Lirette May 28, 1919; John J. Guidry loan document from Bourg State Bank July 9, 1924; J. A. Guidry invoice to Leo Lirette October 26, 1921

SOURCES
1. A Century of Lawmaking for a New Nation: U.S. Congressional Documents and Debates, 1774-1875, American State Papers, Senate
2. Edmond Guidry family genealogical records courtesy Sheri Guidry Bergeron
3. Helen Wurzlow, *I Dug Up Houma-Terrebonne*, 1984
4. The Historical Encyclopedia of Louisiana, Louisiana Historical Bureau, ca. 1940

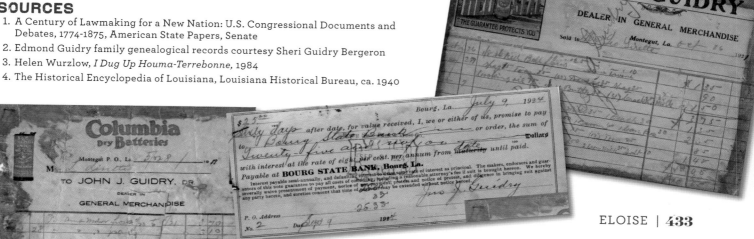

Tobin map 1938

PECAN TREE

At a time after the Civil War when many planters were losing their lands and diminishing their holdings, Jean Louis Cunningham (b. 1840) established Pecan Tree in 1872 along the extreme lower reaches of Bayou Terrebonne. He located his plantation on lands for which Joseph Billiot had received confirmation of ownership by the U.S. government on April 22, 1831.[1]

The modest plantation nevertheless stayed in its founder's hands until 1890 when Robert Clement Cunningham, probably a relative, took over the large farm. Two years later Clovis Dardeau was the owner of record, followed by Edmond Deardon Lapeyrouse, whose second marriage was in Montegut to Louisa Robichaux in 1869. He died in 1899.

Emile Auguste Daigle, owner of the Daigle Barge Line, Ltd., purchased Pecan Tree in 1904. Daigle was one of the parish's most prominent citizens, whose boats the *Harry* and the *Laura* were the first large steamers on Bayou Terrebonne in 1881.[2] The site of the Bayou Terrebonne Waterlife Museum on Park Avenue in downtown Houma was originally developed in the 1880s to serve as a warehouse for Daigle Barge Lines. The building served as a focal point for Houma's growing freight industry.[3]

Waterlife Museum literature added, "Towards the turn of the century, the existing warehouse was purchased by the Cenac family and used in their oyster packing business. It was during this period that Houma assumed the unofficial title of oyster capital of the world. Later (c. 1917), the building was transformed by the A. St. Martin Co., who made it into a labeling and transshipment facility for the Indian Ridge Canning Company, as shrimp began to overtake oysters in economic importance. The building stood empty for years after refrigeration and flash-freezing made canning obsolete."[3]

Another of Daigle's vessels was the *Sadie Downman*,[1] and Daigle's steamer the *Dixie* had its maiden voyage in November 1915 as the largest and most powerful vessel in his fleet and the largest steamboat plying Terrebonne waters.[4] Emile was married to Antoniata Cecile Porche, and the family lived in the 600 block of Main Street.

Emile A. Daigle
(1842-1913)

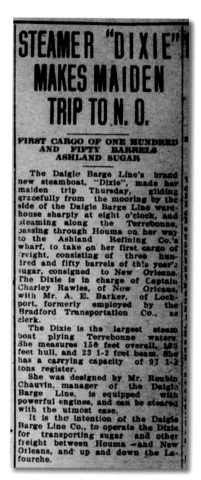

Article on Daigle Barge Line, Ltd.'s steamer Dixie's maiden trip
The Houma Courier *November 13, 1915*

Left, Houma building built by Emile A. Daigle as warehouse, now renovated and used as Terrebonne Waterlife Museum, burned April 18, 1997. Once owned by the Cenac family for use in their oyster packing business, and later by Indian Ridge Canning Co. of the Armand Pierre St. Martin family.

Houma Fish & Oyster Co., Ltd. oyster house, Sea Breeze c. 1910

TEXAS COMPANY YARD

Texas Company Yard boat slip, Lower Montegut 2015

Texas Company Yard dock remnants, Lower Montegut 2015

Opposite page, Pecan Tree in Louisiana Sugar Report *1906-07*

According to an article in the *Houma Daily Courier,* Daigle's boats made trips to New Orleans with local products, and carried groceries, dry goods and other supplies on the return trip. In early steamboat days he dredged Bayou Terrebonne, at his own expense, from Houma to the Company Canal to keep his boats running. He owned several landings and wharfs along Bayou Terrebonne, and at a Houma wharf he built barges and hired a crew of painters and carpenters to maintain his boats. He also was a charter member of Houma Fish and Oyster Company, Ltd. was partial owner of an oyster shop at Sea Breeze.[2]

Daigle's entrepreneurial credentials include his being made a board member for the Lower Terrebonne Refinery and Manufacturing Company in 1891.

The businessman also expanded Daigleville in the late 1890s by building residential homes along Bayou Terrebonne in the proximity of Houma. He was owner of Pecan Tree until 1912, and died the next year. Pecan Tree as a farm seems to have become defunct at that time.

The pre-Civil War house on the property has the current address of 2039 Highway 55, Montegut, Louisiana, on the left bank of Bayou Terrebonne.

SOURCES
1. A Century of Lawmaking for a New Nation: U.S. Congressional Documents and Debates, 1774-1865, American State Papers, Senate
2. *Houma Daily Courier,* September 26, 1971
3. Online information about the Bayou Terrebonne Waterlife Museum, 2015
4. *The Houma Courier,* November 13, 1915

Pecan Tree house 2015

436 | HARD SCRABBLE TO HALLELUJAH

PARISH

Jno. T. Moore Pltg Co. Ltd			Bayou
R. S. Wood & Bro			"
Jno. T. Moore Pltg Co. Ltd		...	"
Jno. T. Moore Pltg Co. Ltd		George	"
Henry Gauthreaux			"
David Levy		le Cuba	"
J. J. Ayo			"
O. E. Pelletier		green	"
J. P. Landry			"
Foret Landry		Way	"
Eno Landry			"
Town of Houma		sh Seat	"
H. W. Connelly		estead	"
Barrow & Duplantis		tle Grove	"
§Gueno Bros	"	Presqu'ile	"
Heirs Alfred Boudreaux	"	Front Lawn	"
Cambon & Champagne	Montegut	Klondyke	"
Chas. J. Champagne	"	St. Agnes	"
Ellender Bros	"	Hope Farm	"
J. L. LeBlanc	"	Aragon	"
Ellender Bros	"	Hard Scrabble	"
Point Farm Pltg. Co. Ltd	"	Point Farm	"
‡L. Terrebonne P. & M Co	"	Central Factory	"
A. A. Saunders	"	Magenta	"
Eug. Fields	"	Angella	"
Mrs. M. J. Saunders	"	Live Oak	"
A. A. Saunders	"	Sunbeam	"
Saunders & Price	"	Eliza	"
Jos. A. Robichaux & Bro	"	Red Star	"
Robichaux & Carlos	"	Argene	"
Leo Lirette	"	Orange Grove	"
Alidor M. Guidry	"	Eloise	"
Emile Daigle	"	Pecan Tree	"
Ernest Guidry	Bourg		"
Ashland P. & M. Co. Ltd	Houma	Ranch	Bayou

* Wells, Fargo & Co. Express.   † Southern Pacific Route, Houma Branch.
§ Now operated by the Lower Terrebonne Planting and Manufacturing Com

Lapeyrouse home in Montegut, Standing Structure survey 1981

Lapeyrouse home 1997

Jean Pierre Junius Lapeyrouse, Sr. (1899-1993), c. 1980

LAPEYROUSE CANAL & STORE

Lapeyrouse Canal was dug by Edmond Deardon Lapeyrouse (1839-1899) from Bayou Terrebonne into the marsh for fur trapping. He was married to Louisa Robichaux (1869-1914), and opened the Lapeyrouse Store sometime in the late 1800s. The original store was constructed from lumber salvaged from a sailing schooner shipwrecked off the Terrebonne Parish coast.

Lineus Joseph Lapeyrouse (1873-1968), besides keeping the store, raised diamondback turtles for commercial sale, an industry that ceased after Prohibition years.

Jean Pierre Junius Lapeyrouse, Sr. (1899-1993) was the last store keeper of the longtime establishment. Jean Pierre's livestock brand was registered as number 1,121 on December 28, 1945 in Terrebonne Parish records. Cause of the store's closing was Hurricane Juan of October 28, 1985, which destroyed the building.

An old oak tree marks the site of the former store, at 2089 Highway 55, Montegut, Louisiana.

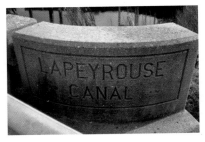

Lapeyrouse Canal bridge marker 2015

Lapeyrouse Store, lower Montegut 1975

Lapeyrouse School, lower Montegut Standing Structures Survey 1981

Cattle brand of Jean Pierre Junius Lapeyrouse, Sr. December 28, 1945

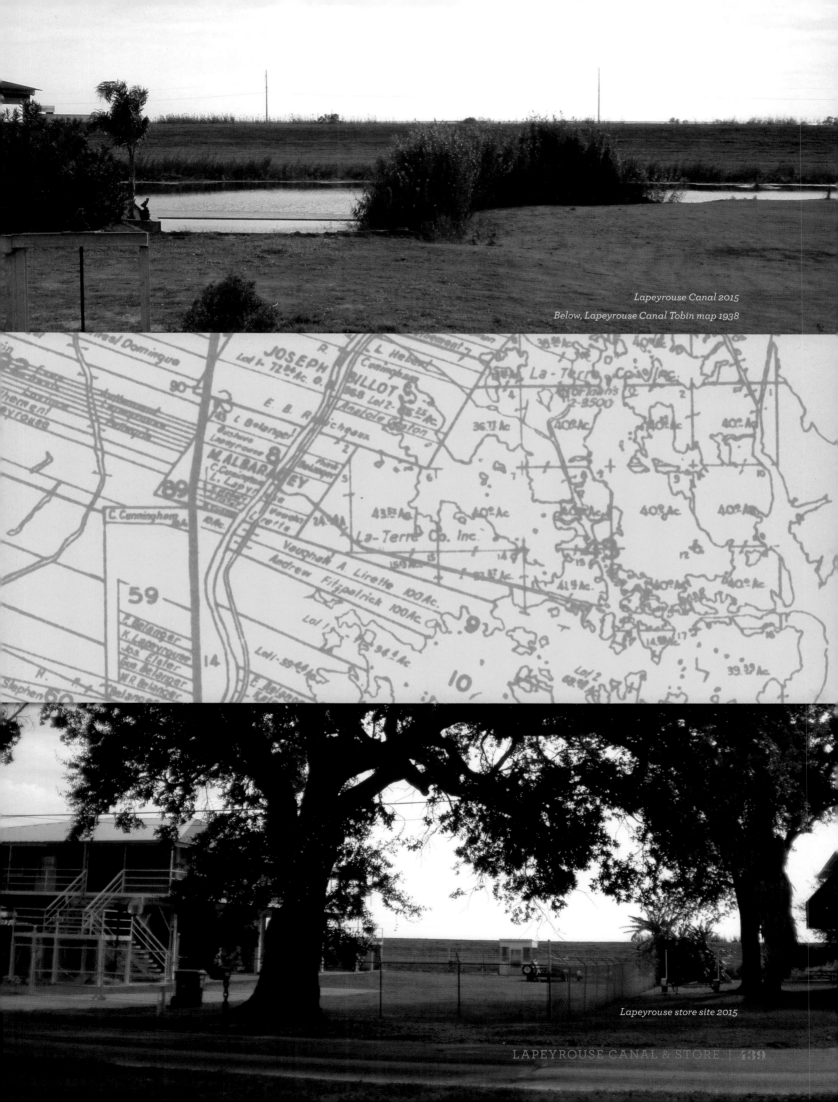

Lapeyrouse Canal 2015
Below, Lapeyrouse Canal Tobin map 1938

Lapeyrouse store site 2015

LAPEYROUSE CANAL & STORE | 439

Boyne Boat Works, last building on Bayou Terrebonne c. 1900

Left to right, Andrew Eugene Boyne (1893-1985); Edward Bennett Boyne (1887-1966); Baby Ruby Boyne (1913-2005) with Onesia Marie Avet Boyne (1894- 1984); John Madison Boyne (1891-1966); John A. Boyne (1858-1924); Elise Marie Belanger (1866-1965), second wife of John A. Boyne

BOYNE BOAT WORKS

John Alexander Boyne (1858-1924) owned land on lower Bayou Terrebonne near Madison Canal upon which his son Edward Bennett Boyne built the Boyne Boat Works.

Edward had two boatways, one on each side of his father and mother's house on the left bank of Bayou Terrebonne. The Boynes, including Edward's brothers Andrew and John Madison, were commercial boat builders using the cypress found in such abundance then in Terrebonne Parish. Among the products of their labors were Clifford Percival Smith's *Helen Snow* and as many as eight boats built at one time for Chinese shrimpers at Manilla Village.[1]

John Alexander's second wife was Elise Marie Belanger (1866-1965) of Montegut, granddaughter of Hubert Madison Belanger, co-founder of Houma. His livestock brand was registered as number 865 on April 29, 1920.

Edward (1887-1966) was a carpenter and boat builder. He was married to Onesia Marie Avet of Cameron, Louisiana; they were parents of Ruby Boyne (1913-2005) who married Earl Anthony Gonsoulin in 1932.

John Alexander's other sons made their living on the water. William Thomas (1889-1976) was a boat builder and boat captain, as was his brother Andrew Eugene (1893-1985). John Madison (1891-1968) was a boat builder and bar pilot.

The boatways were destroyed by the 1926 hurricane, after which the family relocated to Houma and St. Mary Parish.

SOURCES
1. Helen Wurzlow, *I Dug Up Houma Terrebonne*, 1984

Helen Snow skiff at Roberta Grove c. 1900

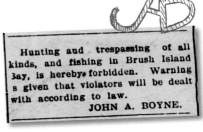

Cattle brand of John A. Boyne April 9, 1920; John A. Boyne trespass notice The Houma Courier June 17, 1916; below, article on size and success of Boyne Shipyard The Houma Courier August 9, 1924

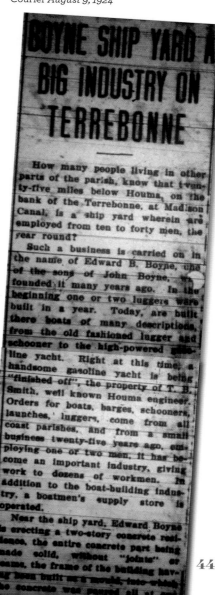

RHODES BROTHERS

Although the Rhodes family did not establish a plantation in Terrebonne Parish, they are notable in that patriarch Thomas came to the area as a surveyor before the founding of the parish, in the early 1800s. In 1812 he married Heloise Bergeron, who was born in Donaldsonville.

Their son John Jackson (also called Jacques Guillaume), born in 1815 at Plattenville in Assumption Parish, was the first Rhodes on record to own land on Bayou Terrebonne, a quarter mile below Bush Canal, which is below Madison Canal. The Rhodes property was on both banks of Bayou Terrebonne. Jackson and his wife Henriette Amanda Hebert married in 1836, and the only other known fact about them is that they were the parents of eight, all of whom settled around their parents' home. John Jackson and Henriette were buried alongside his parents Thomas and Heloise on an Indian mound which no longer exists.

One of John Jackson and Henriette's sons, Thomas, married Victorine Use in 1860 and had eight children also. The family settled near present-day Bush Canal among many Rhodes families in the vicinity. The Rhodes clan lived as subsistence farmers and fishermen. Thomas was a boat builder, and his son Gustave perfected the craft. Thomas and Victorine were also buried on the Indian mound cemetery.

Another son, Leon Alexis, married Victoria May Price on April 14, 1877. They became parents of 12 children. Leon was a grocer and fisherman as were most of his sons. All the children were born on the family property.

Gustave plied his trade as boat builder, and patented the first V-bottom boat. He and his wife Amanda Henry and family, along with (constable) Robert Rhodes and other Rhodes households, stayed on in lower Terrebonne until the destruction by the storm of 1909 convinced many of the family members to move to higher ground in Terrebonne, St. Mary, and Calcasieu parishes.

Bush Canal was dug through Rhodes family property in 1912 to transport sugar cane from Little Caillou to Bayou Terrebonne to be processed at the refinery in Montegut.

After the devastating storm of 1926, many of the Rhodes families relocated to Houma.

SOURCES
1. Helen Wurzlow, *I Dug Up Houma Terrebonne*, 1984

Thomas Rhodes (1838-1918)

Leon Alexis(1855-1956) and Victoria May Price (1860-1909) Rhodes on their wedding day April 14, 1877

Leon Rhodes trespass notice The Houma Courier *March 7, 1903 and July 1, 1905*

Right, Carroll Rhodes (1924-2011) standing with the cross marking the burial place of his ancestors atop an Indian mound at the end of the road paralleling Bayou Terrebone

Madison Canal facing east from Bayou Terrebonne 2016

MADISON CANAL

The three-mile long Madison Canal was dug from Bayou Terrebonne to Bay Madison in 1870 by Hubert Madison Belanger and John Madison Belanger. John Rhodes and his brothers extended it south in 1872.

The Terrebonne Rod & Gun Club bought 40 acres of land in the area on July 23, 1929 for $5,000 from B. Boyne (Boerne). Members of the club were New Orleans business and professional men. Closure date of the club is unknown. Current address of the site is 2177 Highway 55, Montegut, Louisiana. All that remains on the site are two oaks dubbed the Gustave Rhodes Oaks.

Tobin map 1938 Madison Canal; right, "The Clubhouse" at Madison Canal, c. 1940s; below, Terrebonne Rod and Gun Club, Times Picayune July 23, 1929

MADISON CANAL | 445

Bush Canal
Tobin map 1938

Aerial photo Bush Canal 1953

BUSH CANAL

Bush Canal was dug in 1912 through the Rhodes family properties to connect Bayou Little Caillou to Bayou Terrebonne for transporting sugar cane to the refinery in Montegut. In the early days of the canal, barges were pulled by cordelle to the refinery.

The canal is nine miles above where the Sea Breeze community was located. A floodgate measuring 56 feet wide and rising 18 feet out of the water was completed at Bush Canal in July 2011. Built at a cost of $14 million, the Bush Canal floodgate was constructed to keep floodwater from filling Lake Boudreaux during storms, thus protecting Houma, Chauvin, and Dulac. The floodgate was designed to choke the 350-foot wide canal down to 56 feet, allowing less Gulf of Mexico saltwater into the lake, and helping to curb erosion. On November 22, 2011, the Bush Canal floodgate was dedicated in memory of Willis J. Henry, longtime member of the Terrebonne Parish Council.

Bush Canal Floodgate from the southeast 2016

END OF THE ROAD

Southern Bayou Terrebonne below the end of the road and Bush Canal, toward Sea Breeze 2016

DESTINATION: SEA BREEZE

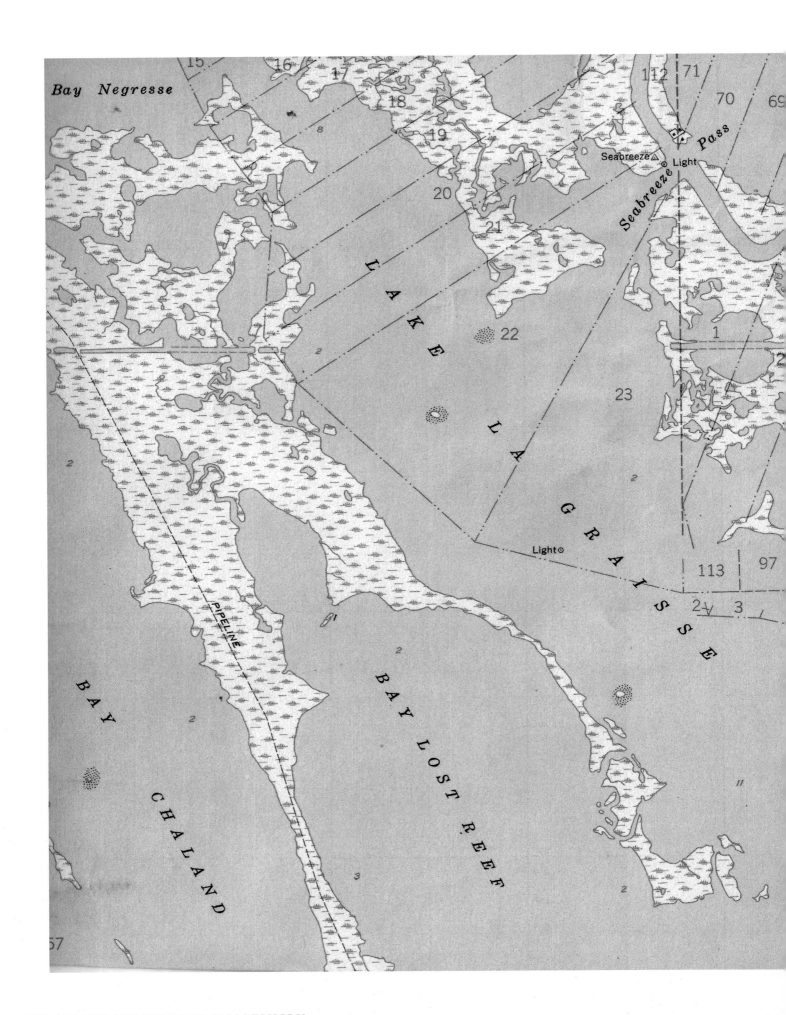

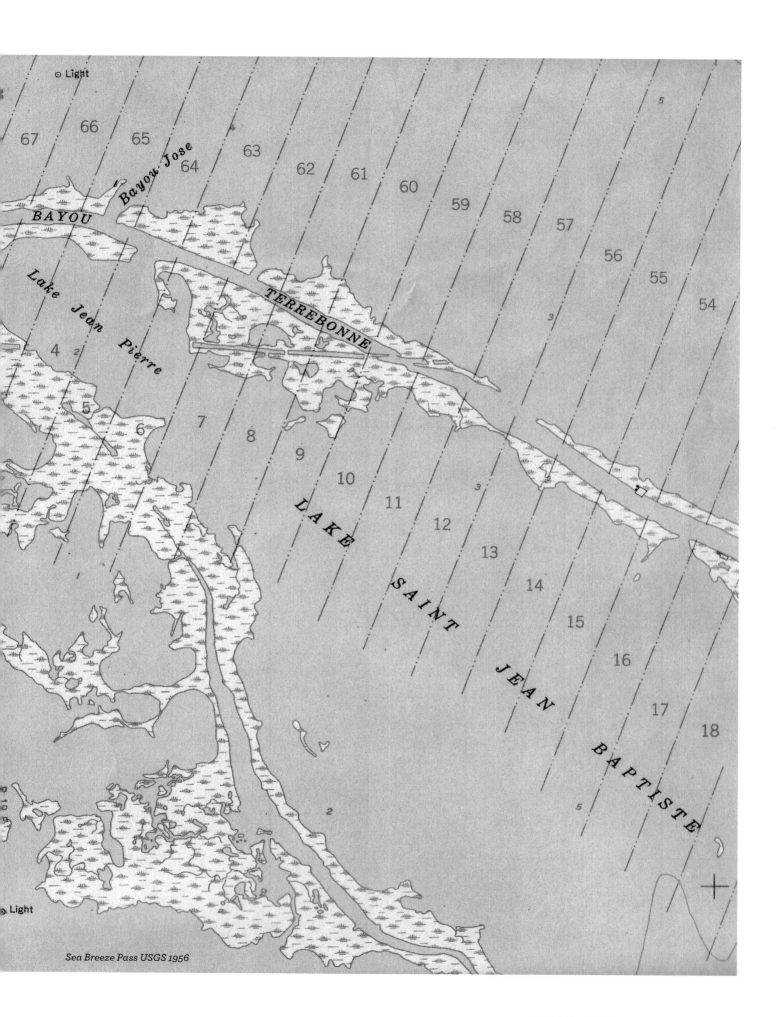

Sea Breeze Pass USGS 1956

DESTINATION: SEA BREEZE | **453**

Dupont camp Sea Breeze c. 1918

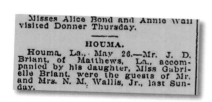

A Trip to Sea Breeze
Times Picayune May 27, 1906

SEA BREEZE

Very close to the Gulf of Mexico, 33 miles below Houma, James Monroe Sanders was the first to dig through the natural levee on the left descending bank of Bayou Terrebonne to landlocked Lake Barré. The work was done with shovels where only portage could transport goods and boats to the lake before Yancey Sanders dug out the land from Bayou Terrebonne to Lake Barré in 1870. A channel on the right descending bank to Lake Lagraisse was completed at a later date, specifics unknown.

This new water route became a competitor to the Barataria & Lafourche Canal (# 2) because it allowed commercial boat transportation from Bayou Terrebonne to Lake Barré to Bayou Lafourche to the Barataria Canal to New Orleans.

In the late 1800s, Sea Breeze was a coastal playground for wealthy citizens of the area. Many families had summer homes there, and annual boat races attracted boaters not only from Terrebonne, but also from surrounding parishes. The August 26, 1926 hurricane destroyed the camps and homes at Sea Breeze, ending the halcyon days on the coast.

Top to bottom, left to right, Houma Fish & Oyster Co., Ltd. Oyster House Sea Breeze c. 1910; A.A. Sanders camp Sea Breeze c. 1918; boat races Sea Breeze c. 1913; Sea Breeze camps and luggers; Elizabeth Stoufflet Viguerie Sea Breeze c. 1924; A.A. Sanders camp Sea Breeze after 1926 storm

Shrimp dryers' camp Sea Breeze c. late 1800s

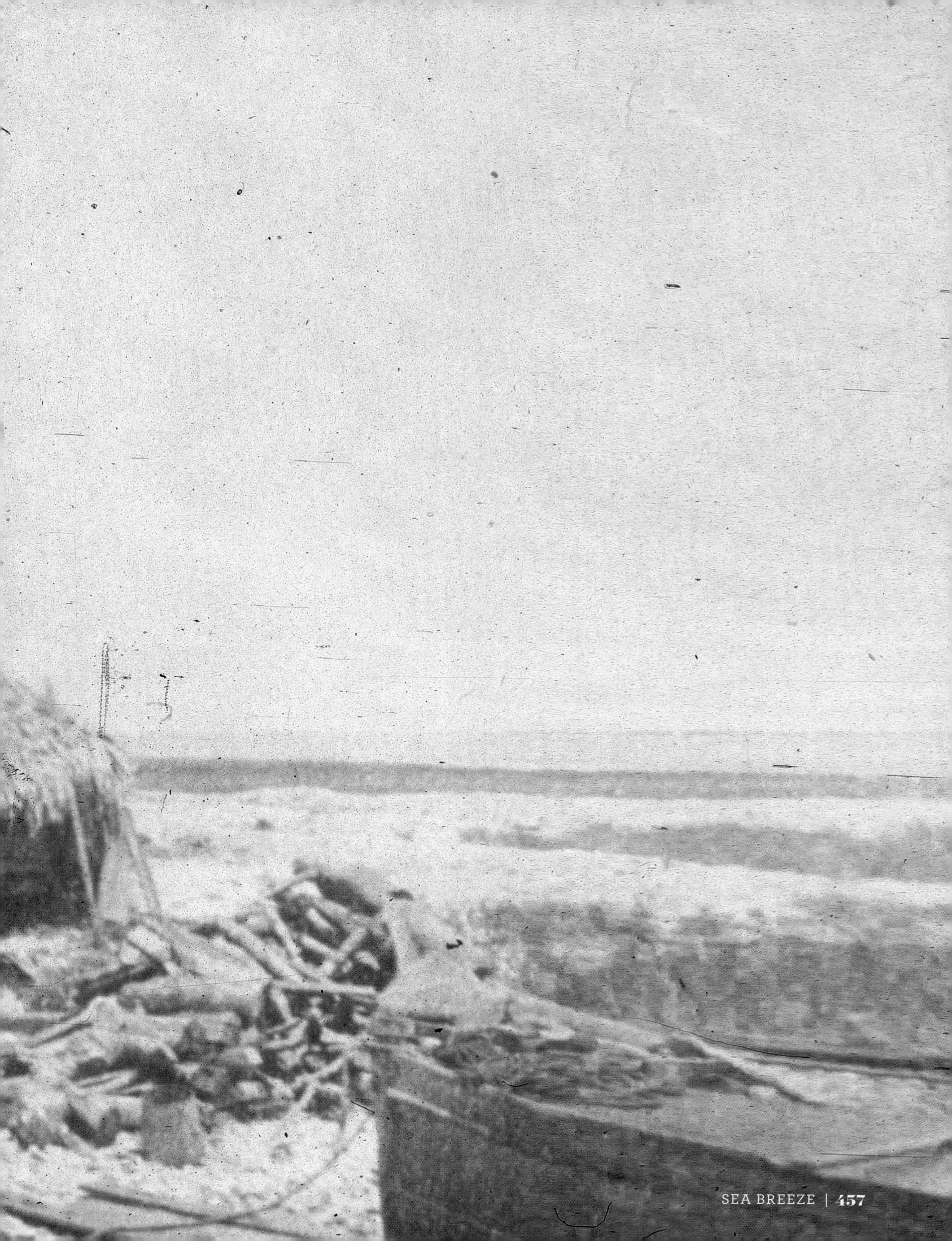

Thaddeus J. Pellegrin's shrimp boat, dock, and shrimp drying platform Sea Breeze, c. 1960

Thaddeus J. Pellegrin's camp and boat dock Sea Breeze, c. 1960

Sea Breeze in 2016 and inset same location in 2010

AFTERWORD

Dr. Chris Cenac might have considered a career as a miner instead of a physician. The man never ceases to astound at the glittering nuggets he is able to unearth.

I say this facetiously. Dr. Cenac, who semi-retired some years ago as a highly respected orthopedic surgeon, has these days turned his considerable passion and skills to history. With this work, he has produced his third (and gorgeous) book, *Volume I: Bayou Terrebonne - Hard Scrabble to Hallelujah*. While focused again on his native Houma and Terrebonne Parish, the new book demonstrates the broader significance of Dr. Cenac's efforts.

He has captured in magnificent detail – using stunning historical pictures, documents and artifacts – a slice of 19th Century agrarian Louisiana that illuminates the rhythms, practices and aspirations of the larger agrarian South. Reading this book, you get a true feel for how our forebears viewed and appreciated their ties to the land and how they worked to preserve and improve their holdings (and sometimes failed and lost them.) It also shows how these legacies survive today in place names and historical homes and buildings, some crumbling, some well-preserved, whose importance might be forgotten save for Dr. Cenac's work. The book is also a poignant reminder of how this fertile but fragile low-country which has nurtured a unique French-influenced culture lies these days diminished and in need of our best restoration efforts.

I grew up in and around Houma and spent my formative childhood years on a small Southdown Sugars farmstead at Waterproof plantation on Big Bayou Black. I had five brothers and a woodsman and fisherman for a father. In the days when "No Trespassing" signs were still rare we roamed freely, squirrel hunting the ridges at Ellendale and traveling often to fish the salt water bayous below Houma, and particularly Sea Breeze. En route we passed through what was a still a rural countryside of seemingly endless sugarcane fields marked now and then by exotic place names – Presquille, Klondyke and Bourg to name a few. The marvel of Dr. Cenac's book is that he traces *all* of these names to their original plantation sources and, indeed, comes to the astonishing conclusion that if you reside in modern-day Houma, you live upon the footprint of an historical plantation. You may also be surprised to learn that before the Civil War, these plantations sprouted indigo, the blue-dye-producing plant, instead of sugar cane.

Volume 1: Bayou Terrebonne is another important book, beautiful enough to decorate your coffee table, packed with so much history that it belongs on the shelf of every library as well. And the good thing – Dr. Cenac is completing *Volume 2: Bayous Black and Buffalo (Dularge)* which will trace land-owning legacies from Bayou Black and modern-day Bayou Dularge. As a proud former Bayou Black resident, I eagerly await its publication.

Ken Wells

Wells is a former editor for Page One of the Wall Street Journal and the author of nine books, including the classic coming-of-age novel, *Meely LaBauve*, and *The Good Pirates of the Forgotten Bayous*, a prize-winning non-fiction work chronicling a bayou people's heroic survival of Hurricane Katrina.

HARD SCRABBLE to HALLELUJAH
LEGACIES
of TERREBONNE PARISH, LOUISIANA

Volume 2: Bayous Black and Buffalo (Dularge)

---EXPECTED RELEASE 2018---

Greenwood plantation, Bayou Black, home of Tobias Gibson, built 1834; painting by Garth K. Swanson 2016 from photo c. 1940

*Greenwood plantation token
Laurent Lacassagne 65 cents 1886*

AFTERWORD | 463

ABOUT THE AUTHORS

Christopher E. Cenac, Sr., M.D., F.A.C.S, grew up in Houma, Louisiana, and graduated from the Louisiana State University system. He attended LSU undergraduate school on academic and athletic scholarships, and completed his residency in orthopedic surgery in 1976. He is a practicing orthopedic surgeon, and has served a term as Terrebonne Parish Coroner.

His intial work, *Eyes of An Eagle*, a 2011 docu-novel, is a biographical account of his great-grandfather's life. The book also provides an illustrated history of early Terrebonne Parish, Louisiana. It was a selected book of the Louisiana Bicentennial Commission, and has been placed in Terrebonne Parish public and Catholic schools as a historical resource. His next book was *Livestock Brands and Marks, An Unexpected Bayou Country History* (published 2013), which was named a 2014 Book of the Year by the Louisiana Endowment for the Humanities.

Dr. Cenac is co-author of the Interprofessional Code approved by delegates of the Louisiana State Bar Association and the Louisiana State Medical Society in 1994. As a member of the Medical-Legal Interprofessional Committee, he was co-author of several published papers including "Subpoenas to Physicians: The Obligations and Consequences"; "Medical Review Panel Process"; and "The Physician as Witness." Professional affiliations are with the American Board of Orthopedic Surgery, as a Fellow in the American Academy of Orthopedic Surgeons, a Fellow in the American College of Surgeons, and a Fellow in the International College of Surgeons. He was appointed in 2003 by the Louisiana Supreme Court to the Judicial Campaign Oversight Committee and was reappointed in 2010. His current professional emphasis is expert testimony in the field of orthopedic surgery in medical-legal litigation at state and federal level judiciaries, and teaching professionalism and ethics to medical and law students, attorneys, and physicians.

He served as king of the Mystic Krewe of Louisianans in Washington, D.C. in 2003. Today his greatest personal interests are history and international travel. He and his wife Cindy reside at Winter Quarters on Bayou Black outside Houma, Louisiana. He has two sons and a daughter—two of whom are also physicians— eight grandchildren, and a widespread family both in and beyond Terrebonne Parish.

Claire Domangue Joller is a national and state award-winning writer. She received the first place award from the North American Catholic Press Association in 2001 for her first year of *Bayou Catholic* newspaper columns in the Arts, Culture, and Leisure category, and followed that with a 2014 second place award in the Association's General Commentary category. Her "Seeing Clairely" column also won a Louisiana Press Association award in 2001 and another in 2012. She was the writer for Dr. Christopher E. Cenac, Sr.'s book *Eyes of An Eagle*, published in 2011 and *Livestock Brands And Marks*, published in 2013, which was named a 2014 Book of the Year by the Louisiana Endowment for the Humanities.

A native of Terrebonne Parish, she first received statewide recognition during Nicholls State collegiate days with a sweepstakes prize from the Louisiana College Writers Society. After a short stint as English and journalism teacher, she variously worked as a newspaper section editor, public relations director for a children's home, technical writer for the environmental section of an engineering firm, development director for a Catholic high school, and editor of *Terrebonne Magazine*. Entries she submitted were selected to be among the 120 nonfiction "minutes" included in Pelican Publishing Company's 2007 anthology *Louisiana in Words*. She lives in Houma, Louisiana, with her husband Emil. She has a daughter, two grandchildren, and a large extended family.

High Water at Whiskey Bay, 1975, Oil on canvas, 30 x 40 inches, George G. Rodrigue

INDEX

1770-1798, Louisiana Census Records 406
1774-1875, American State Papers 135, 139, 147, 148, 161, 181, 200, 208, 233, 241, 279, 283, 285, 287, 289, 291, 308, 315, 319, 327, 357, 369, 385, 399, 401, 402, 405, 412, 416, 433
1785 Land Ordinance 31
1803 Louisiana Purchase 23
1810 U.S. Census 20, 21
1810 Westerman map, Lafourche Interior 20
1830 State Land Grant map 314
1831 Nicholas Leret Claim State Land Grant Map 148
1831 Pierre Menoux State Land Grant map 210, 220
1831 State Land Grant map 54-55, 61, 175, 400, 401
1832 State Land Grant map 286
1832 map, Naquin State Land Grant 335
1832 survey 364
1846 La Tourrette map 4, 93, 102, 105, 132
1846 Richardson map 36
1850 State Land Grant map 36
1850 U. S. Census 17, 18, 362, 409
1853 John La Tourrette map 7
1855 Garlinski survey 289
1856 map, Jean Dupre State Land Grant 361
1856 "Last Island" hurricane 26, 318
1856 Michael Dardar State Land Grant Map 284
1860 U. S. Census 202, 284, 363, 404, 409, 416
1870 U. S. Census 401
1872 Hedgeford Map 125
1877 Act of Partition 93
1880 U.S. Census 95, 284, 320, 337, 364
1884-1885 World's Industrial and Cotton Centennial Exposition
1890 Mississippi River flood 190
1890 U.S. Census 183
1893 Klondyke Map 323
1894 Map of Timbalier Bay 365
1897 Directory of the Parish of Terrebonne 80, 137, 157, 181, 183, 275, 468
1900 U.S. Census 95, 320, 337, 405, 430
1902 Sandersville map 388, 393
1903 A.C. Bell map 147
1907 Isle of Cuba map 107
1909 hurricane 302
1909 map of Morgan's Louisiana & Texas Houma Branch Railroad, Schriever, Waubun plantation, and St. George plantation 75
1909 storm 442
1915 Bocage Map of Terrebonne Parish 13
1921 Map of Aragon 352
1923 Schriever Map 70
1924 Map of Barrowtown 235
1926 hurricane 442, 454
1938 Tobin Map 95, 125, 133, 136, 418, 434, 439, 445, 446
1939 USGS map of Fala - Lake Felicity 343
1940 Map, City of Houma 176–177
1945 Isle of Cuba Map 107
1953 Map of Bourg 294
1956, Sea Breeze Pass Map 453
1960 City Map of Houma 239
1960 Map, Back of Town 182
1981 Standing Structures Survey 351
2010 U.S. Census 181

A

Abendziller, Emil 268
Acadian 100, 427
Acadian Coast 189
Acadians 15, 17, 47, 193, 362
Acadia plantation 50, 103, 143, 145
Acre, defined 31
Adam, Mabel Nellie Adoue 158
Adam, Maturin 67
Adam, Paul Lawrence 158
Adams, Adney Aubin, Sr. 283
Adams, John I. 144
Adams, Rubella Bergeron 283
Adams & Viala 144
Adler, A. & Co. vs. George W. Price 401
Adoue, Mabel Nellie.
 See Adam, Mabel
 Nellie Adoue
African American poet 235
African Americans 48, 108, 132, 157, 186, 229, 235, 256, 275, 281
 See also
 • Christian, Marcus Bruce 235
 • colored men 157
 • Radical Ridge Farm 157
 • The Alley 186
Afton Villa plantation 221
Ag-Lands Investment Co. 357, 369, 395, 399, 401, 402, 405
Agnew, Spiro T.,
 Vice President 412
Agoutte (Aoushe), Catherine
 See Fangui, Catherine
 Agoutte (Aoushe)
Airport Commission 256
Alamo 22
Alcoa Puritan (ship) 244
Alexander, Robert 369
Alford, Sharon A. 7
Alice B. plantation 192
Alidore J. Mahler Addition 217
Allemand, Kevin J. 7
Alley, The 186
Al Saud, Amir Nawaf Ibn Abdulaziz, Prince 244
American Bulb Co. 332
American Legion Baseball Grandstand 264
American Legion Park 188
American State Papers 37, 42, 69, 82, 103, 115, 121, 126, 135, 148, 200, 208, 217, 286, 291, 319, 327, 357, 369, 385, 399, 402
American Sugar Cane League 260
Andrepont, Milton J. 328, 329
Andrew Price School 95, 102, 103
Andrew Price Vocational School 103
Angela plantation 371, 395, 404, 412
Angers, Drew 193
Anglo-Americans 23

Anglo-Saxon sugar planters 23
   See also
- Barrow, R.R. 23
- E. Ogden 23
- Richard Ellis 23
- Thomas Butler 23
- Tobias Gibson 23
- William A. Shaffer 23
- W.J. Minor 23

A.N. Guidry & Sons General Store 433
Antoine, Caesar, La. Lt. Gov. 123
Antoine Naquin Grocery Store and Dance Hall 339
Antoniata Cecilia Porche. *See Daigle, Antoniata Cecilia Porche*
Arab dignitaries visit Houma 244
Aragon home 325, 353
Aragon plantation 11, 19, 102, 199, 325, 326, 352, 353, 356, 357, 367, 371
Arcement, Louis 183
Arcement, Richard 472
Arcement, Richard Anthony 7
Arcemond, Annette Rose *See Guidry, Annette Rose Arcemond*
Arcenaux, Widow Urbain 40
Arceneaux, N.J. 217
Architects Charles Allen Favrot & Louis A. Livaudais Ltd. 311
Architectural Styles / Elements
- *briquette-entre-poteaux* 126
- Colonial Revival 17
- Eastlake 17, 230
- *faux bois* 42
- Greek Revival 17
- heart cypress 126
- "lightning" joint 409
- La. Raised Cottage 17
- mortise/tenon joinery 44, 199
- Queen Anne 17, 187
- Queen Anne Revival 230
- Rococo Revival 53
- Victorian 17

Ardoyne plantation 33, 68, 362
Argene plantation 416
Argyle plantation 157, 192, 284
Arlen B. Cenac, Jr. Foundation 7
Armitage plantation 39 - 43
   See Sargeant/Armitage
Armitage, Charles 40, 41
Armitage, Ellen Rebecca Sargeant 40
Armitage House 41, 47, 48
Armitage, Lucy Dean Foster 41
Army base 216, 217, 471
Arpent 30
Arpenteur 30
Arsenaux, Mrs. Gabriel 40
artillery shells 246
Ashland Branch Railroad 186
Ashland Extension 186
Ashland plantation 283, 287, 373, 383
Assessor's Office 32
Assumption Church 406
Ascension Parish 135, 154, 291
Assumption Parish 53, 66, 93, 130, 132, 144, 225, 291, 317, 395, 418, 442

Atchafalaya River 14, 26, 71, 164
Attakapas District 164
attic water system 79
Aubert, Justine
   See Thibodaux, Justine Aubert
Aubert, Marguerite Barras 93
Aubert, Pierre 93
Aucoin, Alphonse 95
Aucoin, Camille A. 319
Aucoin, Elvina Bergeron 318
Aucoin, Eve Regina Robichaux 319
Aucoin, Rudy R. 7, 472
Aucoin, Silvani Numa 318
Audubon, John James 78
Audubon Sugar School 81
Augustine, Louis, Jr. Rev. 48
Authement, Horace J. 217
Authement Subdivision 217
Autin, Evelyn Bergeron 126
Avet, Onesia Marie 441
Aycock, Armogene F. 281
Aycock, F.A. 285
Aycock, Onezime Theophile 192
Ayo, Abel B. 143, 144
Ayo, Adele. *See Darcey, Adele Ayo*
Ayo, Charles E. 143
Ayo, Donatille Zimorie Olivier 144
Ayo, Eddie 143
Ayo, Ella Williams 144
Ayo, Ernest 143
Ayo, Ivy G. 143, 144
Ayo, J.J. 145
Ayo, Joseph 143
Ayo, Joseph John 144
Ayo, Joseph Odressi, Col. CSA 143, 144
Ayo, Joseph Odressi, Corp. 145
Ayo, Mary 143, 144
Ayo plantation 143
Ayo, Rose.
   See Trahan, Rose (Roselia) Ayo
Ayo, Sidney J. 143
Azucar Rio Lindo Compania 22

B

Babin, Auguste 100, 105, 116
Babin, Augustin 123
Babin, Dexter A. 7
Babin, Elijah (Lule) 265, 272
Babin, L.H., III 7
Babin, Logan H., III 468
Babin, Logan H., Jr. 7
Babin's Grocery Store 212
Babin, Teles 155
Back of Town 168, 182, 183
Badeaux, Udalize.
   See Robichaux, Udalize Badeaux
bagasse 130, 258, 379
Bagley, R.H. 288
Baker, Caroline Louise. *See Bisland, Caroline Louise Baker*
Baker, Joshua Judge 325
Ballard, M. 50
Ballard, V. G., Mr. and Mrs. 147
Ballenger (Belanger), Hubert 93, 100
Balsamine plantation 46, 92, 93, 94, 101, 108

Bank of Terrebonne and Trust Company 301, 362, 433
bankruptcy 123, 236, 275
Banks, Nathaniel P., Gen. 326, 356
Baptiste, Valery Jean 47
Barataria and Lafourche Canal Company 164, 299
Barataria Bay 427
Barataria Canal 80, 161, 164, 166, 168, 169, 173, 186, 299, 454
Barataria-Lafourche Canal Company No. 2 168, 230
Barataria & Lafourche Company No. 1 297
Bardeleben, Skye G. 7
Bardons, P 64
barges 245, 275, 370, 372, 373, 378, 436, 447
bar pilot 441
Barras, Antoine 44, 105
Barras, Brigitte Emilie Thibodaux 105, 114, 139
Barras, Charlotte
   See Bourgeois, Charlotte Barras
Barras House 44
Barras, Jean Valery 44
Barras, Julie Patin 105
Barras, Leufroy (Ludfrois) Emile, Hon. 44, 46, 57, 101, 103, 105, 114, 115, 119, 139, 296, 298
Barras, Marguerite
   See Aubert, Marguerite Barras
Barras, Susann L. 46
Barrios, Harry 208
Barroso, Thomas Villanueva 53
Barrow & Duplantis 275
Barrow, Bartholomew 221
Barrow family child's crib 223
Barrow family china 223
Barrow, Hallette Mary 230, 237
   See Cole, Hallette Mary Barrow
Barrow, Irene Felicie 230, 237
Barrow, Jennie Lodiski Tennent 230, 237
   See Dawson, Jennie Tennent Barrow
Barrow landowners 235
Barrow, Roberta 225
Barrow, Robert Jr. 237
Barrow, Robert Ruffin 237
Barrow, Robert Ruffin, II 230
Barrow, Robert Ruffin, Jr. 168, 169, 225, 227, 230, 233, 234, 236, 237, 275
Barrow, Robert Ruffin, Sr. 23, 132, 135, 164, 217, 221, 225, 228, 230, 232, 233, 236, 237, 245, 274, 275, 278, 281, 284, 298, 352, 361, 408
Barrow, R.R. 23, 173
Barrow, R.R., Collection 469, 471, 472, 474
Barrow, R. R., Jr. 240
Barrow's Canal 186.
   See Barataria Canal
Barrow Subdivision 11, 238, 239, 240
Barrowtown Subdivision 235, 240
Barrow, Volumnia Hunley 228, 230, 236
Barrow, Volumnia Roberta 225, 230, 236, 275. *See Woods, Volumnia*

Roberta Barrow Slatter; See Slatter, Volumnia Roberta Barrow;
Barrow, Volumnia Washington Hunley 221, 230, 236, 237, 275
Barrow, William B. 221
Barrow, Zoe Gayoso 230, 237. *See Topping, Zoe Gayoso Barrow*
Bascle, Ada Lester 368
Bascle, Irene 368. *See Viguerie, Irene Bascle*
Bascle, Richard (Dickie) 303
Bascle, Ruby 303
Bascle, Sam 82
base chapel 256
Batey, Beattie, Beatty, or Beaty plantation 10, 132, 150, 154
Batey, Jesse, Dr. 132, 135, 137
Batey plantation 10, 128, 132, 135, 225
Bateyville 128, 132
Battle of Bisland 326
Battle of New Orleans 53, 154, 308
Battle of Shiloh 193
Baudoin, Ella Mae 130. *See Bernard, Ella Mae Baudoin*
Bay Junop 245
Bayou Barré 415, 427
Bayou Black 4, 14, 15, 17, 23, 26, 48, 61, 77, 108, 116, 152, 164, 168, 186, 225, 299, 319, 363, 422, 462, 464
Bayou Blue 8, 14, 15, 146, 147, 275, 303, 342, 424
Bayou Blue Presbyterian Church 146, 147
Bayou Bomber 247
Bayou Buffalo 15
Bayou Caillou 161
Bayou Cane 10, 15, 146 - 155, 175, 303
Bayou Cane Elementary School 155
Bayou Chacahoula 15
Bayou Derbonne 100, 298
Bayou Dularge 18, 61, 192, 199, 225, 272, 348, 462
Bayou Fala (Faleau) 342
Bayou Grand Caillou 15, 17, 57, 225, 286
Bayou Grand Coteau 18
Bayou History Center 7, 472, 473
Bayou Jean Charles 334, 339
Bayou Lacache 302
Bayou LaCarpe 183, 186
Bayou Lafourche 23, 36, 37, 53, 61, 295, 299, 345, 398, 406, 454
  See dam 190
  See source from 15
Bayou LaGresse 211
Bayou Little Caillou 15, 17, 225, 275, 280, 288, 427
Bayou Little Caillou, origin, map 280
Bayou Little Terrebonne 57
Bayou Petit Caillou 409
Bayou Petit Terrebonne 57.
  See Little Bayou Black
Bayou Pointe-aux-Chênes 11, 15, 18, 328–345, 334, 427
Bayou Salé 181
Bayou Segnette 299
Bayou St. Charles 334
Bayous Lafourche and Terrebonne, early history 14
Bayous, other
  • Big Bayou Black 103, 189, 192
  • Black bayou 15
  • Daspit bayou 19
  • Petit Black bayou 14
Bayou Teche 164
Bayou Terrebonne, head of 14
Bayou Terrebonne Waterlife Museum 435, 436
Bay St. Elaine 245
Bazerque, Bertrande 362. *See Viguerie, Bertrande Bazerque*
Bazet Collection 468, 469, 471, 473, 474
Bazet, Inza 473
Bazet, Inza May (Mae) Sanders 240, 398, 405, 473
Bazet, R. 473
Bazet, Randolph A., Jr. 7
Bazet, Randolph A., Mrs. Sr. 240
Bazet, Randolph August 398
Bazet, Sr., Randolph A. 398
Beary, Thomas E., Col. 78
Beasts of the Southern Wild (movie) 399
  See Magenta plantation
Beattie, Charlton 81
Beattie, Fannie E. Pugh 132
Beattie, John C., Dr. 132, 135
Beattie, S. 146
Beattie, Taylor & Co. 146
Beattie, Taylor, Judge 93, 121, 132, 134, 146
Beaty [Batey], Robert, Sr. 135
Beauchamp, Michael Kelly 69
Beauvais, Mollie Brigitte 95. *See Olive, Mollie Brigitte Beauvais*
Becemis, John 143
Becnel, Thomas 169, 279, 299, 308, 326, 327, 369
Becnel, Thomas A. 225, 230, 233, 237, 241, 410, 412
Belanger 17.
  See First Settlers
Belanger, Brigitte Emelie 100, 116, 297.
  See Thibodaux, Brigitte Emelie Belanger
Belanger, Celeste 297, 299. *See Grinage, Celeste Belanger*
Belanger, Elise Marie. *See Boyne, Elise Marie Belanger*
Belanger, Elmire. *See Guidry, Elmire Belanger*
Belanger, Elmire Irma Azel. *See Guidry, Elmire Irma Azel Belanger*
Belanger, François 175
Belanger/Grinage tract 175
Belanger house 299
Belanger, Hubert 11, 105, 295, 298, 299, 357, 432
Belanger, Hubert M. 175
Belanger, Hubert Madison 298, 299, 314, 432, 441, 444
Belanger, John Madison 444
Belanger, Marie Agnes Celeste 116, 297. *See Tanner, Marie Agnes Celeste Belanger; See Tanner, Marie Agnes Celeste Belanger*
Belanger, Marie Celine Mars 299.
  See Hotard, Marie Celine Mars Belanger
Belanger plantation 298
Belanger, Sophie 299
Belanger, Sophie Comeau 357
Belanger, Sophie Marie Comeau 297, 432
Bell, A.C., map of 1903 147
Bellanger Canal 296
Bellanger, Hubert (Belanger) 296, 297, 298, 299
Belle Farm 103, 363
Belle Grove plantation 42
Bellenger (Belanger), Hubert 408
Bellenger, Hubert 295
Bellevue plantation 66, 317
bell
  Ducros plantation 60
  Hope Farm 327
  Residence Baptist 229
  St. Ann Catholic Church 301
  St. Francis de Sales 211
  St. Matthew's Episcopal 229
  St. Peter's Church 359
Bellview Dairy 234
Bellview Dairy milk bottle cap 234
Bellview Place 10, 217
Bellview Theatre 211
Benedict, W. 284
Bennett, Carl M., Jr. 7
Bennett, Tom 221
Bennie, Robert W. 149
Benoit, Debra S. 7
Berger, John 196
Bergeron, Charles 143
Bergeron, Elmo P., Jr. 7, 126
Bergeron, Elmo P., Sr. 126
Bergeron, Elvina. *See Aucoin, Elvina Bergeron*
Bergeron, Heloise. *See Rhodes, Heloise Bergeron*
Bergeron, Joseph Jess 468
Bergeron, Joseph J., Jr. 7
Bergeron, Michael, Rev. 472
Bergeron, Pierre 146
Bergeron, Rev. Michael A. 7
Bergeron, Rubella 283. *See Adams, Rubella Bergeron*
Bergerons 17. *See First Settlers*
Bergeron, Sheri Lee Guidry 7, 468, 473, 474
Berger, Peter 90, 196, 468
Bermuda lillies 328 *See lillies*
Bernard, Ella Mae Baudoin 130
Bernard, Elvine Leonard 130
Bernard, Leonard , Sr. 130
Bernard's Open Kettle Syrup Mill 130.
  See open kettle and la cuite
Bernard, Valcour (Valcourin) John (Jean) 130
Berry (road) 256
Bertaud, Charles (Berto) 161
Berthelot, Gustave 67
Bertin, Jean 416
Bethancourt, Arthur J. 193
Bethancourt, Emily. *See LeCompte, Emily Bethancourt*
Bethancourt, Joseph 302
Bethancourt, Luke 302
Bienville 157
Big Bayou Black 189, 192, 462
"Big Bopper, The" 384. *See Richardson, J.P.*
Billiot, Alexander 336
Billiot, Alexandre 340, 342
Billiot, Celestin 340
Billiot, Charles 280, 291, 352
Billiot Claim 11, 314, 315
Billiot, Etienne 291, 296, 297, 298, 314, 316, 324

Billiot, Genevieve Armelise 342
Billiot, Jacques 297, 340
Billiot, Jean 340, 422, 427, 432
Billiot, Jean Baptist 300, 316
Billiot, Joesph 340
Billiot, Joseph 435
Billiot, Marianne Marie Nerisse Iris 299
Billiot, Maryland 336. *See Naquin, Maryland Billiot*
Billiot, Michel 280, 297
Billiot, Milton 336, 340
Billiot, Pierre 286, 297
Billiot, Roger 340
Billiot, Rosalie Courteau 340
Billiot, Temé [Barthelemy] 340
Biloxi-Chitimacha-Choctaw, Band of 7, 340, 341
Biographical and Historical Memoirs of Louisiana 419
Bisland and Shields 356
Bisland, Caroline Louise Baker 325
Bisland Cemetery 356
Bisland, Connely 471
Bisland family 199
Bisland, Fannie 471
Bisland, Fannie Ashton Brownson 356
Bisland, Frances (Fannie) 199, 325
Bisland, Jack 183, 353
Bisland, John Brownson 356
Bisland, John R. 289
Bisland, John Rucker 325, 352, 356
Bisland, Joshua Baker 326
Bisland Landing 324
Bisland, Leonora Goode 356
Bisland, Margaret Cyrilla Brownson 325, 353
Bisland, Mary Louisa Lavinia Witherspoon 324, 352
Bisland mill 357
Bisland Oaks 353, 354
Bisland, Ralph 410
Bisland & Shields 327
Bisland, Thomas Shields 325, 353, 356
Bisland, T.S., Capt. 371
Bisland & Watson 102
Bisland, William 324, 325, 352
Bisland, William Alexander 325, 352
Bisland, William, Sr. 324, 325
Bivouac plantation 116
Blackburn, John N., Rev. and Mrs. 147
blacksmith 57, 213, 232, 285, 378
blacksmith shop 77, 145
Blackwater plantation 33
Blake, Adam Eugene 115
Blake, Adam Eugene William, Judge 93, 102
Blake, Cecile Adele Thibodaux 93, 102
Blanchard, C.P. 311
Blanchard, Lydia 208
Blanchard, Marguerite Olympe Hotard 315
Blanchard, Teles P. 315
Blanchard, T.P. 311
Bland, William H. 119
B&L Canal Company No. 1 299
B&L Canal Company No. 2 168, 299
B & L canal system
 See Barataria and Lafourche Canal Company
blimps (LTAs)

 Lighter Than Air 279
blimps 242, 244, 245, 253, 256, 257
blimp names 253
blimps explode 253
Boardville 10, 217
boat builder 441, 442
boats and water craft
 • *Alcoa Puritan* 244
 • Banana boat 169
 • Boyne Boat Works 440
 • *Dixie* (steamer) 435
 • *Dr. Batey* (steamer) 135
 • flatboat 14, 300
 • *Delatour* (dredge) 169
 • *Harry* (steamer) 211, 435
 • *Helen Snow* 441
 • *Houma, The* 167
 • *Hunley* (submarine) 225
 • *Ida Handy* 79
 • *John J. Roe* (riverboat) 79
 • *Laura* 435
 • *Munger T. Ball* (tanker) 244
 • *Norlindo* (tanker) 244
 • *Ohio, The* 166
 • *Pioneer* (submarine) 225
 • pirogues 305, 338, 339
 • *Red Rover* 306
 • riverboats 79
 • *Sadie Downman* 435
 • *St. Remi* (ship) 334, 406
 • *S.S. Benjamin Brewster*
 • *Sun* (ship) 244, 245
 • steamboats 14, 77, 79, 168
 • *Talma* (ship) 362
 • *Terrebonne, The* 165
 • U-boats 244, 245, 246
 • *U.S. Delatour* 297
 • *USS Housatonic* 225
 • *USS KIDD* 245, 257, 272
 • *Victoria* (ship) 362
 • V-bottom 442. *See Rhodes, Gustave*
Bocage, Charles William 13
Bon Ami plantation 143
Bonapartiste 362
Bonvillain 192
Bonvillain, A.A. 155
Bonvillain, Aglaé. *See Burguières, Aglaé Bonvillain; See Denis, Aglaé Bonvillain*
Bonvillain, Alcide J., Sen. 193, 284
Bonvillain, Alphonse 192
Bonvillain, Arthur H. 193
Bonvillain, Bannon 186
Bonvillain Brothers 284
Bonvillain, Felix 60, 284, 368
Bonvillain, Felix A. 193, 284
Bonvillain, Jacques 137
Bonvillain, Louise. *See Breaux, Louise Bonvillain*
Bonvillain, Martial 284
Bonvillain, Martial J. 193, 284
Bonvillain, Norbert 137
Bonvillain, Pauline Camilla Burguières 192
Bonvillain's Grocery 213
Bonvillian, Felicite 100. *See Thibodaux, Felicite Bonvillian*
Boquet, Adam 285, 322
Boquet, G. P. 10, 217
Boquet Subdivision Addendum 217
Bordelon, Ivern, Fr. 428
Bordelon, Ivern M., Rev. 428

Bosque, Bartholome 287
Bouchereau 27, 28, 32, 80, 385, 410
Bouchereau, Alcee 27
Bouchereau, Alcée 385
Bouchereau, L. 32
Bouchereau, Louis 27, 410
Boudraux, Joachin 281
 See coopers
Boudraux, Zerphirin 281
 See coopers
Boudreau, Pelegrain 297
Boudreaus 15. *See First Settlers*
Boudreaux, Alfred 285
Boudreaux, Charles 95
Boudreaux, Francois 297
Boudreaux, Isidore 297
Boudreaux, J. Bte. 295
Boudreaux, Joe 272
Boudreaux, Joseph 297
Boudreaux, Joseph Ulger 126
Boudreaux, Lake 447
Boudreaux, Louisiana Aimee Fields 285
Boudreaux, Marie Evelina/Velina 41. *See Comeaux, Marie Evelina Velina Boudreaux; See Roundtree, Marie Evelina/Velina Boudreaux*
Boudreaux, Renaud 297
Boudreaux, Rosalie Malbrough 126
Boudreaux, Theodule 285
Bourg 284, 298, 300, 316, 410, 433, 462
Bourg Agricultural School 11, 310, 311
Bourg, Angele Laurentia. *See Redmond, Angele Laurentia Bourg*
Bourg, Aubin 149, 157, 199, 234, 240, 299, 319
Bourg, Claude J. 7, 308, 332, 333, 472
Bourg, Claude J., Jr. 328
Bourg Elementary School 310, 311
Bourgeois Carriage House 44, 46
Bourgeois, Charles 47
Bourgeois, Charlotte Barras 46
Bourgeois, H. L. 141
Bourgeois, H.L. 430
Bourgeois, Jacques Jacob 47
Bourgeois, Julia Helene Thibodaux 46
Bourgeois, Lester Charles, Jr. 47, 468
Bourgeois, Lester Charles , Sr. 47
Bourgeois, Lester C., Jr. 7
Bourgeois, Lester C., Sr. 7
Bourgeois, Lester, Sr. 47
Bourgeois, L. Vernon, Jr. 149
Bourgeois Meat Market 46, 47
Bourgeois, Michel 47
Bourgeois, Mrs. Ulysse Barras 44
Bourgeois, Nicholas Grandejean Jacques 46
Bourgeois, Paul 47
Bourgeois, Peter G. 149
Bourgeois, Richard J. 7, 468, 472
Bourgeois, Richard Joseph 47, 69
Bourgeois, Rita Ethel Trahan 47
Bourgeois, Thibodaux, Barras house 44, 45
Bourgeois, Ulysse Ursin 46, 47
Bourgeois, Valerie 44
Bourgeois, Valery Jean Baptiste 46
Bourgeois, Valery, Jr. 47
Bourg, Felonise Celeste 395
Bourg, F.X. 149
Bourg, Harry 217

Bourg, Harry J. 217
Bourg High School 310
Bourg, Houman J.C. 234
Bourg, J.C. 234, 240
Bourg, Lloyd 332
Bourg, Lyes J., Sr. 308, 328, 330, 332, 333
Bourg, Pierre 416, 418
Bourg Post Office 298, 322
Bourg sawmill 301
Bourg School Association 311
Bourg School District 311
Bourg State Bank 433
Bourg State Bank & Trust Co. 275
Bourque, Ch. 297
Boutelou plantation 363
Bowie, James 22, 221
Bowie lands 23
Bowie, Rezin 22
Bowie, William Taft, Rev. 48
Bowman, Michael 272
Boykin plantation 186, 192
Boyne, Andrew 441
Boyne, Andrew Eugene 441
Boyne (Boerne), B. 444
Boyne, Edward Bennett 441
Boyne, Elise Marie Belanger 441
Boyne, John Alexander 441
Boyne, John Madison 441
Boyne, Joyce *See Lasseigne, Joyce*
Boyne, Onesia Avet Edward 333
Boyne, Onesia Marie Avet 441
Boyne, Ruby 333. *See Gonsoulin, Ruby Boyne*
Boyne Shipyard 441
Boyne, William Thomas 441
Bradford, James 326
Bragg, Braxton, Gen. CSA 116
Bragg, Eliza Ellis 116
Bragg, Luther F., Jr. 471
brands 188, 192, 199, 200, 233, 327, 413, 471, 474
  Images of brands included for
  Alcide Carlos 416
  Alexander J. Lirette 148
  Alexandre Billiot 340
  Arthur W. Connely 200
  Aubin Benoni Thibodaux 103
  Aubin Bourg 199
  Bannon Goforth Thibodaux 103
  Bannon Goforth Thibodaux, Sr. 94
  Bisland & Shields 327
  Caleb B. Watkins, 154
  Carl Ellender 413
  Corp. Joseph Odressi Ayo 145
  C.P. Smith 188
  Edmund Fanguy 286
  Elmire Marie Thibodaux 103
  Emile Hebert 318
  Ernest and Wallace Ellender 327, 412
  Eugene Carro 315
  Ernest Louis Guidry 300
  Eugene Fields 399
  Eugenie Thibodaux 103
  Euphrosin Hotard 314
  Fannie A. Bisland 356
  Fay Chauvin 413
  Henry Claiborne Thibodaux 132
  Henry Michael Thibodaux 103
  Henry Schuyler Thibodaux 103
  Hyman Ellender 413
  J. Paul Landry, Sr. 137
  Jean Baptiste Robichaux 421
  Jean Pierre Junius Lapeyrouse, Sr. 438
  Joachim Gueno 282
  John A. Boyne 441
  John Calvin Potts 126
  Joseph A. Gagné, Sr. 202
  Joseph Ellender 327, 413
  Joseph Narcisse Robichaux 421
  Joseph Odressi Ayo 145
  Jules Martial Burguières 192
  Leandre Bannon Thibodaux 103
  Leufroy Barras 115
  Paul Ellender 413
  Philippe Sargeant 126
  Pierre Chaisson 327
  Ray Anthony Guidry 301
  Robert Ruffin Barrow, Jr. 233
  Robert Ruffin Barrow, Sr. 233
  Silvani Numa Aucoin 318
  Soloman C. Lawless 145
  Thomas Elinger 413
  Thomas Ellender (Elinger) 326, 409
  Thomas Ellender Jr. 327
  Thomas Ellender, Jr. 413
  Wallace R. Ellender 413
  Wilson E. Ellender 413
  Van Perkins Winder 69
Brantley, Robert S. 7, 42, 478
Brashear (Morgan City) to Lafayette Railroad 71
Brashear, Walter, Dr. 164
Brasher, Morgan 321
Brasseaux, Carl A. 1, 3, 13, 472
Breaux, Basile 285
Breaux, Becky L. 7
Breaux, Louise Bonvillain 186
Breaux, Marie Madeleine 418. *See Robichaux, Marie Madeleine Breaux*
Breaux-Morrison Addition 186
Briars, The, plantation 39
Bridge, Huey P. Long 260
Brigitte Emelie Thibodaux Barras 105
briquette-entre-poteaux 126
Broussard, Alexandre 362
Broussard, Alice 190
Broussard, Ida Laperle 193, 362. *See Burguières, Ida Laperle Broussard*
Broussard, Joseph (*Beausoleil*) 193, 362
Brown, Eugene R., Rev. 50
Browning, Laura A. 7, 159, 468, 369, 469, 471 - 474
Brown, Richard Paul "Dickie" 7
Brown & Root 415
Brownson, Fannie Ashton. *See Bisland, Fannie Ashton Brownson*
Brownson, Frances (Fannie) 325, 352. *See Bisland, Frances (Fannie)*
Brownson, Margaret Cyrilla 325, 353. *See Bisland, Margaret Cyrilla Brownson*
Brownson, Marianne Musgrave 353
Brunet, Marie Francisca 314
Brunet, Pierre 295
Brush Island 364, 398, 405
Budd, John B. 149
Buddy Holly 384
Buford, C.H. 135
Bull Run plantation 68
Buquet, Joseph 150, 154
Burguières, Marguerite. *See Theriot, Marguerite Burguières*
Burguières Crest 190
Burguières, Ernest Denis 192
Burguières, Ernestine Louisiana 192. *See Viguerie, Ernestine Louisiana Burguières*
Burguières, Eugène 192
Burguières, Eugène Denis 189, 362
Burguières, Ida Laperle Broussard 193
Burguières, J.M. 190
Burguières, J.M., Co., Ltd. 193
Burguières, Jules Martial 188, 192, 193, 362
Burguières, Marie Corinne Patout 192, 193
Burguières, Marie Marianne Verret Delaporte 189, 192
Burguières, Pauline Camilla 192. *See Bonvillain, Pauline Camilla Burguières*
Burguières plantation 188
Burguières, Smith 10, 187, 189, 191, 193, 195
Burke, M. Gene 468, 471, 473
Burke, Michael Gene 7, 32, 145
Buron, Marguerite 119
Buron, Victor 132, 364, 371
Bush and Maginnis refinery 368
Bush Canal 11, 80, 442
Bush, Harry Garland 367, 373
Bush, Harry G., house 368, 383, 384
Bush & Maginnis Well 378
Bush, Reuben 149
Bush's Canal 80
Butler, Anna. *See Ellis, Anna Butler (tombstone)*
Butler, Ann Madeline Ellis 57
Butler, Biby 135
Butler, Biby (child) 135
Butler, Gabe 135
Butler, Henry 135
Butler, Joseph T., Jr. 272
Butler, Martha Anne 135
Butler, Mary 135
Butler, Nace 135
Butler, Thomas 23, 57
Butler, Tom (slave) 135
Butler, W.E. 372, 385, 468, 472, 473
Byrd, Matthew, Rev. 48

C

Cage, A.G. 27
Cage, Albert G. 149
Cage, Duncan S. 275, 278, 289
Cage, James 27, 33, 77, 102
Cage, Kathryn Nolan 326
Cage, Thomas 275
Cage, Thomas A. 149
Cahn, Leon 322
Caillier, Mrs. Albert 137
Caillou, Bayou Little 447
Caillouet, L. Philip 7, 159, 469, 474
Caillouette, Charles 411
Caillouette, Walterine Ellender 411
Caillou Field 364, 408, 409
Caillou Grove 225, 228, 287

Caillou Island 101, 245, 364.
    See Islands
Cajuns, first 425
Calcasieu Parish 364, 442
Calder, David 371
Calder, David E. 367
Calder, David R. 356
Calder, D.R. 367
Caldwell Brothers 102
Caldwell Middle School 108
California 246
Callahan, Francis Deoma 7
Callais, Erin 7
Calumet 326, 356
Calumet Island 101.
    See Islands
Cambon and Champagne 281
Cambon Brothers 282, 322
Cambon, Ferdinand 322
Cambon, Maurice 322
Cambon, Sylvester 322
Campanella, Richard 31
Campbell, Frank 275
Camp Bisland 326, 356
Camp Livingston 260, 264
Camp Livingston POW script 269
Campo, John Mary (Jean Marie) 116, 123
Camp Plauché 260
Camp Polk 260
Camp Ruston 260
Canada 297
Canal Belanger 11, 295, 296, 297, 298,
    299, 302, 311, 315, 415
Canal Belanger Fair 298
Canal Belanger Society 295
Canal Bellanger 286
Canal, Bush 442, 447, 450
Canal, Madison 442
Cane Brake plantation 286
cane hoist 121, 373
cane train 67
Cantrelle, A.P. 275
Cantrelle, Jacques 189
Captain Hicks 154
Caracristi, Prof. 421
carbide acetylene gas plant 62-63
Caribbean Islands 22
Carl F. Dahlberg & Co. 283, 405
Carl F. Dahlberg & Company 395, 399
Carl F. Dahlbert & Co. 401
Carlos, Alcide 416
Carlos, Alcide Ferjus 416
Carlos, Ambrose 416
Carlos, Argene 416
Carlos, Aurelia Letitia Gautreaux 416
Carlos, Charles 416
Carlos, Clement 416
Carlos, Clement Charles 416
Carlos (family at Myrtle Grove) 278
Carlos, Joseph Maurice 416
Carlos, Louisa 416
Carlos, Margueret Naise 416
Carlos, Marguerite Anaise Landry 416
Carlos, Marie Louise Molinere 416
Carlos, Marius 416
Carlos Oak 416
Carlos, Virginia 416
Carlos, Wycliffe 416
Carnton plantation 61
Carolina Biological Supply Company 7, 82
Caro, Linda 472

Caroline Winder. See McGavock,
    Caroline Winder
Carondelet, Baron de,
    La. Governor 100
Carothers 225
Carothers, James L. 296, 298
carpenters 18, 26, 378, 436
    See Jerome Guidry 300
Carradine, Mary Semple 288
Carrere, Dominique J. 407
Carrere, Dominique Jules 407
Carrere, James R., Sr. 7, 474
Carret a Manzanillo 260
Carro, Eugene 315
Carro, Inezida Hotard 315
Carroll, Scott LaPée 7
Cartier, Jacques 100
Casse-Tete Island 101. See Islands
Catfish Lake 342
Catholic chapel 78
Catholic missionary priests 390
Cavalier, Dean 7
Caya, Arthur 257
Cazeau, Pierre 281, 295, 324
C. Cenac & Co. 275
Cedar Grove 225, 286
Celeste, Marie Agnes 297. See Belanger,
    Marie Agnes Celeste
Celotex Company 379, 383
Celotex Corporation 370, 379
Cemetery, Halfway 137
Cenac, A.B., Jr. 7
Cenac, Arlen B., Sr. 7, 472
Cenac, Brent J. 474
Cenac, Cheryl.
    See Cronan, Cheryl Cenac
Cenac, Chris 462
Cenac, Christopher E., Jr. 474
Cenac, Christopher E., Sr. 464, 465, 468,
    469, 471, 472, 473, 474
Cenac, Christopher Everette, Sr. 4, 9, 13,
    28, 72, 169, 287, 468
Cenac, Cindy T. 7
Cenac, Eva Dover. See Keene,
    Eva Dover Cenac
Cenac family 435
Cenac, Jacqueline G. 473
Cenac, Jacqueline Guidry 7, 472
Cenac, Jean-Pierre 28, 181, 287, 301, 348, 351
Cenac, Marie Clothilde Klingman 393
Cenac oaks 181
Cenac, Ovide J. "Jock" 471
Cenac, Paul Michel 393
Cenac, Philip L., Sr., Dr. 196
Cenac, Victorine 301
Cenac, Victorine Aimée Fanguy 287
Cenac, Victorine-Aimée Fanguy 287
Cenac, William Jean-Pierre, Sr. 171
Centenary Methodist College 202
Center for Louisiana Studies 272
Central store 68
Cevin, Valentin 297
Chabert, Viona Lapeyrouse 278
Chacahoula 14, 192, 225
Chaisson, Pierre 322, 324, 327, 352
Chambers, Henry E. 369
Champagne, Albert 323
Champagne, Albert P. 320, 322, 323
Champagne, Albert Philip 320, 322
Champagne, Bruce Philip, Sr. 323
Champagne, Charles Dalferes 320

Champagne, Charles J. 378
Champagne, Charles Joseph 322
Champagne, Charles Kevin 7
Champagne, Cloyd 323
Champagne, Eve Cecile Guidry 320
Champagne, Laura. See McComiskey,
    Laura Champagne
Champagne, Leonise Elodie. See Hotard,
    Leonise Elodie Champagne
Champagne, Marie Evelina Hotard 320,
    322
Champagne, Mary B. 7
Champagne, Odelia Marie (Lucy) 300.
    See Guidry, Odelia Marie (Lucy)
    Champagne
Champagnes house 321
Champagne, Vaniola Duplantis 320
Champagne, Venus Vivian Eudy 323
Champagne, Walton 305
Champagne, Walton E. 305
Champagne, Walton Ernest 306
Champomier, P.A. 27, 28, 32, 40, 57, 77,
    102, 147, 200
Chapel of Saint Sophie 303
Charity Hospital 157, 191
Charpentier, Jean 297
Chatagnier, Lawrence C. 7, 468
Chauvin 17, 268, 413, 428, 447. See First
    Settlers; See Shuvin
Chauvin, Charles David 90
Chauvin, Charles J. 279
Chauvin, Charles L. 275
Chauvin, Elmire 395
Chauvin, Elodie. See Viguerie, Elodie
    Chauvin
Chauvin, Emily. See Ellender, Emily
    Chauvin
Chauvin, Fay 412
Chauvin, Grace Aloysia 300, 301.
    See Guidry, Grace Aloysia Chauvin
Chauvin, Jessie. See Ellender, Jessie
    Chauvin
Chauvin, J. Farquhard 173, 378
Chauvin, Josephine Ellender 412
Chauvin, Leonard J. 7, 32
Chauvin, Lewis 193
Chauvin, Mary Elizabeth "Bettie".
    See Wurzlow,
    Mary Elizabeth "Bettie" Chauvin
Chauvin, Merril Elaine Tucker 90
Cheramie, Angela 69
Cheramie, Angela M. 7, 468, 472
Cheramie, Brian 7, 472
Cherry, Rachel E. 7, 471
Chicago World's Fair 190
Chinese 441
cholera 228, 236
Christ, Charles J. 7, 246, 257, 471
Christ Church Cathedral 221
Christian, Ebed 235
Christian, Emanuel Banks 235
Christian, Marcus Bruce 235, 240
cistern 77, 94, 222
Citizens Bank of Louisiana 143, 157, 234,
    240, 288, 323
City Hotel 158, 159
City Map of Houma 1960 239
Civilian Air Patrol 244
Civil War 18, 26, 27, 40, 50, 62, 64, 71, 78,
    102, 105, 114, 123, 193, 198, 199, 202,
    225, 228, 236, 240, 282, 300, 317, 325,

INDEX | 473

326, 353, 356, 357, 363, 395, 405, 410, 432, 435, 436, 462
Claiborne, William C.C., Gov. 154
Clara (train) 378
Clark, J.J., Rear Admr. 256
Clark Place subdivision 186
Clemens, Samuel Longhorn 79.
 See Twain, Mark
Clemmon, Frank 126
Clerk of Court's Office 32
Cleveland Demolition Company 257
Cleveland, Grover, President 157
Club, Terrebonne Rod & Gun 444
Coats, Sherri 233
Cobb, Thomas Blum 471, 472, 474
Cocodrie 19
Cole Addition 238, 239
Cole, Christian Grenes 237, 279
Cole, Christian Grenes, Mrs. 240
Cole, Hallette Mary Barrow 237, 240, 275, 279
Cole Subdivision 11, 240
College of William and Mary 143
Colley Switch 373
Collins, Gerald 268
Collins plantation 282, 409
Colonial Revival 17, 187.
 See Architectural Styles
Comeau, Sophie Marie 297, 357.
 See Belanger, Sophie Marie Comeau;
 See Belanger, Sophie Marie Comeau
Comeaux, Alfred 41
Comeaux, Marie Evelina/Velina Boudreaux 41
Committee of Arrangement 311
communities
 Acadian Coast 189
 Alley, The 186
 Back of Town 182
 Bateyville 128
 Bayou Blue 424
 Bourg 410
 Canal Belanger 298
 Chacahoula 192
 Cocodrie 19
 Côte de Malbrough 211
 Crochetville 387
 Daigleville 210
 Daspit 19
 Deweyville 183
 Donaldsonville 31, 408, 418
 Dulac 19, 424
 East Houma 230
 Forest Grove 192
 Gibson 26, 424
 Grand Caillou 27
 Grand Coteau 420
 Gray 128, 422
 Hallelujah 19
 Ile a Jean Charles 427
 Klondyke 170
 Lafourche Crossing 300
 Levy Town 19
 Little Caillou 157
 Maringouin 135
 Mechanicsville / Mechanicville 229, 234
 Montegut 158, 416, 424
 Newport 298
 New Ridge 48
 Newtown 183
 Old Ridge 48
 Peterville 19
 Plattenville 287, 395, 406, 418
 Pointe-au-Barré 424, 427
 Pointe-aux-Chênes 19, 424
 Presqu'ile 245
 Plaquemine 410
 Sandersville 387
 Schriever 36
 Silver City 186
 Smithland 188
 Thibodaux 406
 Tigerville 26
 Williamsburg 175
Company Canal 11, 164, 275, 295, 297, 299, 300, 303, 304, 432, 436
Compting, Sarah 408. See Elinger, Sarah Compting
Concannon, Jasper 405
Concord 19, 23
Concordia Parish 60
Confederate Army 199
Confederate Congressman 86
Confederate forces 326, 356
Confederate States Army 300
Confederate States of America 116, 363, 369
Congregation of St. Joseph 102. See St. Bridget Catholic Church
Connell, William 326
Connely, Arthur W. 200
Connely, Arthur Warren 149, 183, 196, 198, 199, 200, 217
Connely, Clara Himel 198, 199.
 See Himel, Clara
Connely, Clerville Himel 199
Connely, Edmund McCollam 199
Connely, Flora Bowdoin 199
Connely, Frances Bisland. See Bisland, Frances (Fannie)
Connely, Georgia M. 201
Connely, Georgia Mallard 199
Connely, Gilmore 199, 383
Connely, Katherine Easton 199
Connely, Lavinia 201
Connely, Lavinia May 199
Connely, Lucy 198
Connely, Lucy Ella 199
Connely Row 217
Connely, Ruth 199
Connely's Row Subdivision 217
Connely Subdivision 217
Connely, William Alexander 199
Consalle, Joseph 364
Consalle, Marie Viguerie 364
Cook Family Corp. 357, 369, 395, 401, 405
Cook Family Corporation 383, 399, 402
Cook, Herman Albert 170
coopers 18, 26, 80, 281, 378
 See Joachin and Zerphirin Boudraux
cooperage 77
Cope, Gayle B. 7, 72, 469
Cora's Restaurant 194
Corbin, Charles 90
Corday, Sandra 272
Cormier, A.J. 311
Cortez, Huey 112
Cortez, Martin 345
Cortez, Ruth Naquin 112
Côte de Malbrough 36, 211.
 See Malbrough Coast

cotton 17, 23, 60, 61, 137, 264, 292
Coulombe, Joseph M., Rev. 431
Council, Terrebonne Parish 447
Courteau, Houma 427
Courteau, Rosalie 427. See Billiot, Rosalie Courteau
Courteau, Tacalobe 427
Courteaux, Ferris 426
Courteaux, Rosalie 300
Courthouse Square 181
Coxen, Augustus 105
Cragin, George D. 78, 79
Cragin, Samuel 78, 79, 90
Crawford plantation 192
Creoles 17, 69, 157, 193
Crescent-Magnolia & Mfg. Co, Ltd. 114
Crescent plantation 33, 152, 186, 192, 225, 284
Crochet, Elise 126. See Sonier, Elise Crochet
Crochetville 387
Crochet, Walter "Bud" 388
Crockett, Carlos B. 7
Cronan, Cheryl Cenac 369
Cronan, Tim 369
Crosier, Bibolet 90
Crosier, Oscar 90
Crowell, Richard B. 31
Cuba 10, 139, 260
Cuba (country of) 114
Cuban sugar 22
Cunningham, Celestin 193
Cunningham, Jean Louis 435
Cunningham, Robert Clement 435
Cypremort plantation 192
cypress, heart 126

D

Dahlberg, Bror G. 379
Dahlberg, Carl 378, 379, 383
Dahlberg, Carl F. & Co. 368, 378
Daigle, Antoniata Cecilia Porche 211, 435
Daigle Barge Line 210, 217, 435
Daigle, Elphege 146, 147
Daigle, Emile 372
Daigle, Emile A. 210, 211, 229, 353, 364, 372, 435
Daigle, Emile Auguste 210, 435
Daigle, Ferdinand 137
Daigle, Marie Molly Josephine Pitre 146, 147
Daigle, Mrs. Oscar 44

Daigleville 10, 210, 211, 213, 215, 216, 217, 229, 436
Daigleville Baptist Church 213, 216
Daigleville School 211
dairies 19, 200, 232 - 235, 295, 303- 304
 Bellview Dairy 234
 Matherne Dairy 295, 303
 Residence Dairy 232
 milk shed 62
Danks, Catherine P. Gwin 284
Danks, John W., Dr. 284
Danziger, Alfred P. 81
D'Arbonnes 17. See First Settlers
Darce, Joseph P. 297
Darce, P.H. 149
Darcey, Adele Ayo 143

Dardar, Albert 339
Dardar, Augustine 339
Dardar, Harry 336, 339
Dardar, Helen Verdin 337
Dardar, Henry 335
Dardar, Loney 336
Dardar, Marie Eve Naquin 337
Dardar, Marion 337
Dardar, Michael 284, 288, 291
Dardar, Rita Antoinette. *See Falgout, Rita Antoinette Dardar*
Dardar, Theodore 337
Dardeau, Alexandre André 395, 397
Dardeau, Clovis 435
Dardeau, Marie Augustine Tasset 397. *See Fields, Marie Augustine Tasset Dardeau*
Darsey, Peggy G. 7
Daspit, Agnes 196. *See Kennedy, Agnes Daspit*
Daspit bayou 19
Daspit-Breaux Addition 186
Daspit, Edna 278
Daspit, Evelida 279. *See Duplantis, Evelida Daspit*
Daspit, Isaac 183
Daspit, Noah J. 183
Daspit, Oscar 10, 149, 183
Daspit, Salvador 279
Daunis, Thomas J. 116
Davidson, Gladys. *See St. Martin, Gladys Davidson*
Davis, Bernard B. 246
Davis, Bernard Bazet 178
Davis, Dan H. 7
Davis, Daniel H. 468, 471
Davis, Donald W. 1, 3, 13, 468, 472, 473, 474
Davis, Eliza. *See Semple, Eliza Davis*
Davis, Jody A. 7, 468
Davis, R.V. 278
Davis, Sarah E. 282
Dawson, Harris P. 237
Dawson, Jennie Tennent Barrow 237
De Agesta, Don Philipe 287
de Bazet, "Lafayette" Bernard Filhucan 181
De Boré, Etienne 17, 22
DeClouet Regiment 297
Defelice, Eva 278
Defelice, Kiss 278
Defelice, Vincent 278
DeFraites, Arthur A. 32
DeFraites, Arthur A., Jr. 7
de Galvez, Don Bernardo, General 363
Delaporte, A.J. 202
Delaporte, Alfred 362
Delaporte, Joseph 363
Delaporte, Marianne Verret 363
Delaporte, Marie Elvire 362, 363. *See Viguerie, Marie Elvire Delaporte*
Delaporte, Marie Marianne Verret 189
De la Ronde, Adelaide Adèle 53
De la Ronde, Marie Elizabeth Eulalie Guerbois 53
De la Ronde, Pierre Denis 53, 54
De la Ronde, Pierre Denis, Maj. Gen. 53, 56, 61
DelClaire plantation 320, 321
De Lucinge, Pierre Adolphe Ducros 53

Dénécé, Jean Marie Joseph 406
Dénécé, Jean Marie Joseph, Father 390
Dénécé, Jean Marie Joseph, Fr. 302
Denis, Aglaé Bonvillain 192
Denis, Alice. *See Dupont, Alice Denis*
Denis, Ernest 192
Derbigny, Charles, Judge 164
Deroche, Albert 348, 350
Deroche Brothers 348
Deroche Brothers Syrup Mill 348, 348–351, 350
Deroche, Colbert 348
Deroche, Joyce 333
Deroche, Mr. 333
Deroche, Odelia Marie 357
Deroche, Robert 350
Deroche, Telly Mae 326. *See Ellender, Telly Mae Deroche*
Deroche, Walter 348
Deroche, William 348, 350
DeSantis, John 135
de Vargas, Don Jose 287
De Villiers, Charles Jumonville 57
Devil's Swamp Path 50
Devin, Lavinia 416
Deweyville 183
Dianne, Joseph 324
Dickson, Charlean St. Martin 7, 471
Diefenthal, Edward L. 7
Diefenthal, Nancy C. 7
Dillard University 235
Dillsworth, John S. 78
Dillsworth, William P. 78
Dinsmore, James 23, 61
Diocese of Houma-Thibodaux 7
Dixie plantation 121, 130
Dixie Shipyard 211
Doescher, Montella 7
Dog Lake 245
Domangue, Armand J., Jr. 321
Domangue, Della Roberts 321
Domangue, Dolly 7, 279. *See Duplantis, Dolly Domangue*
Domangue, Edward 298
Donaldsonville 31, 68, 101, 135, 154, 190, 291, 408, 418, 442
Doré, Emile 378
Doucet, John P. 31, 257, 357
Douglas, S.A. 64
Douglas, Veranese E. 7, 469, 471
Douglas, Veranese Evans 241
Downing, D.D. 137, 139
Downing, William V. D. 137
Downs Hall 202, 208
Downs, J.W. 208
Dr. Batey (dredge) 135
Druealing, R. J. 249
Dubreuil, Claude Joseph 22
Dubuys, Pierre 286
duck decoy carvings 304
Ducros 53, 56, 57, 58, 61, 62, 63, 66, 67, 68, 69, 318
Ducros house 62
Ducros, Joseph Marcel 56
Ducros, Pierre 62
Ducros, Pierre Adolph 56, 61
Ducros plantation 53, 58, 77, 81, 90, 116, 123, 318
Ducros plantation bell 60
Dugas Cemetery 406, 407, 416, 419, 420, 427, 432

Dugas, Jean-Baptiste Placide 407. (Hermit of Terrebonne)
Dugas, Jean Batiste 395
Dugas, Jean Pierre 396, 404, 406, 408, 409
Dugas, Mary Ellen Walker Feyerabend 285
Dugas, Renée/Reynalda Naquin 396, 406
Dugas, Theodiste Ludivine 395, 396. *See Fields, Theodiste Ludivine Dugas*
Dug Road 230
Dulac 19, 22, 23, 33, 77, 208, 424, 447
Dularge plantation 15, 199, 225
Duma, Luis 274
dummy locomotives 378
Duncan, Stephen, Dr. 352
Dunn, J.B. 202
Dunn, Sarah Anne 202
Dupart, Francoise 157
Duplanti, Bte. 295
Duplantis, Ada Daigle 150
Duplantis, Adley J. 150
Duplantis, Agapi 314
Duplantis, Annette. *See Stoufflet, Annette Duplantis*
Duplantis, C.C. 155
Duplantis, Dolly Domangue 7, 279, 472
Duplantis, Evelia 326. *See Ellender, Evelia Duplantis*
Duplantis, Evelida Daspit 279
Duplantis, Filican Alexis (*Tecon* or *Tican*) 150, 153, 154, 348, 351 *See Mr. Carnival*
Duplantis, Gayle 303
Duplantis Grocery Store 212
Duplantis, Henry Clay 275, 279
Duplantis, Iola 303
Duplantis, Jean Baptiste 314
Duplantis, Jean Felix 150
Duplantis, Omer 275
Duplantis, Ronnie 149
Duplantis, Trasimond 367
Duplantis, Vaniola. *See Champagne, Vaniola Duplantis*
Duplantis, Wallace, Mrs.
Leo Lirette Family 422
Dupont, Albert M. 372
Dupont, Alice Denis 193
Dupont, Alphonse 193
Dupont, Claire Marie Pauline 187, 193. *See Smith, Claire Marie Pauline Dupont (Aunt Clara)*
Dupont, Ernest 191
Dupont, Ernest A. 149
Dupont, J. Cyrille, Rep., Mayor 193
Dupont, Jean Marie 193, 364, 372, 395, 397
Dupont, Joseph Cyrille (J.C.) 193
Dupont, Jules S., Sr., Dr. 191
Dupont, Lydie Marie (Astugue) Tasset 397
Dupont, Yvette St. Martin 191
Dupre, Alcee 336. *See Naquin, Alcee Dupre*
Dupre, Alexandre 154
Dupre, Charles 297
Dupre, Gerome 295
Dupre, Grant J. 7, 32, 472, 473
Dupre, Isadore 432
Dupre, J.C. 378

Dupre, Jean 361
Dupre, Jean, Jr. 297
Dupre, Jean, Sr. 297
Dupre, J.M. 410
Dupre, Wiley J. 340
Dupri, Lucius J. 86.
Durand Hotel 158
Duthu, Isidore 362
Duthu, Virginia 362
duty-free sugar and rice 123
Duval, Berwick, Dr. 201

E

Easter Lilies 328–333
East Houma 230
Eastlake 17, 230. *See Architectural Styles*
Eastonia plantation 319
Easton, Tris 319
Easton, Tristram Shandy, Jr. 319
East St. Bridget Plantation 93
East Street School 234
Edward, Onesia Avet 333. *See Boyne, Onesia Avet Edward*
Edwards, Jonathan 257
E. Gajan, Inc. 81
Eichhorn, Mary Lou 7, 72, 319
Elinger 326. *See Ellender*
Elinger, Catherine Roddy 326, 408
Elinger, David 326
Elinger, Elizabeth 409
Elinger, Ernest 326
Elinger, George 326
Elinger, Henry 409
Elinger, James 326
Elinger, Joseph 326, 408, 409
Elinger, Sarah Compting 408
Elinger, Thomas 409, 412
Elinger, Thomas Jr. 326
Elinger, Thomas Sr. 326
Elinger, Wallace 326
Elinger, William 326
Elizabeth Place 10, 217
Eliza plantation 401, 412
Elks organization 368
Ellendale plantation 77, 199, 462
Ellender Act of 1937 323
Ellender, Albert J. 474
Ellender, Albert P. 7, 472
Ellender, Allen J. 474
Ellender, Allen J., Sen. 169, 173, 315, 323, 410, 411, 412, 413, 468
Ellender Brothers 357, 410
Ellender, Carl 412
Ellender, Catherine Roddy 410
Ellender, Clara Crossman 268, 270
Ellender, Claude J. 411
Ellender, Claude J., Mrs. 240
Ellender Collection 272
Ellender, Coralie 7
Ellender, Coralie Hebert 468, 472, 473, 474
Ellender, David 326, 409, 410, 412
Ellender, David (Tabin) 305, 307
Ellender, Dennis 332
Ellender, Dr. Willard A. 411, 412
Ellender, Elda Jarveaux 326
Ellender (Elinger), Thomas 364, 408, 409
Ellender, Elizabeth 326, 410. *See Savoie, Elizabeth Ellender*

Ellender, Ellen Mary 268, 285. *See Walker, Ellen Mary Ellender*
Ellender, Emily Chauvin 326
Ellender, Ernest 409, 410, 412
Ellender, Ernest and Wallace 412
Ellender, Ernest Filmore 378
Ellender, Evelia Duplantis 326
Ellender, George 326, 409, 410
Ellender, Henry 326, 410
Ellender, Henry Jefferson 411
Ellender, Henry Thomas, Sr. 411
Ellender, Hyman 412
Ellender, James 326, 409, 410
Ellender, Jessie Chauvin 411
Ellender, Joseph 323, 326, 327, 357, 410, 412
Ellender, Josephine 412. *See Chauvin, Josephine Ellender*
Ellender, Jr., Thomas 412
Ellender, Jr., Wallace R. 411
Ellender, Lloyd 412
Ellender, Mary Torr Lynch 326
Ellender Memorial Library 7
Ellender, Paul 412
Ellenders 170
Ellender, Allen J., Sr. 411, 412
Ellender, Steven Ernest, Sr., Dr. 357
Ellender, Telly Mae Deroche 326
Ellender, Thelma 240
Ellender, Thomas 409
Ellender, Thomas 409
Ellender, Thomas Anthony 323
Ellender, Thomas (Elinger) 410
Ellender, Thomas J. 103
Ellender, Thomas, Jr. 326, 327, 410
Ellender, Thomas Sr. 326, 327, 364, 413
Ellender, Timothy C., Sr., Judge 7
Ellender, Victoria Jarveaux 326, 327, 411
Ellender, Wallace 409, 410
Ellender, Wallace J., Jr. 7, 323
Ellender, Wallace Joseph, II 323
Ellender, Wallace Richard 411, 412
Ellender, Wallace R., III 468, 472, 473, 474
Ellender, Wallace R., Jr. 268, 270
Ellender, Wallace R., Sr. 327
Ellender, Walterine 411. *See Caillouette, Walterine Ellender*
Ellender, Willard A., Dr. 357
Ellender, William 326, 409, 410
Ellender, Wilson E. 412
Ellerslie 56
Ellis, Anna Butler (tombstone) 116
Ellis, Ann Madeline 57. *See Butler, Ann Madeline Ellis*
Ellis Causeway 57
Ellis, Eliza 116. *See Bragg, Eliza Ellis; See Bragg, Eliza Ellis*
Ellis, Margaret (tombstone) 116
Ellis, Mary Jane Towson 116, 119
Ellis, Odette 91. *See Moore, Odette Ellis*
Ellis, R.G. 40, 57
Ellis, Richard 23
Ellis, Richard G. 123
Ellis, Richard Gaillard 116
Ellis, Richard Gaillard (Gaillain) 56, 57, 116
Ellzey, Bill 327
Eloise plantation 301, 432
End of the Road 449–450
Engeron, Theo 272
Engeron, Theogene 217
Engeron, Theogene J. 217
Engeron, Theogene,

Subdivision 217
Erin plantation 90
erosion 447
erosion, Sea Breeze 460
estates (*vacheries*) 30
Eudy, Venus Vivian. *See Champagne, Venus Vivian Eudy*
Eureka 19
Evans (carpenter) 64
Evergreen Junior High School 143
Evergreen plantation 116-121, 143, 363
Evergreen plantation kitchen 117
Evergreen plantation tenant houses 120 *See tenant houses*
Evergreen School 143
Evergreen's sugar house 146
Executive Order 10501 272

F

Fabre, Curt 305
Fabre, Eula Lebouef 419
Fairfax plantation 325, 326, 353, 356
Faisans, Paul 364
Fala-Lake Felicity map, USUG 1939 343
Fala Village 342, 345
Falgout, Heloise 287. *See Fanguy, Heloise Falgout*
Falgout, Loris 305
Falgout, Lucien (Yak) 305
Falgout, Rita Antoinette Dardar 337
Falgout, Sarahlene 304. *See Whipple, Sarahlene Falgout*
Falterman, Dennis A. 144
Falterman, Dennis Adam, Sr. 144
family, Rhodes 447
Fangui, Antoine 287
Fangui, Catherine Agoutte (Aoushe) 287
Fangui, Jean 287
Fangui, Marie Magdeleine Laporte 287
Fangui, Pierre 287
Fangui, Vincente 287
Fanguy, Charles 287
Fanguy, Clodamire 287
Fanguy, Edmund 11, 286, 287
Fanguy, Heloise Falgout 287
Fanguy, Victorine Aimée. *See Cenac, Victorine Aimée Fanguy*
Fanguy, Vincent 287
Farber, Glyn V. 7
farmers 442
Farm Security Administration of the United States 67, 81, 90, 104, 108, 114
Father of Terrebonne 23, 100. *See Thibodaux, Henry Schuyler*
faux bois 40, 42, 44
Favrot, Charles A. 140
Favrot, Charles Allen 311, 393
Favrot, Charles Allen & Louis A. Livaudais Ltd. 393
Fazende, Paul 371
Federal Writers' Project 235
Feitel, Augusta 68. *See Polmer, Augusta Feitel*
Feitel, Stella. *See Polmer, Stella Feitel*
Feliciana Parish 53
Fernandez, Wayne M. 7
Feyerabend, Angela 268
Feyerabend, Mary Ellen Walker 285
Fields, Ben, Rev. 48

Fields, Celeste Fidelise Pitre 397
Fields estate 372
Fields, Eugene 159, 371, 387, 395, 399
Fields, Eugene Pierre Dalmace 395, 397
Fields, Eugene Pierre Dalmace, Sr. 394
Fields, Henry J. 407
Fields, John H. 149, 202
Fields, Louisiana Aimee 285.
    See Boudreaux, Louisiana
    Aimee Fields
Fields, Marie Augustine Tasset Dardeau 394
Fields, M. Neuville 364
Fields, Neuville 371
Fields, Neuville Demetrius 395
Fields, Neuville Demetrius Henry 395
Fields, Rev. Elie, S.J. 395
Fields School 390
Fields, Theodiste Ludivine Dugas 396
Fields, William 395, 396
Field, William 295
fire 208
First Presbyterian Church 42, 228, 229
First Settlers
  Belanger 17
  Bergerons 17
  Boudreaus 15
  Chauvin 17
  Curtis Rockwood 17
  D'Arbonnes 17
  James B. Grinage 17
  LeBoeufs 17
  Marlboroughs 17
  Marsh, Royal 15
  Prevost 17
  R.H. Grinage 17
  Trahans 17
Fischman, Debra J. 7, 69, 468
Fischman, Diana Lynn 69
Fischman, Jacob L., Dr. 69
Fischman, Kathy Rose 69
Fischman, Nathan Harvey, Dr. 69
flood 26, 67, 173, 190, 340
flood of 1890 190
floodgate, Bush Canal 447
Flora plantation 189, 225
Flores, Michael 321
Florida 244, 245, 246, 260, 383
Floyd, Stella R. Matherne 304
Floyd, William Barrow 233
Flynn, Peter P. 364
Folse, Marcellite. See Hotard, Marcellite
    Folse
Folse, Marie Marcellite 314. See Hotard,
    Marie Marcellite Folse
Fonseca, Peter W. 471
Foolkes, Edmond J. 222
Forest Grove 192, 225
Fort Bisland 325, 326, 356
Fortier, Alcee 15, 19, 42, 468
Fortmayer, Claire. See Roberts, Claire
    Fortmayer
Fort Quitman 363
Foster, Lucy Dean 41. See Armitage,
    Lucy Dean Foster
Four Deroche Brothers Pure Cane Syrup 349
Four Isle 245
Franck, Charles L. 165
Franklin POW camp 260
Freedmen's Bureau 18, 102, 326

Free, Gregory 478
free people of color 286
    See
    Marianne Marie Nerisse
    Iris 288, 316
    Marie Jeanne 286
    Pierre Billiot 286
French families 15
French Opera 366
French Revolution 362
French West Indies 22
Frontlawn plantation 11, 284, 285
Front Lawn plantation 225, 268, 284, 285
Fryou, Adrianne. See Richardson,
    Adrianne Fryou
Fryou, Joseph 268
Fryou, Richard 384
Fugatt, Jean J. 7
Funerals in lower Terrebonne 431
Fuqua, Phebe Olds. See Pierce, Phebe
    Olds Fuqua Woods; See Woods, Phebe
    Olds Fuqua
fur trapping 438

G

Gage, D.S. 27
Gagné, Joseph A. 202, 203, 205, 207, 209, 275
Gagné, Joseph A., House 10, 202, 203, 205, 207, 209
Gagne, Joseph Auguste 149
Gagné, Joseph Auguste, Sr. 202
Gagné, Sarah Anne Dunn 202
Gaidry, Adolph 216
Gaidry, Agelee Cadiere 216
Gaidry, Arnold 216
Gaidry, Clara Katherine Slatter 232
Gaidry, Edward J., Sr., Judge 7
Gaidry, Harold Langdon 232, 233
Gaidry, Laura 229
Gaidry, Laura V. 216
Gaidry, Lillie Lea McKnight 232
Gaidry, Lowell 216
Gaidry, Mrs. W.J. 217
Gaidry, Paul E. 216
Gaidry, Roberta L. 232. See Piel, Roberta
    L. Gaidry
Gaidry, Serule 216
Gaidry, Wanda Faye Claire Ledet 233
Gaidry, Wanda L. 7
Gaidry, Wilfred 216
Gaidry, Wilson J. 216
Gaidry, Wilson J. "Doc", III 233
Gaidry, Wilson J., III 7, 232, 233
Gaidry, Wilson J., Jr. 232
Gaidry, Wilson Joseph, Jr. 232
Gaidry, W.J. "Doc", III 471
Gaidry, W.J., Jr. 232
Gaidry, W.J., Sr. 229
Gallay, Francois 407
Galtier, Etienne V., Rev. 302, 303
Gandy, Miss 208
Garden Society, 240
Garlinski, Joseph 281, 294
Garlinski survey of 1855 289
Gartwohl, Willy 268
Garver, C.F. 311
Gary, Leon, Mayor 256
Gaudin, Paul J., Rev. 302
Gautier, Pierre 296, 298

Gautier, Pre. 295
Gautreaux, Aurelia Letitia 416.
    See Carlos, Aurelia Letitia Gautreaux
Gautreaux, Emile 126
Gautreaux, Henry 126
Gautreaux, Joseph Mrs. 363
Gautreaux, Marie Angela Angeline 422.
    See Lirette, Marie Angela Angeline
    Gautreaux
Gay, Anna
    See Price, Anna Gay
Gay, Edward James 387
Gay, Edward James, Rep. 159
Gayoso, Manuel, Gov. 230, 241, 356
Gayoso, Margaret Cyrilla Watts 356
Gazeaux, Pierre 274
Gazzo, Lottie 473
Gazzo, Lottie Winnifred Sanders 399
Geneva Convention 264
Geneva Convention of 1929 264
George Pitre Lane 217
George Rodrigue Foundation of the Arts 4, 7
Georgetown Memory Project 135
Georgetown University 135
Georgia 246
Georgia plantation 383
Georgia [state of] 77
Georg, Irvin 268
German POWs 268
Germans 245, 256, 260, 272
Germans (settlers) 17
German sub 244
Gibbens, J. Louis, Sr. 468, 471
Gibson 18, 26, 193, 229, 363, 424
Gibson, Ambrose 102
Gibson, Randal 28
Gibson, T. 28
Gibson, Tobias 23, 26, 463
Gilbert, Walker 77, 85
Gilmore, A.B. 473
gin house 40
Giovagnoli, Madeline Sanders 321
Giovagnoli, Paul M. 320
Gisclair, Adelaide 291. See LeCompte,
    Adelaide Gisclair
Glenwood plantation 146, 147
Godchaux Sugar 372
Goforth, William, Judge 101
Golden Meadow 245, 342
Gonsoulin, Earl Anthony 441
Gonsoulin, Pat 333
Gonsoulin, Ruby Boyne 333, 441
Good Templars Hall 182
Goodyear Tire and Rubber Company 253
Gorman, Carolyn Portier 7, 268
Goucheaux, Alexandra 95
Goucheaux, Edna 95
Goucheaux, Louise 95. See Hobert,
    Louise Goucheaux
Gould, Jeffrey 121
G. P. Boquet Subdivision 217
Grabert, Loney J. 7
Grand Bois 431
Grand Caillou 17, 18, 27, 33, 56, 228, 229, 252, 268, 287, 326, 359, 363
Grand Caillou Tract 57
Grand Coteau 420
Grand Fair 311
Grand Isle 244, 256, 364, 471

Gravois, Celeste. *See* Tieken, Celeste Gravois
Gravois, Earl 383
Gray 128, 130, 132, 422
Gray, D.F. 275
Gray, Mary Anne Hanagriff 128
Gray school 103, 128
Gray, Walter J. 132
Gray, Walter Jasper, Jr. 129
Gray, Walter Jasper, Jr. D.V.M. 128–129
Gray, Walter Jasper, Sr. 128
Great Western Railroad mail routes 298
Greek Revival 17, 40, 42, 53, 78, 90, 208. *See* Architectural Styles
Greenwood plantation 18, 264, 463. *See* plantations
Gregoire, Bt. 297
Gregory, Douglas A. 173
Grey, Mary Anne Hanagriff 128
Griffin, Nancy Sarah 236, 274. *See* Wright, Nancy Sarah Griffin
Griffin, Ramona A. 7
Grinage, Celeste Belanger 297, 299
Grinage, James B. 17, 181. *See* First Settlers
Grinage, Richard H. 17, 175, 299, 432. *See* First Settlers
Grisman, Mr. 392
Grivot Guards 326
Groebel, Henry 116, 119
Groebel, Marioneaux 116
Gros, Eula Mae 44
Gros, Percy J., Sr. 44
Grundy, Felix 60
Guatemala sugar mill 383
Gueno, A.E. 282
Gueno, Amelesie Robichaux 282
Gueno, A.O. 282
Gueno Bros' refinery 285
Gueno Brothers 281, 282
Gueno, Delmas 280, 281
Gueno, J. 288
Gueno, Jh. 295
Gueno, J.O. 282
Gueno, Joachim 282
Gueno, Joachin 281
Gueno, Joseph 280, 281
Gueno, Joseph P. 281, 282
Gueno, Pierre 280
Gueno & Smith 282
Gueno & Bush 281
Guerbois, Marie Elizabeth Eulalie . *See* De la Ronde, Marie Elizabeth Eulalie Guerbois
Guiberteau, Bert A. 7
Guidroz, Eugene L. 285
Guidroz, Eugene Louis 285
Guidroz, J.L. 311
Guidroz, Terry P. 7, 468
Guidroz, Terry Philip 82
Guidry, Adele 422. *See* Lirette, Adele Guidry
Guidry, Alidore Noel 433
Guidry, Anazile Helena Robichaux 433
Guidry, Annette Rose Arcemond 300
Guidry, Charles 433
Guidry, Claudia 433
Guidry, David A. 7
Guidry, E. 288

Guidry, Edith 301, 332
Guidry, Edith Guidry 433
Guidry, Edmond 357, 432, 433
Guidry, Elmire Belanger 357
Guidry, Elmire Irma Azel Belanger 432
Guidry, Elodie 411. *See* Jarveaux, Elodie Guidry
Guidry, Ernest Louis 300
Guidry, Eve Cecile. *See* Champagne, Eve Cecile Guidry
Guidry, Farquard 433
Guidry, Farquard P. 300, 301, 303, 378
Guidry, Gillis 433
Guidry, Grace Aloysia Chauvin 300, 301
Guidry, Horace 433
Guidry, Horace "Jack" 433
Guidry, Horace (Jack) 301
Guidry, "Hurry Hurry" 216
Guidry, Iris 433
Guidry, J. A. 433
Guidry, Jean Baptiste 300
Guidry, Jerome 300
Guidry, John 393
Guidry, John, store 393
Guidry, John Joseph 301, 433
Guidry, Loretta 301
Guidry, Loretta Guidry 433
Guidry, Louella 433
Guidry, Ludovia Lirette 393
Guidry, Ludovia Valerie Lirette 433
Guidry, Neze 279
Guidry, Odelia Marie (Lucy) Champagne 300
Guidry, Ray Anthony 301
Guidry, Robley 433
Guidry, Robley "P.K." 301, 433
Guidry's Grocery 213
Guidry, Sherwin 348, 351, 393, 398, 409, 419
Guidry, Sherwin "Chabbie" 340, 393, 433, 474
Guidry, Sidney, Mrs. 422
Guidry, Velma 433
Guidry, Wenzel 433
Guidry, Wenzel A. "Manan" 433
Guilfoil (ship) 244
Guillot, Marculus 67
Guillot, Sylvere 67
Guion, Caroline Lucretia Winder 39
Guion, Caroline Zilpha 40. *See* Nicholls, Caroline Zilpha Guion. *See* Zilpha, Caroline
Guion, George Seth, Judge 39
Gulf Intracoastal Waterway 168, 169, 200, 202, 211
Gulf of Mexico 14, 15, 80, 242, 244, 246, 252, 454
Gulf Sea Frontier 245
Gunn, Ralph 42
Gustine, Rebecca 123. *See* Minor, Rebecca Gustine
Gustine, Sarah 123. *See* Potts, Sarah Gustine
Gwin, Catherine P. 284. *See* Danks, Catherine P. Gwin

H

Haché grant 21–22, 161, 183
Haché, Joseph 186

Hackett, A.J. 48
Hagen, Melissa A. 7
Half of Richland 324
Halfway Cemetery 77, 94, 100, 135, 136, 137, 139
Halfway plantation 137–139, 143, 157
Hallelujah community 19
Hall, Thomas C. 318
Hamilton, Alexander 100
Hamilton, Willie 213
Hanagriff, Mary Anne 128. *See* Gray, Mary Anne Hanagriff
hangar demolition 257
Hanks, Wilbert J., Rev. 229
hardscrabble (defined) 412
Hard Scrabble plantation 326, 327, 408-412
Hard Scrabble store 411
Hardy, Dr. Florent, Jr. 7
Hare & Baker 33
Harper's Weekly 15, 17, 23, 25, 28, 132
Harris, H.M., Lieut. Commander 253
Harris, Rebecca 235
Harris, T.H. 311
Harry Bourg Corp. 7
Harry Bourg Subdivision 217
Harry J. Bourg Corporation 469, 471
Hartnett, Valerie 154
Hart, William A. 399
Hatch, Winslow 193, 278
Hawaiian Treaty 123
Headquarters Army Service Forces 272
heart of cypress 62
Hebert, Augusta Marie. *See* Verret, Augusta Marie Hebert
Hebert, Carey Francis "Buddy" 32
Hebert, Cary R. "Buddy" 7
Hebert, Cora 420
Hebert, Coralie. *See* Ellender, Coralie Hebert
Hebert, Emile 318
Hebert, Gail . *See* Trahan, Gail Hebert
Hebert, Gustave 419, 420
Hebert, Henriette Amanda. *See* Rhodes, Henriette Amanda Hebert
Hebert, Hypolite 161
Hebert, Louis 419
Hebert, Marguerite Camilla. *See* Robichaux, Marguerite Camilla Hebert
Hebert, Olivia 149
Hebert, Peter H., Sr. 173
Hebert, Tim 181
Hedgeford map 1872 125
Hedgeford plantation 119, 122, 123, 126
Hedgeford plantation map of 1872 122
Heffermann, Herbert 268
Heirs, Gaubert 300
Hellier, Harry 217
Hellier, J. H. 10, 217
Hellier, J.H. 217
Hellier, J Harry 193
Henri, Jean Baptis 297
Henry, Amanda. *See* Rhodes, Amanda Henry
Henry, Ch. M. 297
Henry Dardar 335
Henry, Edith 332
Henry, Guillaume 297
Henry, J. Bt. Theodore 297

Henry, Jean Baptiste Theodore 356, 396
Henry, Toby 393
Henry, Willis J. 447
Hermitage, The 61
Hermit of Terrebonne 407
Hesse & Vergez 85
H. G. Bush Petroleum Company 399
H. Groebel & Co. 119
Hicklin, Margaret 40. *See Mead, Margaret Hicklin*
Hickman, Aannis "Annie" Slatter 230
Hickman, Thomas Smith 232
Hidalgo, Frank 67
Himel, Clara 199. *See Connely, Clara Himel*
Himel, J.E. 137
Historic New Orleans Collection 7, 32, 72, 319, 469, 472, 473, 478
H.L. Bourgeois High School 128, 139
Hobert, Edgar B. 90, 95
Hobert, Louise Goucheaux 95
Hobert store 95
Hobertville 95
Holden line 71
Holland 302
Hollinshead, John 321
Holly, Buddy 384
Hollywood POW facility 264
Holy Rosary Catholic Church 217
Home Building Supply Company 170
Home Place 221
Homestead plantation 10, 197, 199, 200, 201, 202
Honduras 18, 19, 186, 225, 230. *See Plantations*
Honduras Addition 186
Honduras sugar 22
Honolulu Plantation Company 260
Hooper Cottage 208
Hooper, Ella Keener 202, 204, 205, 208, 471
Hooper, Wilhemina 202, 208
Hope Farm 11, 170, 272, 324, 325, 326, 327, 334, 353, 356, 357, 359, 371, 410, 412
Hope Farm mill 327
Hope Farm Road 412
Hope Farm Sugar House 324
Hornsby, Holton J. 144
Hornsby, Samuel 409
hospital 77
Hotard, Agapi 314
Hotard, Athalie. *See Oubre, Athalie Hotard*
Hotard, Emma 338
Hotard, Estelle. *See Jacobs, Estelle Hotard*
Hotard, Euphrosin 302, 314, 315, 322
Hotard, Euphrosin, Jr. 315
Hotard, Inez. *See Patterson, Inez Hotard*
Hotard, Inezida 315. *See Carro, Inezida Hotard*
Hotard, James P. 298
Hotard, Leonise Elodie Champagne 315
Hotard, Marguerite Olympe 315. *See Blanchard, Marguerite Olympe Hotard*
Hotard, Marie Celine Mars Belanger 299, 314
Hotard, Marie Evelina. *See Champagne, Marie Evelina Hotard*

Hotard, Marie Marcellite Folse 314
Hotard, Noah 338
Hotard's General Store 338
Hotard, Sydney. *See Sheffield, Sydney Hotard*
Houma #1 (Prisoner of War Camp) 260, 264. *See Woodlawn POW camp*
Houma #2 (Prisoner of War Camp) 260, 264, 268
Houma Academy 42, 181, 228, 229, 235, 366
Houma Academy/St. Francis School 42
Houma, aerial photo of 178–179
Houma Airport 245, 247
Houma Area Convention and Visitors Bureau 7
Houma Branch Railroad 71, 181
Houma Brick and Box Company, Inc. 173
Houma Canal 168
Houma City Council 256
Houma, City of 174–181
Houma Colored High School 186
Houma Colored School 186
Houma Consolidated Association, The 202
Houma Country Club 253
Houma Courier 433, 435, 436, 441, 442, 468, 469, 471, 472, 473, 474
Houma Cypress Co. 323
Houma Cypress Co. Ltd. 10, 170, 171, 173
Houma Fire Co. No. 1 181
Houma Fire Department 285
Houma Fish and Oyster Company, Ltd. 196, 436, 455
Houma Golf Club 253
Houma Heights 10, 217
Houma Indians 264, 338, 430
Houma Indians Baseball stadium 264
Houma Motor Co. 273
Houma Naval Air Station 244, 245, 246, 247, 248, 251, 253, 255, 256, 272
Houma Oil Co. Ltd. 421
Houma Oil Company 420
Houma Oil & Mining Company 378, 421
Houmas Indians 427
Houmas tribe 15. *See Native Americans*
Houma-Terrebonne Lumber Company 170
Houma-Terrebonne Regional Airport 253
Houma Tribe 431
Hubbard, John 196
Hubbard, Josephine Loevenstein 196
Huey P. Long Bridge 260
Humble Canal 415
Hummel, Amanda Ellender 328
Hummel, George Anthony 328
Hunley, H.L. 224, 233
Hunley, Horace Lawson 225
Hunley (submarine) 225, 471
Hunley, Volumnia W. *See Barrow, Volumnia Washington Hunley*
Huntington, C.P. 72
hurricane 237
Hurricane Juan 438
hurricane Katrina 399
Hurricane Katrina 69, 462
hurricane of 1909 302
Hutchinson, Cynthia Trosclair 357
Hutchinson, Jamie, Dr. 308

Hutchinson, Jamie Ellis-John 7
Hutchinson spring stopper bottle 68

I

Iberia Parish 192
Iberville Parish 170
Idlewild plantation 225
Ile a Jean Charles 11, 334, 335, 337, 339, 340, 342, 362, 427
Ile de Cuba 114
Ile de Jean Charles 334. *See Ile a Jean Charles*
Indian mound 442
Indian Ridge Canning Company 435
Indian School 11, 208, 428, 430
indigo 22, 189, 462
Indonesia sugar cane factory 260
Infante, J. 114
Intracoastal Canal 168, 186, 202, 211
Iris, Jean Baptiste 299
Iris, Marianne Marie Nerisse 11, 288, 295, 299, 300, 316, 418. *See Billiot, Marianne Marie Nerisse Iris; See Nerisse, Marianne Marie*
Island of Cuba 105, 114
Island of Cuba map, 1907 107
Island Road 339
Islands
  Caillou 101
  Calumet 101
  Casse-Tete 101
  Labrosse 101
  Timbalier 101
Isle of Cuba plantation 44, 46, 95, 104–115, 105, 107, 108, 114, 115, 119, 139
Isle of Cuba Plantation mill site and quarters 109
Isle of Cuba store 114. *See Levy's General Merchandise*
Italian POWs 268
Ivanhoe plantation 192

J

J4F Widgeon (aircraft) 245
Jackson, Andrew, President, General 53, 61, 143
Jacobs, Estelle Hotard 315
Jacobs, Louis E. 315
James A. Robichaux & Co. 421
James Martin & Breaux 405
Janvier, Charles 385
Janvier, Eugene 297
Japan 333
Japanese POWs 268
Japan lillies 328
Jarveaux, Auguste 411
Jarveaux, Elda 326. *See Ellender, Elda Jarveaux*
Jarveaux, Elodie Guidry 411
Jarveaux, Victoria 326, 327. *See Ellender, Victoria Jarveaux; See Ellender, Victoria Jarveaux*
Jastremski, L.H., Dr. 217
Jaubert, David 348
Jaubert, George J. 7, 351, 473
Jean Dupre State Land Grant Map 361
Jeanne, Marie 286
Jefferson Parish 164, 244

Jefferson, Thomas 31
Jenkins, Bernard F., Cmdr. 246
Jenkins, Bernard F., Commander 246
Jesuits 22, 135
Jesuit Plantations Project 132
Jewish 69, 114, 115
Jim Crow 431
Johnson, Henry 135
Johnson Ridge 48, 50, 51
New Ridge 48
Old Ridge 48
John T. Moore Planting Co. 90
Jolet, A. 183
Jolet, A., Jr. 234, 240, 393
Jolet survey 240
Joller, Claire Domangue 1, 3, 13, 28, 72, 169, 287, 465
Joller, Emil W. 7, 465
Jones-Costigan Act 323
Jones, Francis W. 181
Joseph Haché Grant 174
Joseph, Henry 50
Joseph, J.R. 50
Joseph, Lillian 7, 50, 468
Joseph M. Cudahy (tanker) 244
Julia plantation 53, 67, 80, 81, 90, 114
Justine Aubert. *See Thibodaux, Justine Aubert*

K

Kamus, Willy 268
Kane, Ferdinand 372
Keene, Eva Dover Cenac 399
Keene, Mike 399
Keener Cottage 208
Keener, Ella. *See Hooper, Ella Keener*
Keener, Ella 202
Kelley, David B. 7
Kelly, David B. 468
Kennedy, Agnes Daspit 196
Kennedy, Alfred 149
Keystone Loening (Texaco plane) 244
Keys, William H. 149
King, Cecile Salome (Lilly). *See Watkins, Cecile Salome (Lilly) King*
King, Grace Elizabeth 69
Klingman, Henry J. 387
Klingman, Marie Clothilde. *See Cenac, Marie Clothilde Klingman*
Klingman, Mr. 159
Klondyke 11, 170, 319, 322, 323, 462
Klondyke map 1893 323
Klondyke River 322
Klondyke store 323
Knox, Frank 255
Kraemer, Vincent 7
Kraemer, Ying 7
Krajewski crusher 378
Krupa, Kimberly 431
KTIB radio 48
Kunstler, Mort 468
Künstler, Mort 19
Kurtz, Frank 155

L

Labat, Narcisse 302, 407
Labit, Raphael C., Monsignor 301, 302, 303
Labit, Rena Verrett 420
Labrosse Island 101. *See Islands*
Lacache 409
LaCarpe plantation 286
Lacher-Feldman, Jessica 7
Lac (Lake) Barré 427
la cuite 130, 348
Lafont, Lee 345
Lafourche and Terrebonne Canal Co. 123
Lafourche Crossing 300
Lafourche Heritage Society 31, 257, 357
Lafourche Interior 15, 20, 21, 36, 42, 100, 101, 297
Lafourche Parish 396, 401, 418
Lagarde, Geraldine L. 7
Lake Barré 245, 334, 398, 405, 427, 454
Lake Field[s] 299, 396
Lake Fish & Oyster Company 272
Lake Lagraisse 454
Lake Long 299, 304
Lake Okeechobee 383
Lake Pelto 244, 245
Lake Pelto Field 244
Lake Salvador 299
land claims
  Auguste Babin 116
  Charles Bergeron 143
  Charles Bertaud 161
  Charles Billiot 280
  Constant Pitre 132
  David Ellender 412
  Edmund Fanguy 286
  Esteban Rodríguez Miró y Sabater 361
  Estevan Miro 408
  Etienne Billiot 316
  François Malbrough 39
  Front lawn 284
  Henry and Thomas Roddy 408
  Henry Schuyler Thibodaux 288
  Henry S. Thibodaux 100
  Hubert Bellanger (Belanger) 297
  Jean Baptiste Theodore Henry 396
  Jean Batiste Dugas 395
  Jean Billiot 422, 432
  Jean Charles Naquin 396
  Jean Naquin 401, 402
  Jean Pierre Dugas 396, 404, 408, 409
  John Mary (Jean Marie) Campo 116, 123
  John Mason 137
  Joseph Gueno 280
  Joseph Haché 175
  Joseph Malbrough 39
  Laurent Pichoff 123
  Louis Sauvage 418
  Luis Duma 274
  Magenta 396
  Michael Dardar 284, 288
  Michel Billiot 280
  Nicholas Leret 148
  Peter Watkins (Williams) 143
  Pierre Bourg 416, 418
  Pierre Bergeron 146
  Pierre Gazeaux 274
  Pierre Gueno 280
  Pierre Menoux 202, 236
  Rosalie Courteau 427
  Walker Gilbert 85
land grants
  Alexander Verdin 427
  Auguste Babin 105, 116
  Charles Bertaud 161
  Charles Billiot 280, 352
  Charles Jumonville de Villiers 57
  Constant Pitre 132
  Edmund Fanguy 286
  Etienne Billiot 316
  Haché grant 161
  Henry Schuyler Thibodaux 100, 288
  Hubert Ballenger (Belanger) 105, 297
  Jean and Charles Naquin 335
  Jean Baptiste Theodore Henry 352, 396
  Jean Billiot 422, 427, 432
  Jean Charles Naquin 370, 396
  Jean Naquin 400, 401, 402
  Jean Pierre Dugas 396, 404, 408
  John Mary (Jean Marie) Campo 116, 123
  John Mason 137
  John D. Smith 61
  Joseph Gueno 280
  Joseph Haché 21-22, 186
  Joseph Mollere 77
  Laurent Pichoff 123
  Louis Sauvage 418, 427
  Marianne Marie Nerisse Iris 418, 288, 299
  Michael Dardar 284, 288
  Michel Billiot 280
  Phebe Olds Fuqua Woods 317
  Pierre Bergeron 146
  Pierre Bourg 416, 418
  Pierre Chaisson 322, 352
  Pierre Denis de la Ronde 54
  Pierre Gueno 280
  Pierre Menoux 210, 236
  Spanish Grant 295
  State Land Grant Map 284
  Tacalobe Courteau 427
  Thomas Villanueva 54
  Thomas Villanueva Barroso 53
  Walker Gilbert 77, 85
land measurement terms 30
Land Ordinance of 1785 31
land patents 31, 286-288, 290, 337, 352
  See Land Patent of William Bayard Shields 288
Landreneau, Dean 7, 44, 468
Landry, Burnelle P. 173
Landry, E.N. 217
Landry, Eno 137
Landry, Foret T 137
Landry, Gregory J. 173
Landry, J. Paul 137
Landry, J. Paul, Sr. 137, 139
Landry, Marguerite Anaise 416. *See Carlos, Marguerite Anaise Landry*
Lapeyrouse Cana 438
Lapeyrouse Canal 415, 439
Lapeyrouse, Edmond Deardon 435, 438
Lapeyrouse, Jean Pierre Junius, Sr. 438
Lapeyrouse, Louisa Robichaux 435, 438
Lapeyrouse Public School 411
Lapeyrouse store 438-439
Lapeyrouse, Viona 278. *See Chabert, Viona Lapeyrouse*
Laporte, Marie Magdeleine 287. *See Fangui, Marie Magdeleine Laporte*
Larne, Henry 363
Larose, Brian W. 7, 472
Larose, F.G., Jr. 474

Larpenter, Jerry J. 149
La Savanne 30
Laskey Cottage 208
Lassage, Emile 67
Lasseigne, Joyce Boyne 333
"Last Island" hurricane 26
La Tourrette, John 7
Laura plantation 192. *See Laurel Farm plantation*
Laura (train) 378
Laurel Farm plantation 192
Lauret, Debi L. 7
Lauton, Richard 305
Laux, Dorothy 154
Laver, Tara Z. 7
Lawford, Pat 208
Lawless, R.C. 143
Lawless, Richard C. 143
Lawless, Soloman C. 143, 145
Lawson, Isaiah, Bishop 359
Lay, Jean 364
Lay, Jean Baptiste 364
Lay, Jeanne Viguerie 364
L.C. Smith Commercial College 420
Leach's Machine and Boilerworks 246
Leagues (ligas) 30
Leaneaux 150
LeBlanc, Christian J. 469
LeBlanc-Doré, Edward A. 357
LeBlanc, Edward 357
LeBlanc, J.A. 137
LeBlanc, J.L. 378
Leblanc, J. Leonce 378
LeBlanc, J. Leonce 356, 357
Leblanc, Leo 119
Leblanc, Prosper 119
Leblanc, Theotiste 407
LeBoeuf, Eulice A. 144
LeBoeuf, L.C. 217
Lebouef, Lela 153. *See Picou, Lela Leboeuf*
LeBoeuf, Magdeleine 334, 337. *See Naquin, Magdeleine LeBoeuf*
LeBoeuf, Mary Cosper 7
LeBoeufs 17. *See First Settlers*
LeBoeuf Subdivision 217
Lebouef, Eula. *See Fabre, Eula Lebouef*
LeBouef, Gabriel 297
Lebouef, Horace "Petch" 426
Lebouef Subdivision 217
LeCompte, Adelaide Gisclair 291
LeCompte, Anna 315
LeCompte, Athalie 315
LeCompte, Bailey 315
LeCompte, Clifford 315
Lecompte, Elise. *See Lirette, Elise Lecompte*
LeCompte, Emily Bethancourt 315
LeCompte, Emma 315
LeCompte, Estelle 315
LeCompte, Florence 315
LeCompte, François 290, 291
LeCompte, Inez 315
LeCompte, Joseph 290, 302, 315
LeCompte, Joseph François 291
LeCompte, Joseph G. 315
LeCompte Property 314
LeCompte, Rudolph 315
LeCompte, Sydney 315
Le Danois 422
LeDanois, P. 423

Ledet, Aurelie 233
Ledet, Billy 44, 45
Ledet, C.A. 169
Ledet, Charles A. 173
Ledet, Js. Pre. 297
Ledet, Laise Marie 334, 337, 340, 345, 362
Ledet, Leo 119
Ledet, Renee L. 7
Ledet, Wanda Faye Claire 233. *See Gaidry, Wanda Faye Claire Ledet*
Lee, Dayna Bowker 345
Leeville 245
Left Bank 217
Leggett, Anita Gravois 384
Leggett, Delbert L. 384
Legion Pool 256
Lehmann (prisoner of war) 262
Leighton plantation 61
Lejeune, Clement 305
Lejeune, Elsie Whipple 304
Lemuel Dyer (ship) 363
Leneaux 154
Lenox Hotard Post No. 31 264
Leonard, Elvine 130. *See Bernard, Elvine Leonard*
Le Petit Presbytere 301
Lepine & Buron 137, 288
Lepine & Ferry 137
Lepine, Jules 132, 135, 137
Leret, Nicholas 10, 148. *See Lirette family*
Lester, Ada. *See Bascle, Ada Lester*
Levees 26
Leverich, Henry S. 123
Levert, John Baptiste 90
Levert, John Baptiste, Col. 67, 81
Levron, Al J. 7
Levron, Jules 50
Levron & Peltier Butchers 119
Levy, David 108, 114
Levy, Marcus M. 114
Levy, Max Mayer 108
Levy's General Merchandise 114. *See Isle of Cuba store*
Levy's General Store 115
Levy Town 19, 107, 108, 115
Lewis, Frances N. 420. *See Robichaux, Frances N. Lewis*
Ligas (leagues) 30
"lightning" joint 409
Lilac plantation 105, 114. *Also known as Lylac Avenue plantation*
lilies, Easter 328
Linear measure 30
Lipham, Gary D. 7
Lirette, A. 474
Lirette, Adele Guidry 422
Lirette, Alex 422
Leo Lirette Family 422
Lirette, Alexander 148, 150, 154, 422
Lirette, Alexander J. 148
Lirette, Alexander, & Sons 422
Lirette, Alexandre 149
Lirette, Alfred 422
Lirette, Arthur 422, 428, 431
Lirette, Bagley 422
Lirette, Debbie 305
Lirette, Elise Lecompte 428
Lirette, Ella 422
Lirette, Emile 422
Lirette family 148

Lirette Field 415
Lirette, John Peter 283
Lirette, Leo 149, 372, 422, 423, 433
Lirette, Ludovia. *See Guidry, Ludovia Lirette*
Lirette, Ludovia Valerie 433
Lirette, Marie Angela Angeline Gautreaux 422
Lirette, Marie Celestine Louviere 422
Lirette Oil Field 19
Lirette, Oscar 422
Lirette, Volcar 148
Little Bayou Black 56, 57, 61, 69, 77, 108
Little Caillou 15, 17, 33, 157, 225, 275, 303, 372, 390, 431, 442
Little Caillou Catholic chapel 390
Little Zion Baptist Church 217, 229, 235, 359
Livas, Thomas 50
Livaudais, Louis A. 311, 393
Live Oak plantation 11, 22, 23, 26, 33, 102, 371, 398, 404, 405, 420, 427
livestock brand 103, 115, 137, 202, 300, 301, 340, 356, 399, 416, 438, 441
Lloyd, Richard 123
Locust Grove plantation 225
Loevenstein, Herman 196
Loevenstein, Herman H. 211
Loevenstein, Josephine. *See Hubbard, Josephine Loevenstein*
Logan H. Babin, Inc. 283
Longfellow House 90
Long, Huey P., Gov. 68
long lot 30
Lorio, N. 90
Lorio plantation 320
Lorvin, Walker 337
Lottinger, Lee P. 169, 173
Louisa plantation 363
Louise (farm) 157
Louisiana and Texas Railroad 72
Louisiana and Texas Railroad and Steamship Company 71, 75
Louisiana Celotex Co. 376
Louisiana Celotex Co. mill 380
Louisiana Census Records: 1770-1798 406
Louisiana Comprehensive Statewide Survey 208
Louisiana Conference of the Methodist Church 424
Louisiana Historical Association 94, 235, 272
Louisiana Historical Bureau 32
Louisiana Infantry Regiment, 18th 193
Louisiana Militia 53, 77, 154
Louisiana Purchase 23, 139, 297, 361
Louisiana Raised Cottage 17. *See Architectural Styles*
Louisiana Secretary of State 143, 275
Louisiana Military Heritage 245, 246, 272
Louisiana State Archivist 7
Louisiana State University 199, 464
Louisiana State University Hill Memorial Library 7, 279
Louisiana State University Library 468, 469, 472, 473
Louisiana Sugar Belt 362
Louisiana Sugar Exchange in 1883 356
Louisiana Sugar Planter's Association 81

INDEX | 481

Louisiana Sugar Report 27, 32, 80, 385, 402, 410, 436
Louisiana Territory 30, 189
Louisiana Western Extension Railroad 71
Louis Miller Terrebonne Career and Technical High School 264
Louviere, Marie Celestine 422. *See Lirette, Marie Celestine Louviere*
Lower Ducros plantation 68
Lower Terrebonne Refinery 364, 375, 378, 387
Lower Terrebonne Refinery and Manufacturing Company 33, 282, 285, 364, 372, 387, 436,
Lower Terrebonne Refinery Train 373
Lower Terrebonne Refining Company 301, 433
Luke, Adruel B. 7
Luke, Lois Hebert 472
Luke Subdivision 217
Lunny, James 275
Lutecia plantation 85, 90
Lylac Avenue plantation 105, 114. *See Lilac plantation*
Lyman, E. 137
Lynch, Mary Torr 326. *See Ellender, Mary Torr Lynch*
Lyons, Daniel 221
Lyons, Dwayne 7
Lyons, General 149
Lyons, Jennie 221
Lyons, Warren, Sr. 7

M

Mabile, Carolyn Walker 7
Maccabees organization 368
MacDonell French Mission School 202, 208
MacDonell, Tochie 202
MacDonell United Methodist Children's Services 202 - 209
Madewood plantation 132
Madison, Bay 444
Madison Canal 11, 298, 441, 444
Magenta 371, 395, 396, 401, 402, 404, 412
Magenta plantation 193, 397, 398, 399
Maginnis, Charles B. 373
Maginnis, Charles Benjamin 356
Magnolia Grove plantation 116
Magnolia Methodist Episcopal Church 48
Magnolia plantation 48, 51, 56, 57, 62, 77, 78, 79, 90, 102, 108, 115, 116, 196, 225, 298
Magnolia United Methodist Church 48, 50, 51
Mahalick, Claire Moreau 7, 468
Mahler, Alidore 217
Mahler, Alidore J. 10, 217
Maier, Douglas 217
Maier, Emile 283
Main Project Road 114
Main Street POW facility 264
"maisons de terre" 339
Malbrough Coast 36
Malbrough, Eulalie 41
Malbrough, Francois 56
Malbrough, François 36, 39
Malbrough, Joseph 36, 37, 39
Malbrough, Rosalie 126. *See Boudreaux, Rosalie Malbrough*
Malbrough Settlement 36
Mandalay 19
Manilla Village 441
Manning, Keith Julian 7, 42
Manning, Susan Poché 7, 42
manufacturing of sugar 260
maps —————————
  1810 Lafourche Interior 20
  1830 State Land Grant 314
  1832 State Land Grant 286
  1856 Michael Dardar State Land Grant 284
  1831 Nicholas Leret Claim State Land Grant 148
  1831 Smith property 61
  1831 State Land Grant 175
  1831 State Land Grant 54
  1831 Joseph Haché Grant 174-175
  1846 Bayou Terrebonne origin, 36
  1846 La Tourrette Map 4
  1846 Richardson Map 36
  1850 State Land Grant map, Bayou Terrebonne to Schriever 36
  1853 John La Tourrette 7
  1870 Newtown 184
  1872 Hedgeford plantation 125
  1894 USGS Map Timbalier Bay 365
  1899 Newtown 183
  1903 Moore Planting Co. 71
  1905 Smithland 188
  1907 Isle of Cuba 107
  1909 Schriever railroad, Waubun, St. George, 75
  1915 Bocage Map of Terrebonne Parish 13
  1917 Pilié and Glenwood plantations, 147
  1921 Aragon 352
  1923 Schriever 70
  1924 Barrowtown 235
  1938 Hedgeford plantation 125
  1938 Tobin 95, 125, 133, 136, 418, 434, 439, 445, 446
  1939 Fala - Lake Felicity, USGS 343
  1940 City of Houma 176-177
  1945 Isle of Cuba 104
  1945 Julia Plantation 67
  1947 Ellender Brothers 413
  1953 Bayou Cane 152
  1953 Bourg 294
  1960 Back of Town 182
  1960 City of Houma 174
  Detail of Johnson Ridge 49
  Map of the Lafourche Country 30
  Prisoner of War camps in Louisiana 268
  Property map of the northern Lafourche 31
  Soil Conservation map 334
Mardi Gras 150, 151, 153
Marie, Frederick 149
Maringouin 135
Marion, Joseph J. 119
Marion, L.E. 423
Marionneaux, Olivier 119
Marlboroughs 17. *See First Settlers*
Maronge, Philip 47
Marsh, Royal 15. *See First Settlers*
Martial, Jules 193
Martin, Bertha 208
Martin, Litt 468
Martin, Philip E. 7
Martin, Polly Broussard 208
Martin, Teles 305
Martin, Terral J., Jr. 7, 32, 469, 471, 472, 473, 474
Martin, Whitmell P. 68
Martin, William Littlejohn 61
Masonic Lodge Number 5984. 359
Mason, John 137
Mastero, Miche & Co. 146
Masters, Sylvia 399
Matherne, Aaron W. 304
Matherne, Anatole 291
Matherne, Bruce C. 304
Matherne, Celeste Savoie 291
Matherne Dairy 295, 303, 304
Matherne, Della 303
Matherne, Ernest 303, 326, 327, 372, 378
Matherne, Farmand J., Jr. 7, 304, 472
Matherne, Farmand J., Sr. 303 -304
Matherne, Floyd H. 304
Matherne, Irvin P. 304
Matherne, Kenneth 303
Matherne, Stella R. 304. *See Floyd, Stella R. Matherne*
May, Elizabeth 404
Mayfair Club 264
May, Mary Jane 404. *See Sanders, Mary Jane May*
Mayor of Mechanicville 234
  See Thompson, Joseph, Sr. (Kingfish)
McCarthy Family Corp. 357, 369, 395, 401, 405
McCarthy Family Corporation 383, 399, 402
McCollam, Edmund Slattery 199
McCollam family 77
McCollam, William 196
McComiskey, Laura Champagne 323
McCoy Dining Hall 208
McDonald, John 361
McEnery, John, Gov. 157
McFarland, Baldwin & Company 67
McGavock, Caroline Elizabeth "Carrie" Winder 61
McGavock, Caroline Winder 61
McGavock, John 61
McGee, Donna. *See Onebane, Donna McGee*
McGee, Victor 42
McGehee, L.R., Mrs. 473
McGowen Hall 208
McKnight, Lillie Lea 232. *See Gaidry, Lillie Lea McKnight*
McMillan, Betty 153
McNeill, Angus 57
Mead, F. 93
Mead, Francis L. 40, 41, 123
Mead, Margaret Hicklin. *See Nicholls, Caroline Zilpha Guion*
Mechanicsville 11, 157
Mechanicsville / Mechanicville 229, 234, 235, 240

Melodia plantation 46
Menard, Jean Marie Charles, Fr. 406
Menoux-Bowie 221
Menoux, Pierre 199, 202, 210, 221, 236
Menoux, Pierre, State Land Grant Map 1831 210, 220
Menuet, A., & Co. 114
Mess Hall 255, 256
Metcalf, Georgianna Marie. *See* Viguerie, Georgianna Marie Metcalf
"metes and bounds" measurements 30
Methodist church in Gibson 424
Methodist Mission Chapel 424
Methodist Woman's Board of Home Missions 202
Metro Life Insurance Company 305
Mexico, Gulf of 447
Michel, S.P., & Co. 146
Michot, Stephen S. 31, 257, 357
Mid-South Morgage Co. 405
Mid-South Mortgage Co. 357, 369, 395, 401
Mid-South Mortgage Company 399, 402
mill crane foundation 121
Miller, Mark Edwin 427
mill, St. Brigitte sugar 102
Milome, Joseph 297
Minerva school 103
Minor, Henry Chotard 364, 372
Minor, Rebecca Gustine 123
Minor, William John 23, 28, 61, 123
Miro, Estevan 408
Mission of St. John's Church 384
Mississippi 216, 274, 284
Mississippi Department of Archives and History 472
Mississippi River 14, 26, 80, 101, 164, 181, 299, 478
Mississippi River flood in 1890 190
Mississippi Springs 410
Mississippi [state of] 23, 39, 60, 123, 137
Mississippi, Vicksburg 317
Moen, Janelle M. 7, 472, 473
Moffet (road) 256
Molinere, Craven 338
Molinere, Levest 338
Molinere, Louis 338
Molinere, Marie 430
Molinere, Marie Louise 416. *See* Carlos, Marie Louise Molinere
Mollere, Joseph 77
Monmouth plantation 23
Monsan, J. 297
Monsan, Joseph 314
Montegut 387-393 *See* Gabriel Montegut
Montegut Elementary School 392, 393
Montegut, Gabriel 159, 368, 387
Montegut, Gabriel, Col. 157
Montegut, Gabriel, Dr. 157, 157–159, 159, 234, 240, 289, 319
Montegut, Gabriel, II 157–159
Montegut Insurance Agency 157, 159
Montegut, Lizzie Willis 157
Montegut Post Office 159, 387
Montegut POW camp 260, 264
Montegut School 339, 393
Moore Cane Hook 80
Moore, Charles 67
Moore, Charles V. 91

Moore, Charles Verhagen 81, 90
Moore, John Thomas, Jr. 67, 80, 81, 91
Moore, John Thomas, Sr., Capt. 67, 79, 90
Moore, John T., Jr. 90
Moore, John T. Jr. & Co. 91
Moore, John T., Planting Co., Ltd. 67, 71, 80, 81, 90
Moore, John T., school 103
Moore, Julia M. Freret 80
Moore, Mary Frances 284. *See* Robertson, Mary Frances Moore
Moore, Odette Ellis 91
Morgan, Charles 71
Morgan City 50, 299
Morgan, David J. 7
Morgan's Louisiana & Texas Houma Branch Railroad map of Schriever, Waubun plantation, and St. George plantation 1909 75
Morgan Steamship Company 71
Morning Star Baptist Church 48, 50
Morrison, Chester A. 173
Morrison, Chester F. 7, 173
Morrison Home Center 173
Morrison Terrebonne Lumber Center 173
mortise and tenon joinery 44, 199
Mosaic sugar cane disease 18, 333
Mount Repose 353
Mount Repose plantation 324
"Mr. Carnival" 153
Mulberry plantation 192, 199, 225, 272, 284
Mulledy, Thomas, Rev. 132, 135
Munger T. Ball (tanker) 244
Munson, J.J. 383
Munson Mechanical Sugar Cane Harvester 383
Murphy, Tegwyn 91. *See* Weigand, Tegwyn Murphy
Musgrave, Marianne 353. *See* Brownson, Marianne Musgrave
Mutual National Bank 146
Myles, Flossie 186
Myrtle Grove 11, 225, 228, 232, 236, 237, 246, 274, 275, 278, 279
Mystic Krewe of Louisianians 464

N

Naquin, Adam 336
Naquin, Albert P., Chief 7, 340, 341, 472, 473
Naquin, Carroll "Chock" 340
Naquin, Charles 335, 396
Naquin, Christopher 335
Naquin, Demé, Chief 337, 339, 340
Naquin, Denicia 336
Naquin, Evest 330
Naquin, Frances 337
Naquin, Gabriel 339
Naquin, Gabriel F. 340
Naquin, Henry 330, 332
Naquin, Hippolite 397
Naquin, J. Charles 297
Naquin, Jean 335, 400, 401, 402, 403
Naquin, Jean Charles 334, 337, 370, 396
Naquin, Jean M. 297
Naquin, Jean Marie 337
Naquin, Jean Victor, Chief 339
Naquin, Jean (widow of) 297
Naquin, Jeff 340

Naquin, Joseph, Mrs. 339
Naquin, Levest 335
Naquin, Lillian (Naquin) 340
Naquin, Magdeleine LeBoeuf 337
Naquin, Marie Eve. *See* Dardar, Marie Eve Naquin
Naquin, Maryland Billiot 336
Naquin, Noelia 336, 337
Naquin, Pauline 337
Naquin, Pierre Alexander 336
Naquin, Renée/Reynalda 406. *See* Dugas, Renée/Reynalda Naquin
Naquin, Roch Raphael, Rev. 472
Naquin, Rosalie *See* Pitre, Rosalie Naquin
Naquin, Royce 345
Naquin, Simon 337
Naquin, Ulysse 302
Naquin, Victor Chief 334
Naquin, Vital 339
Narcisa Sugar Company token 260
narrow gauge railroad 146, 373, 410
NAS Houma 246, 247, 253, 255, 256
Nast, Thomas 132
National Dragster Magazine 469
National Register of Historic Places 42, 69, 90, 91, 208, 221, 232, 390, 391
Native Americans 15, 211, 342, 427, 430
See
Pierre 281
Rosalie Courteaux 300
Indian mound 442
Potawatomi 79
Waubunsee 79
natural gas well 367, 388, 405
Naval Airship Association, Inc. 246
Naval Air Station 11, 19, 241 - 259, 272, 279
Naval Air Station POW facility 264
Negro huts 40, 57
Negro quarters 371
Nerisse, Mariane Marie 295. *See* Iris, Marianne Marie Nerisse
See Sauvage, Marianne Marie
Neubaurer, Warren A. 90
Nevers, Clarence 235
Neville, Lusignon 50
Newell, Emily 288
New Hope Cottage 208
New Magnolia Baptist Church 108, 115
New Mount Pilgrim Baptist Church 235
New Orleans Bank 318
New Orleans, Opelousas, & Great Western Railroad 71, 77
Newport 295, 296, 298, 299
New Port 296, 298
Newport and Company 295
New Ridge 48
Newtown 183, 184, 186
New Zion Baptist Church 217, 229, 359
Nicholas, Robert Carter, Col., Sen. 143
Nicholas, Susan Adelaide Vinson 143
Nicholas, Wilson Cary, Gov. 143
Nicholls, Caroline Zilpha Guion 40
Nicholls, Francis T. 40
Nicholls State University Archives 7, 19, 37, 103, 121, 272, 327, 468, 469, 471, 472, 473, 474, 478
Nixon, Mrs. Richard M. 412
Nixon, Richard M., President 412
Nolan, Kathryn 326. *See* Cage, Kathryn Nolan

Nolan, Lynn F. 7, 469, 472
Nolan, Torre 102
Nolan, Wiliam T. 472
Nolan, William 7
Nolan, William T. 72, 102, 155, 468, 469
Norlindo (tanker) 244
North Thibodaux Prisoner of War Camp 266, 267
Nova Scotia 15, 17, 47, 193, 362

O

Oak Forest 19
Oak/Gem theatre 392
Oak Grove plantation 225
Oaklawn plantation 383
Oak ridge plantation 431
Oakwood plantation 288
Oak Wood plantation 193
Odozgoyti, Vincent 287
Odessa's Place 182
Office of State Lands 468, 469, 471, 472, 473, 474
Office of the Provost Marshal General 264, 272
Ogden, E. 23
Ohio, The (paddlewheeler) 166
Old Ridge 48
Old State Capitol 42
Oleander Subdivision 11, 239, 240
Olive, Frank 95, 102
Olive, Mollie Brigitte Beauvais 95, 102
Olivier, Donatille Zimorie 144. *See Ayo, Donatille Zimorie Olivier*
O'Neal, Albert Deutsche, Sr. 161
O'Neal, Mary Sanders 161
Onebane, Donna McGee 7, 193, 369, 471, 473
Opelousas 71
Opelousas convention of sugar planters, 1855 123
open kettle 130, 348, 370
Opera House 181, 366
Orange Grove 422, 423
Orange Grove plantation 10, 11, 146, 192, 363, 423
Ostheimer, Barbara K. 7
Ostheimer, William A. 7, 200
Otis, William 78
Oubre, Athalie Hotard 315
Oubre, N.J. 315
Our Lady of Prompt Succor chapel 303
Our Lady of the Most Holy Rosary Catholic Church 211, 229
overseer 18, 57, 69 , 77, 80, 155, 264, 268, 275, 279, 281, 283, 285, 367, 369, 383
  See
  • Aycock, F.A. 285
  • Benedict, W. 284
  • Breaux, Basile 285
  • Duplantis , Trasimond 367
  • Rodrigue, Raymond 283
  • Watkins, John 155
Ovide J. "Jock" Cenac Collection 471
"Oyster Capital of the World" 435
Oyster House, Sea Breeze 455
oyster shipping 181

P

Paquet, Peter C., Father 302
Paquet, Pierre C., Rev. 302
Paradis, M.R., Rev. 147
Parker, Faye 7
Pass, Sea Breeze 453
Patout family 192
Patout, Marie Corinne 193, 362. *See Burguières, Marie Corinne Patout*
Patterson, Charles R., Sr., Mrs. 240
Patterson, Douglas P. 7, 472
Patterson, Dudley 315
Patterson, Inez Hotard 315
Patterson, Ruth 240
Patterson, Susan O. 7
Patty A.Whitney 472
Paul, Lydia Semple 288
Paulsen, Bethany C. 7
Paulsen, Eric A. 7
Peabody College 420
Pecan Grove 11, 291
Pecan Tree plantation 217, 435, 436
Pelegrin, Oliva 373
Pellegrin, Allie. *See Robichaux, Allie Pellegrin*
Pellegrin, Bruce 419
Pellegrin, Gertie 278
Pellegrin, Michael T. 474
Pellegrin, Thaddeus J. 458-459
Pelletier, Gerard, Rev. 428
Peltier, Ozemé Euzelien 119
Pelton, John 77, 102
People's Bank, The 157, 322, 421
People's Bank & Trust Company of Houma 420
People's Sugar Company 420
Peoples Sugar Co. Stores 421
Perché, Napoléon-Joseph, Archbishop 302
Persac, Adrien 92, 99, 101, 366, 473
Peterville 19
Petit Black bayou 14
petit presbytere 302
Petit Terrebonne tract 57. *See Magnolia plantation*
Pettigrew, Anita 240
Pettigrew, Ashby W., Jr., Mrs. 240
Pharr, A.J., Rev. 48
Phyllis, The 168
Pichoff, Laurent 123, 300
Pichof, Laurent 295
"Pick" (Indian game) 340
Picou, Lela Leboeuf 153
Piedra, Jonathan J. 114
Piedra, Manuel A. 114
Piel, Peter 232
Piel, Roberta L. Gaidry 232
Piel, Roberta Louise Gaidry 232
Pierce, Phebe Olds Fuqua Woods 317
Pierce, Phebe Woods 318
pigeon house 222
Pilié 422
Pilié, Armand 146
Pilié plantation 10, 143, 146, 146-148, 147, 148
Pinel, Bruce 332
Pinel, Charlie 332
Pinel, Lloyd 332
Pinel, LouLou 332

Pinel, Philip 332
Pinel, Wilma 332
Pioneer (submarine) 225
pirogue 305, 338, 339
Pitre, Celeste Fidelise 397. *See Fields, Celeste Fidelise Pitre*
Pitre, Constant 132
Pitre, George 10, 217
Pitre, Glen 393
Pitre, Gwen A. 472
Pitre, Hippolite 397
Pitre, Rosalie Naquin 397
Pitre, Marie Molly Josephine. *See Daigle, Marie Molly Josephine Pitre*
Pittman, John Bradford 236, 275
Pitt Street Bridge 123
Pizzuto, Greg 233
Plaisance 283
plantations *See listings by name*
  Acadia
  Acadia
  Afton Villa
  Alice B.
  Angela
  Aragon
  Ardoyne
  Argene
  Argyle
  Ashland
  Ayo
  Balsamine
  Batey
  Bayou Place
  Belanger
  Belle Grove
  Bellevue
  Bivouac
  Blackwater
  Bon Ami
  Boutelou
  Boykin
  Briars
  Bull Run
  Burguières
  Caillou Grove
  Cane Brake
  Carnton (Tennessee)
  Carothers
  Cedar Grove
  Collins
  Concord
  Crawford
  Crescent
  Cypremort
  DelClaire
  Dixie
  Ducros
  Dularge
  Eastonia
  East St. Bridget
  Eliza
  Ellendale
  Ellerslie
  Eloise
  Erin
  Eureka
  Evergreen
  Fairfax
  Flora
  Frontlawn or Front Lawn
  Georgia

Glenwood
Greenwood
Halfway
Hard Scrabble / Hardscrabble
Hedgeford
Honduras
Homestead
Idlewild
Ivanhoe
Isle of Cuba / Island of Cuba
Julia
Lacache
LaCarpe
Laurel Farm
Leighton
Lilac
Live Oak
Locust Grove
Lorio
Louisa
Lower Ducros
Lutecia
Madewood
Magenta
Magnolia Grove
Magnolia
Mandalay
Melodia
Monmouth
Mount Repose
Mulberry
Myrtle Grove
Oak Grove
Oak Forest
Oaklawn
Oak ridge
Oakwood
Oak Wood
Orange Grove
Pecan Tree
Pilié
Point / Pointe Farm
Poverty Flat
Porche Hermitage
Presqu'ile
Ranch
Red Star
Residence
Richland
Ridgeland
Rienzi
Roberta Grove
Rosedown
Rural Retreat
Sarah
Sargeant/Armitage
Semple and Shields
Smith & Barrow
St. Agnes
St. Brigitte/Bridget
St. George
Stormy Point
Sunbeam
Terrebonne
Upper Ducros
Versailles
Waterproof
Waubun
Waverly
Westover
White Hall or Whitehall

Wilkerson
Windermere
Woodlawn
plantation tokens 115
  See "tokens"
Plaquemine 410
Plaquemines Parish 300
Plater, David D. 7, 50, 103, 145
Plater family 143
Plattenville 287, 395, 406, 418, 442
Poche, Stanley Joseph 283
Poché, Susan. See Manning, Susan Poché
Pointe-au-Barré 11, 342, 424, 426, 427, 428, 429, 431
Pointe-au-Barré definition 427
Pointe-Aux-Chênes 11, 15, 19, 308, 328, 329, 330, 333, 334, 339, 340, 371, 424, 427, 428
Pointe-Aux-Chênes Indian Tribe 345
Pointe-Aux-Chênes sawmill 308
Pointe-Aux-Chênes School 339
Pointe Coupee 100, 105, 297
Pointe / Point Farm 11, 192, 221, 225, 334, 361 - 371, 393, 397, 409, 422
Pointe Viguerie 362
Point Farm Planting Co. 368
Point Farm Store 384
Polk, Leonidas, Bishop 225
Polmer, Augusta Feitel 68
Polmer brothers 81
Polmer Brothers Louisiana Bottling Works 68
Polmer Brothers, Ltd. 68, 69
Polmer, Irvin 68, 69
Polmer, Irvin Feitel 69
Polmer, Leon 68
Polmer, Mervin 68, 69
Polmer, Samuel 68
Polmer, S. Cahlman 68
Polmer, Stella Feitel 68
Polmer store 68, 69
Porche, Antoniata Cecilia 211, 229.
  See Daigle, Antoniata Cecilia Porche
Porche Brothers 348
Porche, Elmire Marie Thibodaux 103
Porche, Evariste 57, 103, 114, 296
Porche Hermitage plantation 363
Porche, Hypolite 296
Porche, Jack 285, 353
Porche, Joachim 103
Porche, Myra 285
Portier, Carolyn 268. See Gorman, Carolyn Portier
Portier, Clovis 268
Portre-Bobinski, G. 193, 357
Port Texaco 245
Posey, H.F. 282
potato farm 199
Potawatomi 79
Potts [Hedgeford], John C. 119
Potts, John Calvin, Maj. 123, 126
Potts, Sarah Gustine 123
Potts (Sonier) house 122–127
Poverty Flat 19
Powell, C. L. 183
Powell, Joseph 82
Powell, Thomas E. 82
Powell, Thomas E., III 7
Powell, William S. 233
POW Germans escape 272
POW letters 270

POW paintings 272
POW passport 272
Prejean, Abel P. 149
Prejean, Leoni 67
Prejean, Pierre 67
Prentiss, Seargent Smith 324
Presqu'ile 11, 245, 272, 280 - 283
Presqui'le mill 280
Presquille 462
Prevost 17. See First Settlers
Price, Andrew, U.S. Rep. 93, 102, 121
Price, Anna Gay 102
Price, Fielding George 401
Price, Forest Jackson 401
Price, Franklin "Frank" 401
Price, George W. 401
Price, George W. vs. A. Adler 401
Price Hine & Co. 71
Price, John Wallace 401
Price, Lee Frederic 401
Price, Marie Eliza Cordilia 401
Price, Mary Louise Ross 401
Price, Victoria May. See Rhodes, Victoria May Price
Price, William 407
Price, William Ernest 401
Price, William Franklin 401
Prisoner of War Camps 11, 19, 256, 260–273, 261, 263, 264, 265, 267, 269, 271, 272, 273
Prisoners of War / sugar cane harvest 265
Prospect Street Bridge 279
Protestant Episcopal Church 384
Pugh, Fannie E. 132. See Beattie, Fannie E. Pugh
Pugh, Fannie Whitmell 317. See Wood, Fannie Whitmell Pugh
Pugh, Margaret. See Woods, Margaret Wood Pugh
Pugh, Mary 121
Pugh, W. W. 284
Purkins, Charles 295

Queen Anne 17. See Architectural Styles
Queen Anne Revival 230
Queen Anne style 187
Queen Bivalve
  See Connely, Lucy
Quick, Walton James 393
Quimby, John 119
Quitman, John A. 23

Rabin, Estelle Polmer Slipakoff 68
Radical Ridge Farm 157
Railroad bridge at Thibodaux 66
railroad, mill 378
railroads
  Ashland Branch 186
  Holden line 71
  Houma Branch 71
  Louisiana and Texas & Steamship Co. 71, 75
  Louisiana Western Extension 71
  New Orleans, Opelousas, & Great

INDEX | 485

Western 77
    Price Hine & Co. and Berger lines 71
    Southern Pacific 71
Ramel, Antoine 280, 281
Ranchos (*sitios*) 30
Ranch plantation 287
Randolph, James 50
Realty Operators 264
Reconstruction 150, 235
Recreation Hall 255
Red Cypress Lumber and Shingle Company, Ltd. 173
Redmond, Angele Laurentia Bourg 393
Redmond, Clara A. 7, 472, 473
Redmond, Enos 393
Redmond, James 384
Redmond, Jim 383, 384
Red Star 418, 419, 420, 421
Refinery, Lower Terrebonne 370
Regional Military Museum 7, 257, 471
relocate island residents 340
Rembert, Keneth L. 7, 32, 272, 468, 469, 473, 474
Residence Baptist Church 7, 229, 235
Residence Dairy 232
Residence Dairy milk bottle and cap 233
Residence Dairy truck 233
Residence Dairy Wagon 233
Residence plantation 220, 220-227, 221, 224, 225, 229, 229-237, 230, 232, 233, 236, 237, 274, 275
Residence Subdivision 217
Revolutionary War 15, 39, 363
Rex Carnival Krewe 72
Rex Organization 468
Rhodes, Amanda Henry 442
Rhodes Brothers 442
Rhodes, Carroll 442
Rhodes, Gustave 442
Rhodes, Gustave, Oaks 444
Rhodes, Heloise Bergeron 442
Rhodes, Henriette Amanda Hebert 442
Rhodes, John 444
Rhodes, John Jackson (Jacques Guillaume) 442
Rhodes, Leon Alexis 442
Rhodes, Robert 442
Rhodes, Thomas 442
Rhodes, Victoria May Price 442
Rhodes, Victorine Use 442
rice 17, 123, 137, 264, 310, 340
rice factor 123
Richard, Charles, Fr. 302
Richard, Charles, Rev. 302
Richardson, Adrianne Fryou 384
Richardson, J.P. 384
Richaud, Everette 256
Richland 352
Richland plantation 192, 324
Ridgefield 155
Ridgefield plantation 39, 44
Ridgeland plantation 192, 225, 284
Rienzi plantation 101
Right Bank 217
Rightor, A.F. 286, 287
Rightor, Henry 32, 82, 369
Rillieux, Norbert 26
Rita Ethel Trahan. *See Bourgeois, Rita Ethel Trahan*
Rives, T. 50

Roberta Grove Boehm bird collection 241
Roberta Grove plantation 11, 19, 225, 228, 230, 234, 236, 237, 238, 239, 240, 241, 274, 275, 441
Roberts, Carol Ann 321
Roberts, Claire Fortmayer 321
Roberts, Courtney Claire 321
Roberts, Della Marie 321
Roberts house 321
Roberts, Kay Toups 321
Robertson, Billy Earp 147
Robertson, Farrel Mrs. 147
Robertson, Francis Epes, Dr. 284
Robertson, John P. 284, 314, 317, 325, 326, 352, 353, 361, 364, 371
Robertson, Louis 48, 50
Robertson, Mary Frances Moore 284
Robertson, Thomas B., Gov. 100
Roberts, Preston C. 321
Roberts, Preston Carl, III 321
Roberts, Preston Carl, Sr. 321
Robichaux, Ada Ashley Sanders 420
Robichaux, Ada Marie 420
Robichaux, Albert J., Jr. 406
Robichaux, Alfred 420
Robichaux, Allie Pellegrin 419
Robichaux, Amelesie 282. *See Gueno, Amelesie Robichaux*
Robichaux, Anazile Helena. *See Guidry, Anazile Helena Robichaux*
Robichaux & Carlos 416
Robichaux, Easton 393
Robichaux, Ernest 387
Robichaux, Ernest L. 387
Robichaux, Etienne Paul 419
Robichaux, Eve Regina. *See Aucoin, Eve Regina Robichaux*
Robichaux, Fernand "Mr. Pete" 173
Robichaux, Frances N. Lewis 420
Robichaux, Honore 419, 420
Robichaux, Honorine 419
Robichaux, Jean Baptiste 421
Robichaux, Jean Baptiste Honoré 421
Robichaux, Jean Baptiste Marin 418, 421
Robichaux, J.N. 372, 406
Robichaux, Joseph A. 102, 420
Robichaux, Joseph Alfred 420
Robichaux, Joseph Luby 419
Robichaux, Joseph Narcisse 419, 420, 421
Robichaux, Leo N. 407
Robichaux, Louisa. *See Lapeyrouse, Louisa Robichaux*
Robichaux, Marguerite Camilla Hebert 419
Robichaux, Marie 406
Robichaux, Marie Madeleine Breaux 418
Robichaux, Mr. and Mrs. Etienne 421. *See Robichaux, Marguerite Camilla Hebert*
Robichaux, Narcisse 420
Robichaux, Odelia Stoufflet 393
Robichaux, Paul 407
Robichaux School 390
Robichaux, Theresa A. 7
Robichaux, Udalize Badeaux 407, 419
Robichaux, Ursula 419
Rockwood, Curtis 17, 139. *See First Settlers*
Rockwood-Grinage Store 432

Rockwood, Sarah Semple 288
Rococo Revival 53
Roddy, Catherine 408, 409. *See Elinger, Catherine Roddy; See Elinger, Catherine Roddy*
Roddy, Catherine (elder) 409
Roddy, Henry 408
Roddy, Marie Catherine Thomas 408
Roddy, William 408
Rodrigue, George G. 4, 8, 9, 11, 29, 34, 96, 162, 194, 218, 292, 312, 346, 425, 466, 492, 497, 500
Rodrigue, Jacques George 7
Rodrigue, Mallory Page 7
Rodrigue, Raymond 283
Rodrigue, Wendy Wolfe 7
Rods (*varas*) 30
Rogers, Dorothy "Dot" 7
Rogers, Lily Sonier 126
Rogers, Roy 212
Roman Catholic 301
Rommel's Afrika Corps 260
Rosedown plantation 42
Roselawn 10, 217
Ross, Mary Louise 401. *See Price, Mary Louise Ross*
Ross, William 356
Roundtree, Joseph Darden Hilache 41
Roundtree, Marie Evelina/Velina Boudreaux 41
Roundtree, Willie 44
Roussell, Infonte 90
Rozands, Charlton Peter 149
R.R. Barrow Collection 469
Rucker, John 356
Ruiz, Alfred J. (Freddie) 262
Rummel, Joseph Francis, Archbishop 216, 428
Rural Retreat 157, 316, 317, 318, 319, 320, 322, 371, 404
Rural Retreat plantation 11, 66, 157, 317
Russel, Linda Walker 285
R. M. Parker, Jr. (ship) 244, 245

S

Saadi Sites 217
Sabine River 71
Saboular, J.R. 288
Sacred Heart Catholic Church 301, 302, 390, 428
Sacred Heart Church 366, 369, 431
Sacred Heart (Church) Parish 390
Sainte Anne, la petite Chapelle de 302
Sanders, A.A. 398, 455
Sanders, Ada Ashley 404, 420. See Robichaux, Ada Ashley Sanders
Sanders, Alice 405
Sanders, Allen A. 378, 384, 385, 402, 404, 405, 412
Sanders, Allen Andrew 395, 397, 398, 404
Sanders, Annie E. 404
Sanders, Charlotte 404
Sanders, Eleonora Benton 404
Sanders, Elizabeth "Annie" Armogene 404
Sanders, Ella J. Trahan 397
Sanders estate 372
Sanders, Fannie Mary Eudora 404
Sanders, Hattie 405

Sanders, Hattie Altia 404
Sanders, James Monroe 364, 371, 397, 402, 404, 405, 420, 427, 454
Sanders, Jim 405
Sanders, J.M. 404
Sanders, Laura 405
Sanders, Laura Jane 404
Sanders, Madeline. *See Giovagnoli, Madeline Sanders*
Sanders, Marie Lucretia 404
Sanders, Marthe Alice 404
Sanders, Mary. *See O'Neal, Mary Sanders*
Sanders, Mary J. 405
Sanders, Mary Jane May 404
Sanders, May. *See Bazet, Mae Sanders*
Sanders & Price 401
Sanders, Robert 405
Sanders, Robert H., Mrs. *see Orange Grove* 422
Sanders, Robert Hunley 404
Sandersville 387
Sandersville map of 1902 388, 393
Sanders, Warren James Goode 404, 405
Sanders, Yancey 405, 454
Sanders, Yancey John Lee 404
Santo Domingo 22
Sarah Anne Dunn. *See Gagné, Sarah Anne Dunn*
Sarah plantation 268, 282, 409
Sargeant/Armitage plantation 39, 46, 123
Sargeant, Ellen Rebecca 40
Sargeant, Philippe M. 40, 123, 126
Sargeant plantation 39, 48
Sargeant, Ellen Rebecca. *See Armitage, Ellen Rebecca Sargeant*
Sargeant, widow 40
Sauls, Linda 305
Sauvage, Louis 300, 418, 427
Sauvage, Marianne Marie 418. *See Iris, Marianne Marie Nerisse*
Savoie, Celeste 291. *See Matherne, Celeste Savoie*
Savoie, Elizabeth Ellender 326
Savoie, Jules Homére 326
Savoie, Lucien 281
Savoie, Murphy 384
Savoie, Murphy H. 7
Savoie, Murphy J. 384
Saxon, Lyle 235
Schexnayder, Jeanette F. 7, 468, 469
Schexnayder, Maurice, Rt. Rev. 216
Schneider, Clemens, Rev. 302
schools
   African America high school, first 256
   Andrew Price School 95
   Andrew Price Vocational 103
   Audubon Sugar School 81
   Bourg Agricultural School 310
   Bourg Elementary School 310
   Bourg High School 310
   Caldwell Middle School 108
   Daigleville School 211
   Dillard University 235
   East Street School 234
   Evergreen Junior High 143
   Evergreen School 143
   Fields school 390
   Gray school 103
   H. L. Bourgeois High School 128
   Houma Academy 235
   Houma Colored High School 186
   Houma Colored School 10, 186
   Ile à Jean Charles school 337
   Indian schools 208, 428
   John T. Moore, Jr. school 103
   Lapeyrouse Public School 411
   L.C. Smith Commercial College 420
   Louis Miller Terrebonne Career and Technical High School 264
   MacDonell French Mission School 202
   Minerva school 103
   Montegut Elementary School 393
   Montegut School 366, 393
   Peabody College 420
   Pointe-Aux-Chênes School 339
   Robichaux School 390
   Schriever Graded School 103
   Schriever School 103
   St. Bridget school 103
   St. Charles College 420
   St. Francis de Sales 181, 366
   Stoufflet School 430
   Straight University in New Orleans 105
   Terrebonne High School 339
   University of New Orleans 235
Schott, Matthew 272
Schriever 18, 23, 33, 36, 44, 48, 68, 69, 71, 75, 72, 79, 81, 85, 91, 159, 225, 297, 367
Schriever Graded School 103
Schriever Hotel 72, 81
Schriever, John George 71, 73
Schriever map, 1923 70
Schriever railway depot 164
Schriever School 103
Schutten, Carl J., Rev. 302
Schuyler, Philip, Gen. 100
Scurto, Elizabeth M. 7
Seabreeze 280
Sea Breeze 11, 14, 364, 436, 447, 450, 453, 454, 455, 456, 458, 459, 460, 462
Sea Breeze Cut 398, 405
Second Louisiana Calvary 282
Second Restoration 362
Sefton, Thomas 356
Seminole War 420
Semple and Shields plantation 288
Semple, Eliza Davis 288
Semple, Joseph 288, 289
Semple, Lydia. *See Paul, Lydia Semple*
Semple, Lydia 288
Semple, Mary 288. *See Carradine, Mary Semple*
Semple plantation 157, 289
Semple, Robert H. 289
Semple, Sarah 288. *See Rockwood, Sarah Semple*
Semple & Shields 11, 288, 289
Shaffer, John D., Mr. and Mrs. 385
Shaffer, John D., Sen. 114
Shaffer, M. Lee, Jr. 469
Shaffer, M. Lee, Sr. 468
Shaffer, Thomas R. 318
Shaffer, William A. 23
Shaver, C.P. 68
Sheffield, R.K. 315
Sheffield, Sydney Hotard 315
Shelly, Michel 90
Sheriffs of Terrebonne Parish 149
Shields, Martha Jane Lenora Witherspoon 324, 352
Shields, Rodney 53
Shields, Thomas R. 274
Shields, Thomas Rodney 324, 352
Shields, W.B. 288
Shields, William Bayard 288
Ship's Company 255
Ship Shoal Lighthouse 244
Shrimp dryers' camp 456
Shuvin 17. *See Chauvin; See First Settlers*
Silver City 186
Simms, Amos 149
Simpson, Oramel Hinckley, Gov. 315
Sisters of Charity 364
sitios (lots) 30, 31
sitios (ranchos) 30
Sitterson, J. Carlyle 19, 23, 26, 28, 135
Slade, Ernest 249
Slater, Bessie 232
Slatter, Aannis "Annie" 230. *See Hickman, Aannis "Annie" Slatter*
Slatter, Bessie 230
Slatter, Clara 230, 232. *See Gaidry, Clara Katherine Slatter*
Slatter, David 230, 232
Slatter, Volumnia Louise 230, 232
Slatter, Volumnia Roberta Barrow. *See Barrow, Volumnia Roberta*
Slatter, William J. 230
slave labor 18, 26, 27
slave murdered 361
slave owners, leading
   Barrow, Robert Ruffin, Sr. 230
   Batey, Jesse 132
   Belanger, Hubert Madison 299
   Bisland, John Rucker 352
   Bisland, William Alexander 325
   Butler, Thomas 57
   Cazeau[x], Pierre 281
   Ellis, Richard 56
   Gueno, Delmas 281
   Gueno, Joseph 281
   Ramel, Antoine 280
   Sargeant, Philippe M. 40
   Thibodaux, Henry Schuyler 100
   Thomas Mulledy, Fr. (Jesuit) 132
   Winder, Van Perkins 57
   Woods, William Lackey 316
slaves 40, 48, 50, 57, 101, 115, 132, 137, 221, 230, 240, 299, 325, 352
slaves, origin of 281
   See also
   Adelaide 281
   Alexis 281
   Butler, Biby 135
   Butler, Biby (child) 135
   Butler, Gabe 135
   Butler, Henry 135
   Butler, Martha Anne 135
   Butler, Mary 135
   Butler, Nace 135
   Butler, Tom 135
   Campbell, Frank 275
   Catalina 69
   Christian, Ebed 235
   Coxen, Augustus 105
   ex-slaves 108

Figaro 281
Johnson Ridge 48
Julie 281
Levy Town 115
Lundy 281
Lyons, Daniel 221
Lyons, Jennie 221
Mandinga 281
Negro huts 40, 57
Negro Quarters 371
Pierre 281
quarters 109
Smedes, C.E. 183
Smith & Barrow plantation 157
Smith, Claiborne T. 233
Smith, Claire (Clara) Marie Pauline Dupont (*Aunt Clara*) 187, 190, 193
Smith, Clara Mildred 188, 193, 357, 473
Smith, Clifford Percival 187 - 193, 441
Smith, Clifford T. 32
Smith, C.P., & Co. lumber 188
Smith, C.P., & Company 170
Smith, C.P., Cypress Co. 192
Smith, C.P., Insurance Agency 188
Smith, Ephon, Rev. 48
Smith, Helen Snow 190
Smith house 187, 192
Smith, John Davidson, Dr. 39, 61
Smith, Kenneth W. 7, 187, 189, 191
Smithland 188, 471
Smith, Lois Ruth 188
Smith, T. Baker 186, 190
Smith, T. Baker, Mrs. 190
Smith, William Clifford 7, 190, 468, 471
Soda pop factory 68
Soil Conservation Map 334
Songe, Joseph Bernard 423
Soniat, Judith A. 7
Soniat, Judy A. 468, 472
Sonier, Alcide 126
Sonier, Elise Crochet 126
Sothern, James M. 7
South Coast 245, 260, 264, 283, 357, 368, 369, 370, 378, 379, 382, 383, 395
South Coast Company 260, 283, 357, 368, 378, 379, 383, 399, 401, 405
South Coast Corporation 245, 368, 370, 382, 383, 402
South Coast Presqui'le Store 283
Southdown 7, 17, 22, 23, 33, 61, 123, 128, 155, 264, 356
Southdown Engine 17
Southdown Sugars 462
Southern Defense Command 260
Southern Pacific Railroad 71
Southern Pacific railway 373
Southland Raceway 144
South, Martha R. 7, 471, 473
South Terrebonne Subdivision 11, 238, 239, 240
Spaniards 17
Spanish Christian Pentecostal church 260
Spanish colony 15
Spanish governor 230
Spanish land grant 11, 53, 93, 100, 105, 288, 295, 299, 314
Springer (carpenter) 64
S.S. Benjamin Brewster (ship) 244, 245
Staff Cottage 208
St. Agnes plantation 319, 320, 321

Stahls, Paul F., Jr. 42, 53, 61, 69, 78, 82
Standing Structures Survey of 1981 351
St. Ann Catholic Church 301, 302, 303, 432
State Land Grant Map 1831 175, 400
State Land Grants 396, 403, 409
Statement of Sugar Crop Made in Louisiana 27, 28, 32, 40, 102, 143
Statement of the Sugar and Rice Crops Made in Louisiana 27, 32
State Superintendent of Schools 311
St. Bernard Parish 53
St. Bridget Catholic Church 18, 102, 106, 108, 115
St. Bridget Chapel 103, 112
St. Bridget Church 57, 103, 108, 111, 112
St. Bridget school 103
St. Brigitte / Bridget plantation 23, 37, 46, 85, 93, 95, 98–103, 105, 108, 115, 132, 139, 297
St. Brigitte plantation home 103
St. Brigitte plantation sugar mill 102
St. Charles College 420
St. Charles Parish 164
Steel, David 67, 81, 90
Stewart, Jordan 149
St. Francis Cemetery 157
St. Francis Church 424
St. Francis de Sales bell 211
St. Francis de Sales Catholic Church 211, 216, 228, 229, 327
St. Francis de Sales School 42, 181, 229, 366
St. George 67
St. George plantation 75, 80, 81, 84, 85, 86, 90, 95, 114
St. James Parish 100, 143, 189, 300, 432
St. John Episcopal Church 39
St. John's Chapel 385
St. John the Baptist Catholic Church 432
St. John the Baptist Church 406, 407
St. Joseph Cemetery 95
St. Joseph Church 130, 406
St. Joseph Church (Thibodaux) 103
St. Joseph Congregation 106, 108. See St. Bridget Catholic Church
St. Louis Canal 410
St. Louis Cypress Co. 170, 410
St. Martin (A.) Co. 435
St. Martin, Eugene C., Dr. 191
St. Martin, Gladys Davidson 161
St. Martin house 10, 187, 189, 191, 193, 195
St. Martin, Hugh P., Jr. 191
St. Martin, Hugh P., Sr. 191
St. Martin, Michael J., Sr. 471
St. Martin, Michael X., Sr. 7
St. Martin, Rhea Marie 191
St. Martin, Roy J. 191
St. Martin, Thaddeus Ignace (T.I.), 161
St. Martin, T.I. 474
St. Martin, Virginia R. 7
St. Martin, Yvette 191. See Dupont, Yvette St. Martin
St. Mary Parish 15, 173, 188, 192, 193, 325, 326, 353, 356, 362, 363, 364, 441, 442
St. Matthew's Episcopal Church 42, 228,

229, 237, 398
St. Michel Subdivision 217
St. Patrick Catholic Church 228, 229
St. Peter Catholic Mission Chapel 428, 429
St. Peter's Baptist Church 11, 358, 359
St. Peter's Church bell 359
St. Remi (ship) 334, 406
St. Sophie Chapel 303
St. Tammany Parish 53
St. Therese, Françoise 299
Stokes, Carlton D. 471
storm, Last Island 318
Stormy Point plantation 317, 320
Stoufflet, Alidore 393
Stoufflet, Annette Duplantis 393
Stoufflet, Emery 426
Stoufflet, Ernest 393
Stoufflet, Melice Walker 393
Stoufflet, Odelia. See Robichaux, Odelia Stoufflet
Stoufflet School 430
Stoufflet Store 426
Stoufflet, Wallace 393
Straight University in New Orleans 105
Streams, Louis, Jr., Rev. 48
Stripling, A. 50
Strong storms / hurricanes 26
submarine 225
Sugar Bowl 17, 28, 130
sugar cane and POWs 260
sugarcane process, open kettle 26
sugar house 18, 93, 325
sugar levee 22
sugar mill 77, 102, 114, 121
sugar mills 372
Sunbeam 401, 402, 412
Sundbery House 170
Sundbery, Oscar C. 170
Sundbery Shopping Center 170
Sun (ship) 244, 245
Suthon, Hugh S. 378
Suthon, Lucius 196
Suzanna Roberta (bell) 229
Swanson, Garth K. 7, 463, 474
Swanton, John 19, 427
sweet potatoes 301
Sytko, Glenn 272

T

Talbot, Russell W. 7, 473, 474
Talma (ship) 362
tank batteries (WWII) 245
Bay Junop 245
Bay St. Elaine 245
Caillou Island 245
Dog Lake 245
Four Isle 245
Golden Meadow 245
Lake Barré 245
Lake Pelto 245
Leeville 245
Tanner, Lemuel 56, 77, 102, 139, 161, 298
Tanner, Lemuel P. 116
Tanner, Marie Agnes Celeste Belanger 76, 77, 102, 119, 139
Tanner, Mrs. L. 77, 102
Tasset, François, Fr. 397
Tasset, Lydie Marie 193
Tasset, Lydie Marie (Astugue) 395, 397. See Dupont, Lydie Marie (Astugue)

Tasset
Tasset, Marie Augustine 395, 397.
　　See Dardeau, Marie
　　Augustine Tasset;
　　See Fields, Marie Augustine Tasset
　　Dardeau
Tate, Joseph 469
Taylor, Dick, Gen. 326, 356
T. Baker Smith 7
tenant house 101, 120, 200, 369
Tennent, Charles 230
Tennent, Zoe Gayoso 275
Tennessee 230, 284
Tennessee [state of] 23, 60, 61
Terrebonne Coast 245
Terrebonne Factory 383
Terrebonne Farms 68, 81, 90, 108, 114
Terrebonne General 257
Terrebonne General Medical Center 196
Terrebonne High School 158, 159, 260, 339
Terrebonne Lumber and Supply
　　Company 173, 323
Terrebonne Memorial Park 10, 196
Terrebonne Parish Assessor 7
Terrebonne Parish Clerk of Court Office 7, 295, 469, 472
Terrebonne Parish Consolidated
　　Government 7, 32, 232, 431, 468, 469, 471, 472, 473
Terrebonne Parish Medical Society 256
Terrebonne Parish, origins of 14
Terrebonne Parish Police Jury 105, 256
Terrebonne Parish Public Library 7, 468, 469, 472
Terrebonne Parish School Board 7, 102, 103, 155, 158, 211, 256, 311, 315, 368, 390, 393, 430, 431
Terrebonne Parish Sugar Crop Report 1844-1917 32-33
Terrebonne plantation 383
Terrebonne Project 53, 67
Terrebonne Station 36, 70, 71, 72, 165
Terrebonne Sugar Co. 378, 401
Terrebonne Sugar Company 33, 283, 285, 378, 400
Terrebonne Waterlife Museum 435
Terrell, Robbie E. 321
Texaco 393
Texas 71, 246
Texas Company (Texaco) 11, 244, 245, 246, 256, 436
Texas Company Yard 436
Theogene Engeron
　　Subdivision 217
Theriot, Adoiskia Gaidry 216
Theriot (Terraiu), Augustave 193
Theriot, Clifton 103, 121, 471
Theriot, Clifton P. 7, 471, 478
Theriot, Cyrus J. 471
Theriot, Cyrus J., Jr. 7, 469
Theriot, Edward L. 141
Theriot, Elphege 217
Theriot, Jason P. 257, 474, 478
Theriot Lumber Yard 215
Theriot, Marguerite Burguières 193
Thewes, William (Billy), Ensign 252-253
Thibodaux, Aubane 46
Thibodaux, Aubin Benoni 103
Thibodaux, Bannon Goforth 93, 101, 103
Thibodaux, Brigitte Emelie Belanger 23,
37, 77, 93, 95, 100-103, 115, 137, 139, 175.
　　See Belanger, Brigitte Emelie
Thibodaux, Brigitte Emilie 103, 139.
　　See Barras, Emilie Brigitte Thibodaux
Thibodaux Brothers 102
Thibodaux, Cecile A. 93. See Blake,
　　Cecile Adele Thibodaux
Thibodaux, Elmire Marie 103
Thibodaux, Eugenie 103
Thibodaux, Felicite Bonvillian 100
Thibodaux, Henri Claiborne 100
Thibodaux, Henry (Henri) Claiborne 100, 102, 132
Thibodaux, Henry Schuyler
　　(Father of Terrebonne) 23, 46, 77, 85, 93, 95, 100 - 102, 103, 105, 115, 116, 137, 139, 175, 280, 288, 296 - 298, 324.
Thibodaux, Henry Schuyler (tomb) 137
Thibodaux, Jerry P. 7
Thibodaux, Julia Helene 46.
　　See Bourgeois, Julia Helene
　　Thibodaux
Thibodaux, Justine Aubert 93, 94
Thibodaux, Leandre B. 101
Thibodaux, Leandre Bannon 103, 123
Thibodaux Massacre of 1887 121, 132
Thibodaux, Mathilde M. Toups 100
Thibodaux, Melissa N. 7
Thibodaux, Michel Henri (or Henri
　　Michel) Joseph
　　102, 103
Thibodaux, Mrs. A. 95
Thibodaux, Nettie 208
Thibodaux, P.A. 126
Thibodauxville 100, 137
Thibodaux, Wallace 102
Thibodeaux, B.A. 123, 126
Thomas Ellender Estate 410
Thomas Ellender Jr. 326, 327
Thomas House 208
Thomas, James 50
Thomas, Marie Catherine^. See Roddy,
　　Marie
　　Catherine Thomas
Thompson, Antonio 272
Thompson, J.H., Rev. 359
Thompson, Joseph, Sr. (Kingfish) 234
Tieken, Celeste Gravois 384
Tieken, John 384
Tigerville 26, 225
Timbalier 336, 364
Timbalier Island 101, 364.
　　See Islands
timber, cypress 170
Times Picayune, The 445, 454, 468, 471, 472, 473
tokens
　　Celotex Corp. 379
　　Houma Cypress Co. 173
　　Narcisa Sugar Co. (Cuba) 260
　　South Coast Co. 283
　　South Coast Corp. 382
Topping, Zoe Gayoso Barrow 237
Toupes, George 286
Toups, Leonard L. 81
Toups, Louis 297
Toups, L.S. 68
Toups, Mathilde M. 100. See Thibodaux,
　　Mathilde M. Toups
Toups, Priscilla 264

Toups, Prosper, Jr. 264
Toups, Prosper, Sr. 264
Township system 31
Towson, Jane Ellis 116
Towson, Mary Jane. See Ellis,
　　Mary Jane Towson
Trahan, Angela F. 471
Trahan, Angela Marie Fonseca 7
Trahan, Angie 233, 241
Trahan, Ella J. 397. See Sanders, Ella J.
　　Trahan
Trahan, Gail H. 469
Trahan, Gail Hebert 7, 145
Trahan, Garland A 469
Trahan, Garland Anthony 7, 145
Trahan, Ivy 143, 145
Trahan, Rita Ethel 47
Trahan, Rose (Roselia) Ayo 144-145
Trahans 17. See First Settlers
Train depot 168
Treaty of Fontainebleau 30
Tremoulet, Casimir 363
Trinidad 246
Trosclair, Cynthia. See Hutchinson,
　　Cynthia Trosclair
Trosclair, Elizie 326
Trosclair, Elysée Alfred 357
Trosclair, Odelia Deroche 357
Trosclair, Odelia Marie Deroche 357
Truxillo, Clarence 350
Tucker, Joseph, Capt. 103
Tucker, Merril Elaine. See Chauvin,
　　Merril Elaine Tucker
Tucker, William P. 196
Turner, Louis 126
turtles, diamondback 438
Twain, Mark 79
typhoid fever 81
Tyson, Jean Jn. 295

U

Union Army 326, 356
Union forces 71, 356
Union soldiers 62, 78
Union troops 353
United Houma Nation 7, 338, 340, 427, 431
University of New Orleans 235
University of Southwestern Louisiana 147, 272
Upper Ducros 53, 67, 80
U. S. Army Corps of Engineers 80
U. S. Census of 1850 409
U. S. Census of 1860 404, 409, 416
U. S. Census of 1870 401
U.S. Census of 1880 337
U. S. Census of 1900 405, 430
U.S. Coast Guard 245-246, 253
U.S. Corps of Engineers 168
USDA Agricultural Research Service
　　Sugar Cane Research Unit 61
Use, Victorine. See Rhodes,
　　Victorine Use
USGS Map Timbalier Bay 1894 364
USGS Map Fala - Lake Felicity, 1939 343
Usie, Leryes 7, 323, 472
U.S. Naval Air Station 11, 19, 240, 241, 242, 243, 244, 245, 246, 247, 249, 251, 252, 253, 255, 256, 257, 259, 279
Houma Naval Air Station 244

U.S. Navy 245, 246, 252, 253, 256
USO 256
USS Housatonic 225
USS KIDD 245, 257, 272
USS KIDD Veterans Memorial 245, 253, 257, 272
Uzee, Philip D. 147

V

vacheries (estates) 30
Valens, Ritchie 384
Valentine 266
Valentine Prisoner of War Camp 267
Valentine Sugars 260
VanBergen, John P. 353
VanBuren, I. 50
Van Buren, Martin, President 60
van den Broek, Adrian, Fr. 302
Van den Broek, Adrien, Rev. 302
varas (rods) 30
Verdin, Alexander 427
Verdin, Alexandre 297
Verdin, Felicie 345
Verdin, Forest 345
Verdin, Genevieve Armelise Billiot 342
Verdin, Helen. *See Dardar, Helen Verdin*
Verdin home 345
Verdin, Laurent 345
Verdin, Laurent, Jr. 345
Verdin, Laurent Mitchell, Sr. 342
Verdin, Laurent M., Sr. 342, 344
Verdin, Mayfield 345
Verdin, Octave 342
Vermilion Parish 192, 364
Verret, Augusta Marie Hebert 420
Verret & Company 281
Verret, Jacques 101, 189, 363
Verret, Kirby A. 7, 431
Verret, Marianne 189, 363. *See Delaporte, Marianne Verret*
Verret, Martial 149, 363
Verret, Nicolas 189
Verret, Solomon 297
Verrett, Rena 420. *See Labit, Rena Verrett*
Versailles plantation 53
Veterans Flying School 256
Vice, David G. 173
Vice, Robert J. "Bobby" 173
Vicknair, Inez 208
Victorian 17
Victoria (ship) 362
Victor, W. 50
Viguerie, Albert 410
Viguerie, Albert R. 283, 378
Viguerie, Albert Robert 282, 378, 383
Viguerie, Alexis (Alejandro) 362, 363
Viguerie and Marmande, planters 372
Viguerie, Anthony R. 369
Viguerie, A.R. (Albert Robert) 281, 366, 367, 368, 373
Viguerie, Arthur Camille 367 - 369
Viguerie, Arthur Robert 390
Viguerie, Bertrande Bazerque 362
Viguerie brothers 370
Viguerie Canal 364
Viguerie, Clara 368
Viguerie, Clara (Sister Marie Corinne) 364

Viguerie, Elizabeth Stoufflet 455
Viguerie, Elodie Chauvin 369
Viguerie, Elvire 363
Viguerie, Emma 367
Viguerie, Ernestine Louisiana Burguières 192
Viguerie, F. 372
Viguerie, François 119, 192, 362, 363, 364, 366, 370, 372, 397, 422
Viguerie, François Camille 362, 364, 378
Viguerie, François (Frank) Camille 192
Viguerie, Georgianna Marie Metcalf 364
Viguerie, Guillaume 362, 363
Viguerie, Irene Bascle 368
Viguerie, Jean 362
Viguerie, Jeanne 362, 363, 364. *See Lay, Jeanne Viguerie*
Viguerie, Jean Pierre 192, 361, 362, 363, 364, 366, 370, 395, 397, 408, 409, 410, 422
Viguerie, Joyce 369
Viguerie, Marie 362, 363, 364. *See Consalle, Marie Viguerie*
Viguerie, Marie Elvire 364
Viguerie, Marie Elvire Delaporte 192, 363
Villanueva, Thomas 54
Vinson, Susan Adelaide. *See Nicholas, Susan Adelaide Vinson*
virgin cypress 302, 362
Voisin, Gerald J. 7, 272
Voisin, Keith M. 173
Voisin Place 217
Voisin, Ray C. 173
Volumnia W. Hunley 230
Voris, J.M. 149
Voss, William 217

W

Wade Claim. *See Bertaud, Charles (Berto)*
Wade Estate 157
Walcott, Marion Post 468
Walker, Craig M. 472, 474
Walker, Dr. Craig M. 7, 410
Walker, Ellen Mary Ellender 268
Walker, Glenn 7
Walker, Herman E., Jr. 7, 285, 472
Walker, Herman Ernest, Sr. 268, 285, 333
Walker, Herman E., Sr. 285, 310
Walker, Herman, Jr. 268
Walker, Herman, Sr. 325, 357
Walker, Leslie T., Sr. 285
Walker, Linda 285. *See Russel, Linda Walker*
Walker, Lori R. 7, 472
Walker, Mary Ellen 285. *See Dugas, Mary Ellen Walker Feyerabend; See Feyerabend, Mary Ellen Walker*
Walker, Melice. *See Stoufflet, Melice Walker*
Walker, Mr. 161
Walker, Theophile 311
Walker, Tina H. 7
Walker, Tina Hebert 410, 474
Wallis, Quvenzhané 399
Walter, Jim, Corporation 283, 357, 369, 383, 395, 399, 402
Walter Land Company 283, 357, 369, 399, 401, 402, 405

Walther, Dick C. 471
Walther, Owen 114
War Between the States 132, 210, 326, 356
War of 1812 77, 143, 154, 297
war stamps 186
Waterproof 19, 128
Waterproof plantation 462
Waterproof store 192
water transportation 230
Waties, J.C. 186
Watkins, Caleb B. 149, 154
Watkins, Caleb Baker 154
Watkins, Cecile Salome (Lilly) King 155
Watkins, Gilbert Luke 155
Watkins, John Washington 155
Watkins, Joseph W. 155
Watkins, Judge J. Louis, Sr. 155
Watkins, Marguerite 19, 103
Watkins, Marguerite E. 154
Watkins (Williams), Peter 143
Watkins, William Sternin 154
Watson, John 325
Watson, John C. 314
Waubun 7, 48, 67, 68, 298
Waubun chapel 78
Waubun plantation 75, 77, 114
Waubun refinery 81
Waubun store 68, 81, 82
Waubunsee, chief 79
Waud, A.R. 15
W.A.V.E. 256
Waverly plantation 61
Webber, Felix 193
Weber, Otto 262
Wedgeworth, Celeste St. Martin 7, 471
Wegmann, Anthony J., Rev. 216
Weigand, Joseph John, Jr. 7, 90
Weigand, Tegwyn Murphy 7, 91
Wells, Ken 462
Wesley House 202, 208
Westerman, Audrey B. 20, 21, 42, 319
Westerman maps, Lafourche Interior, 1810 20-21
West Feliciana Parish 221
Westover Plantation 272
West Point 40
Westwego 299
Whipple, C.A. 311
Whipple, Caroline 304
Whipple, Charles Frederick 304
Whipple, Charles (Willie) 304
Whipple, Elsie 304. *See Lejeune, Elsie Whipple*
Whipple, Eugene 304
Whipple, George 304
Whipple, Mark 304
Whipple, Mark McCool 304, 305, 307
Whipple, Mary Augustine Mars 304
Whipple, Roderick (Duck) 305, 307
Whipple, Roy P., Sr. 305
Whipple, Sarahlene Falgout 304
Whipple, Thomas A. 304, 305
Whipple, Tom 304
Whipple, Walter 304, 306
Whipple, Walter, Jr. 304
Whipple, Willie 304
White, Eliza 53
White, George W. 53
White Hall plantation 143
Whitehead, M.W. 137
White, Laura M. 202, 205, 208

White League 132
White, Maunsel 53
Whitney Central Trust & Savings Bank 282, 283, 368, 378, 395, 399, 401, 402, 405
Whitney, Elward P. 7
Whitney, Linwood P. 7, 471
Whitney, Patty 279
Whitney, Patty A. 7, 473
Whitney, W. J. 234
Whitten, Horace L. 82
Wickliffe, Robert C., Gov. 202
Wilbert, Frederick 170
Wilkerson Plantation 264
William C. McTarnahan (ship) 244, 245
Williamsburg 10, 146, 148, 150, 154
Williams, Charles 385
Williams, Ella
  See Ayo, Ella Williams
Williams, Frank B. 378
Williams, James 50
Williams, W. Horace, Co. 246
Willis, Lizzie 157. *See Montegut, Lizzie Willis*
Wilson, H. 352
Wilson, Harry L. 183
Wilson, James D. 387
Wilson, J.D. 48, 367
Winde, Nina 69
Winder, Caroline Lucretia 39, 60, 61.
  See Guion, Caroline Lucretia Winder
Winder coat of arms 57
Winder, Elizabeth 60
Winder, Felix G. 85
Winder, Harriet Handy 60
Winder, John N. 196
Winder, Louise 61
Winder, Martha 62
Winder, Martha Grundy 60, 61
Windermere 19, 80, 85, 90
Winder, Nina 61, 62
Winder, Sallie 61
Winder, Thomas Jones, Dr. 57
Winder, Van 61, 62, 90
Winder, Van Perkins 57, 69
Winde, Thomas L. 90
wine cellar 78
Witherspoon, Martha Jane Lenora 324, 352. *See Shields, Martha Jane Lenora Witherspoon*
Witherspoon, Mary Louisa Lavinia 352. *See Bisland, Mary Louisa Lavinia Witherspoon*
"Witness trees" 30
Wolf, Samuel 363
Wonder Lake 415
Woodlawn plantation 27, 33, 77, 102, 256, 260, 275, 287, 326, 373
Woodlawn POW camp 262, 264, 272. *See Houma #1 (Prisoner of War Camp)*
Woods, Andrew Van 318
Woods, Covington Barrett 66
Woods, Fannie Whitmell Pugh 66, 199, 317, 318
Woods, Maggie 318
Woods, Margaret Wood Pugh 66, 317
Woods, Mildred 318, 471
Woods, Phebe Olds Fuqua 317
Woods, Richard Covington 53, 66, 317, 318
Woods, Rodney Shields 66, 317, 318
Woods, S. 372
Woods, T. Albert 230
Woods, Virginia 318
Woods, Volumnia Roberta Barrow Slatter 135, 230, 234
Woodworth, George W. 78
World's Industrial and Cotton Centennial Exposition 72
World War I 129, 245
World War II 47, 235, 244, 246, 253, 257, 260, 270, 272, 285, 333. *See Prisoner of War camps; See tank batteries; See U-boats*
Wright family 474
Wright, Holden 275
Wright, Holden E. 236, 274
Wright, Jasper K. 157
Wright, Nancy 289
Wright, Nancy Sarah Griffin 236, 274
Wright, Robert R. 144
Wurzlow, August 157
Wurzlow, Calvin, Mayor 193, 298
Wurzlow, E.C. 181, 183, 275, 285, 373, 385, 468
Wurzlow, Francis William "Frank," Jr. 41
Wurzlow, Helen 69, 91, 126, 145, 159, 169, 170, 173, 193, 200, 211, 217, 233, 241, 279, 285, 289, 315, 319, 323, 327, 357, 366, 369, 385, 399, 405, 433, 441, 442, 468, 469, 471, 472, 473, 474
Wurzlow, Margaret Alice Toups Davis (Peggy) 262
Wurzlow, Mary Elizabeth "Bettie" Chauvin 41

Y

Yancey, Patrick H. 7
Yancey, Stanley E. 7, 471
yellow fever 62

Z

Zerger, Mel, Rev. 48
Zeringer, F.X. 193
Zilpha, Caroline 40
Zuber, Jean 84, 86

Evangeline, 1989, Oil on canvas, 36 x 24 inches, George G. Rodrigue

PHOTO CREDITS

Unless otherwise noted, photos and maps are in the Cenac Collection at the Nicholls State University Archives in Thibodaux, Louisiana. Research of these materials may be arranged by contacting the Archives and Special Collections Department at Allen J. Ellender Memorial Library on the Nicholls State University campus in Thibodaux.

Cover: Wallace R. Ellender III and Coralie Hebert Ellender

Page 5: David B. Kelly, Ph.D., Coastal Environments, Inc.

Page 7: Personal collection of Dr. Christopher E. Cenac, Sr.

Page 14: Bazet Collection, Nicholls State University Archives

Page 16: Donald W. Davis

Page 17: Top, Bazet Collection, Nicholls State University Archives; bottom photo, W.E. Butler, *Down Among the Sugar Cane*

Page 18: Top right, Donald W. Davis; bottom left, Bazet Collection Nicholls State University Archives

Page 19: Top, Laura A. Browning; middle, © 1997, Mort Kunstler, Inc.; bottom, Terrebonne Parish Public Library

Pages 20-21: Joseph Jess Bergeron, Terrebonne Genealogical Society

Page 22: Top, J. Louis Gibbens, Sr.

Page 23: Middle left, from Helen Wurzlow, *I Dug up Houma Terrebonne*; bottom left, *Barrow Family History*; bottom right, M. Lee Shaffer, Sr.

Page 26: Articles from Laura A. Browning

Page 27: Right, Nicholls State University Archives; bottom, Bazet Collection, Nicholls State University Archives

Page 29: Estate of George G. Rodrigue

Pages 30-31: *The Lafourche Country Vol. III*

Pages 34-35: Estate of George G. Rodrigue

Pages 36 and 37: Nicholls State University Archives

Page 45: N. Dean Landreneau

Page 47: Lester Charles Bourgeois, Jr.

Page 48: Both photos, Lillian Joseph

Page 49: Map, Keneth L. Rembert; photo, Lillian Joseph

Page 52: Map, Terrebonne Parish Consolidated Government; photo, Richard J. Bourgeois and Angela M. Cheramie

Pages 54-55: Office of State Lands, State of Louisiana

Page 56: Richard J. Bourgeois and Angela M. Cheramie

Page 57: Top, Hill Memorial Library Special Collections, Louisiana State University Library; middle, Helen Wurzlow, *I Dug Up Houma Terrebonne* 1984; bottom right, Richard J. Bourgeois and Angela M. Cheramie

Pages 58 and 59: Bazet Collection, Nicholls State University Archives

Page 60: Top right, Richard J. Bourgeois and Angela M. Cheramie; bottom left, Terrebonne Parish Public Library

Page 61: Top right, Office of State Lands, State of Louisiana; middle left, middle right, and bottom right, The Battle of Franklin Trust; bottom left, *Thibodaux Minerva* September 24, 1853

Page 62: Middle bottom, Richard J. Bourgeois and Angela M. Cheramie

Page 66: Top left, Richard J. Bourgeois and Angela M. Cheramie

Page 67: Top, Richard J. Bourgeois and Angela M. Cheramie; *New Orleans Times Democrat* April 23, 1909; bottom, Terrebonne Parish Consolidated Government

Page 68: Top left and bottom, Richard J. Bourgeois and Angela M. Cheramie; left middle, *Houma Courier* February 15, 1945

Page 69: Top right, Debra J. Fischman; middle, Dr. Christopher Everette Cenac, Sr., *Livestock Brands & Marks*; bottom, Richard J. Bourgeois and Angela M. Cheramie

Page 70: Top, Logan H. Babin III; bottom, Nicholls State University Archives

Page 71: Middle top, Litt Martin Collection, Nicholls State University Archives; middle bottom, *Houma Courier* Magazine Edition 1906; bottom, 1897 *Directory of the Parish of Terrebonne* by E.C. Wurzlow

Page 72: Top, Helen Wurzlow, *I Dug Up Houma Terrebonne* 1984

Page 73: Top, William T. Nolan and the Rex Organization; inset, Laura A. Browning

Page 74: Jody A. Davis

Page 75: Foldout map, William Clifford Smith, PE, PLS

Page 76: Background map, Keneth L. Rembert; bottom, *La Sentinelle de Thibodaux* September 2, 1865

Page 79: Bottom, *Houma Courier* Magazine Edition 1906

Page 80: Top, Judy A. Soniat, Terrebonne Parish Public Library; lower left, Helen Wurzlow, *I Dug Up Houma Terrebonne* 1984

Page 81: Top left, *Houma Courier* Magazine Edition 1906; top right, Alcee Fortier, *Louisiana*; lower right, Terry P. Guidroz; bottom, 1897 *Directory of the Parish of Terrebonne* by E.C. Wurzlow

Page 85: Inset ad, New Orleans *Times Picayune* October 4, 1868

Page 90: Top, middle, and Windermere purchase articles, *The Weekly Thibodaux Sentinel and Journal of the 8th Senatorial District* July 21, 1877, November 26, 1881, and January 12, 1884; Peter Berger photo, Daniel H. Davis; trespass notice, *Houma Courier* September 29, 1900

Page 91: Middle, *Houma Courier* June 16, 1900; bottom *The Louisiana Planter and Sugar Manufacturer*

Page 92: Nicholls State University Archives

Page 93: Top right, *Thibodaux Minerva* April 26, 1856; middle, *Houma Ceres* April 25, 1857; bottom, Nicholls State University Archives

Page 94: Right middle, Dr. Christopher E. Cenac, Sr., Livestock *Brands & Marks*

Page 95: Right, M. Gene Burke, Terrebonne Parish Consolidated Government; bottom, Keneth L. Rembert

Pages 96-97: Estate of George G. Rodrigue

Pages 98-99: Nicholls State University Archives

Page 100: Top, second from bottom and bottom, Nicholls State University Archives; middle, Sheri G. Bergeron

Page 101: Nicholls State University Archives

Page 102: Top, Marion Post Walcott photo June 1940 (Library of Congress); article *The Weekly Thibodaux Sentinel and Journal of the 8th Senatorial District* September 22, 1894

Page 103: Bottom right, Nicholls State University Archives; left, Dr. Christopher E. Cenac, Sr., *Livestock Brands & Marks*

Page 104: Terrebonne Parish Consolidated Government

Page 105: Top, Claire Moreau Mahalick; middle, *The Weekly Thibodaux Sentinel and Journal of the 8th Senatorial District* February 29, 1896; Isle of Cuba notice, *Houma Courier* October 30, 1897

Pages 106-107: Claire Moreau Mahalick donation to Nicholls State University Archives

Page 108: Top and bottom, Claire Moreau Mahalick donation to Nicholls State University Archives

Page 109: Claire Moreau Mahalick donation to Nicholls State University Archives

Pages 110-111: Jeanette F. Schexnayder

Page 112: Top and middle, Jeanette F. Schexnayder; bottom, Lawrence C. Chatagnier *Bayou Catholic* photo in *Schriever Memories* by Gayle B. Cope

Page 113: Jeanette F. Schexnayder

Page 114: Top, Helen Wurzlow, *I Dug Up Houma Terrebonne*; top and bottom articles, *Houma Courier* May 8, 1897 and June 16, 1900

Page 115: Top, Dr. Christopher E. Cenac, Sr., *Livestock Brands & Marks*; Isle of Cuba notice, *The Thibodaux Sentinel* June 11, 1910; store, Col. Joseph Tate Collection, Hill Memorial Library, Louisiana State University Library; invoice, Nicholls State University Archives

Page 116: Bottom right, M. Lee Shaffer, Jr.; map, Keneth L. Rembert

Page 117: Nicholls State University Archives

Page 119: Right center, right middle, right bottom and left top, *Houma Courier* April 11, 1903, October 31, 1896, January 23, 1897, and March 16, 1897; left middle, *Le Sentinelle de Thibodaux* September 2, 1865; left bottom and notice of Evergreen Sale, *The Weekly Thibodaux Sentinel and Journal of the 8th Senatorial District* October 9, 1880 and April 2, 1888

Page 120: Bottom, Nicholls State University Archives

Page 122: Terrebonne Parish Consolidated Government

Page 123: Top, *Thibodaux Minerva* December 23, 1854; middle *Opelousas Patriot* December 8, 1855;

second from bottom and bottom, *New Orleans Republican* December 18, 1868 and March 15, 1868

Pages 124-125: Terrebonne Parish Consolidated Government; right inset, Keneth L. Rembert

Page 126: Left top and middle, *New Orleans Republican* June 4, 1876; bottom left, Dr. Christopher E. Cenac, Sr. Livestock *Brands and Marks*

Page 128: Helen Wurzlow, *I Dug Up Houma Terrebonne*

Page 130: Top, *Houma Courier* November 11, 1962; middle and bottom, *Houma Courier* November 6, 1997

Page 131: *Good Earth Cookbook* 1976

Page 132: Top left, *The Times* (Houma) July 14, 2015; middle, Dr. Christopher E. Cenac, Sr. Livestock *Brands & Marks*; both newpaper items, (ad) *La Sentinelle de Thibodaux* September 9, 1865 and (notice) September 2, 1865

Page 133: Keneth L. Rembert

Page 134: All attributions are in text

Page 135: Laura A. Browning

Page 136: Map, Keneth L. Rembert

Page 137: Article, *The Weekly Thibodaux Sentinel and Journal of the 8th Senatorial District* March 25, 1876; lower left, Dr. Christopher E. Cenac, Sr. Livestock *Brands & Marks*

Page 138: *Houma Courier* October 8, 1972

Page 139: Top right, *Houma Daily Courier* October 8, 1972; ads, left and right, *Houma Courier* November 13, 1897 and January 30, 1897

Page 140-141: Laura A. Browning

Page 142: Bottom, *Houma Courier* March 5, 1978

Page 143: Nicholas death article, *The Feliciana Democrat* January 3, 1857; Nicholas photo, Library of Congress

Page 144: Raceway articles, *National Dragster Magazine*, June 9, 1972

Page 145: Top right and left, Garland A. and Gail H. Trahan; middle and bottom, Dr. Christopher E. Cenac, Sr. *Livestock Brands & Marks*

Page 146: Top, *Houma Courier* October 8, 1972; bottom, *La Sentinelle de Thibodaux* September 9, 1865

Page 147: Top photo, *Houma Courier* October 8, 1972; map, Terrebonne Parish Consolidated Government

Page 148: Article, *Houma Courier* April 11, 1946; bottom right, Dr. Christopher E. Cenac, Sr. *Livestock Brands & Marks*; Office of State Lands, State of Louisiana

Page 149: Terrebonne Parish Clerk of Court Office

Page 150-151: Both pages masthead, *Houma Courier* February 13, 1947 (Terrebonne Parish Public Library); convertible photo, *Houma Courier* October 8, 1972

Page 152: Providence-GSE Terral J. Martin, Jr. PLS

Page 153: Photo, *Houma Courier* February 12, 1972; map, Keneth L. Rembert

Page 154: Brand, Dr. Christopher E. Cenac, Sr., *Livestock Brands & Marks*

Page 155: Upper left, Watkins family photo book by Christian J. LeBlanc; left middle, Houma Courier Magazine Edition 1906; woman in chair, left bottom, and middle bottom, Watkins family photo book by Christian J. LeBlanc; bottom right, William T. and Lynn F. Nolan

Page 156: Center and Montegut photo, Helen Wurzlow, *I Dug Up Houma Terrebonne*; candidacy article, *The Weekly Thibodaux Sentinel and Journal of the 8th Senatorial District*, March 12, 1892; documents, L. Philip Caillouet, Ph.D. and Nicholls State University

Page 157: Photo, Laura A. Browning; documents, L. Philip Caillouet, Ph.D., and Nicholls State University Archives

Page 158: Adam couple photo, Helen Wurzlow, *I Dug Up Houma Terrebonne*

Page 159: Article, Laura A. Browning; letterhead, L. Philip Caillouet, Ph.D. and Nicholls State University Archives; bottom left, Helen Wurzlow, *I Dug Up Houma Terrebonne*

Page 160: Bottom, Terrebonne Parish Clerk of Court Office

Page 161: Bottom, *Houma Courier* March 5, 1978

Pages 162-163: Estate of George G. Rodrigue

Page 164: Barrow Collection Nicholls State University Archives

Page 165: Bottom, Historic New Orleans Collection

Pages 166-167: Both photos, R.R. Barrow Collection, Nicholls State University Archives

Pages 168-169: Background, Nicholls State University Archives

Page 169: Top left and top right, Bazet Collection, Nicholls State University Archives; lower right, R.R. Barrow Collection, Nicholls State University Archives;

Page 170: Top, LSU Postcard Collection, Louisiana State Library; bottom, Standing Structure Survey, State of Louisiana

Page 171: Left top and second at left, *Houma Courier* Magazine Edition 1906; third and bottom left, Helen Wurzlow, *I Dug Up Houma Terrebonne*

Page 172: Background, Helen Wurzlow, *I Dug Up Houma Terrebonne*; tokens, Glyn Farber; document, L. Philip Caillouet Ph.D. and Nicholls State University Archives

Page 173: All photos, Helen Wurzlow, *I Dug Up Houma Terrebonne*

Page 174: William Clifford Smith, PE, PLS, and Keneth L. Rembert

Page 175: Both, Office of State Lands, State of Louisiana

Page 176-177: Cyrus J. Theriot, Jr., Harry J. Bourg Corporation

Pages 178-179: William Clifford Smith, PE, PLS

Page 180: Laura A. Browning

Page 181: Article, Laura A. Browning

Page 182: William Clifford Smith, PE, PLS

Page 183: Top right, *Houma Courier* June 26, 1897; sale ad, *Houma Courier* March 4, 1903; map, Terrebonne Parish Consolidated Government; Daspit, Helen Wurzlow, *I Dug Up Houma Terrebonne*

Pages 184-185: Terrebonne Parish Consolidaed Government

Page 186: Top, Veranese E. Douglas; lower left, *Houma Courier* October 21, 1943 (Terrebonne Parish Public Library)

Page 187: Both, William Clifford Smith, PE, PLS

Page 188: Top corner, left middle, women's photos, William Clifford Smith, PE, PLS; cattle brands, Dr. Christopher E. Cenac, Sr. Livestock *Brands & Marks*; Smithland layout, C. Mildred Smith, *A Southern Neo-Colonial Home*

Page 189: Second photo, Donna McGee Onebane, *The House that Sugarcane Built*

Page 190: Left top, William Clifford Smith, PE, PLS; crest, Donna McGee Onebane, *The House that Sugarcane Built*; letter, Nicholls State University Archives

Page 191: Upper right, Nicholls State University Archives; St. Martin photos, Michael J. St. Martin, Sr., Celeste St. Martin Wedgeworth; Charlean St. Martin Dickson

Page 192: Top, William Clifford Smith, PE, PLS; other photos, Donna McGee Onebane, *The House that Sugarcane Built*; brand, Dr. Christopher E. Cenac, Sr., Livestock *Brands & Marks*

Page 193: Small photos, *Houma Courier* Magazine Edition 1906; bottom photo, William Clifford Smith PE, PLS

Pages 194-195: Estate of George G. Rodrigue

Page 196: Meeting notice, *Houma Courier* October 13, 1917; Connely photo, William A. Ostheimer; Berger photo, Helen Wurzlow, *I Dug Up Houma Terrebonne*

Page 197: William A. Ostheimer

Page 198: All photos William A. Ostheimer

Page 199: Top right, *The Weekly Thibodaux Sentinel and Journal of the 8th Senatorial District* March 5, 1898; brand, Dr. Christopher E. Cenac, Sr. Livestock *Brands & Marks*; Dr. Connely and Fannie Bisland, Martha R. South

Page 200: Brand, Dr. Christopher E. Cenac, Sr., Livestock *Brands & Marks*

Page 201: Bottom left, middle and right, William A. Ostheimer

Page 202: Top, *Houma Courier* July 15, 1854; brand, Dr. Christopher E. Cenac, Sr., Livestock *Brands and Marks*; Ella K. Hooper photo, Clifton P. Theriot, Nicholls State University Archives

Page 203: Top, Clifton P. Theriot, Nicholls State University Archives

Page 204: Bottom, Clifton P. Theriot, Nicholls State University Archives

Page 205: Both, Clifton P. Theriot, Nicholls State University Archives

Pages 206-207: Clifton P. Theriot, Nicholls State University Archives

Page 208: Bottom, Clifton P. Theriot, Nicholls State University Archives

Page 210: Top and Daigle photo, Helen Wurzlow, *I Dug Up Houma Terrebonne*; map, Office of State Lands, State of Louisiana

Pages 214-215: Cyrus J. Theriot, Harry J. Bourg Corporation

Page 216: Photo, W.J. "Doc" Gaidry III; map, William Clifford Smith, PE, PLS

Page 217: New Zion, Carlton D. Stokes; Army base chapel, *Houma Courier* March 2, 2016; Bourg and Mahler photos, Cyrus J. Theriot, Harry J. Bourg Corporation; Engeron photo, Helen Wurzlow, *I Dup Up Houma Terrebonne*

Pages 218-219: Estate of George G. Rodrigue

Page 220: Map, Office of State Lands, State of Louisiana

Page 221: Bowie, J. Louis Gibbens, Sr.; plantation house, R.R. Barrow Collection, Nicholls State University Archives; R.R. Barrow, Sr. photo, Barrow Family book

Page 224: Hunley schematics, Laura A. Browning

Page 225: Bottom, R.R. Barrow Collection Nichols State University Archives

Pages 226-227: Nicholls State University Archives

Page 228: Second row left, Thomas Blum Cobb, deceased; second row right and Presbyterian Church, Bazet Collection, Nicholls State University Archives; St. Patrick, Dr. Dick C. Walther, DVM, deceased; bottom left, St. Matthew's Church Collection Nicholls State University Archives

Page 229: Top left, Carlton D. Stokes; bottom left, Clifton P. Theriot

Page 230: Article, *Houma Courier, July 24, 1897*

Page 232: Milk container, Peter W. Fonseca; cane wagon, *Houma Courier*, October 8, 1972

Page 233: Gaidry photos, Helen Wurzlow, *I Dug Up Houma Terrebonne*; right top and middle, Thomas Blum Cobb, deceased; milk bottle, Peter W. Fonseca; brands, Dr. Christopher E. Cenac, Sr. Livestock *Brands & Marks*

Page 234: Map, M. Gene Burke, Terrebonne Parish Consolidated Government: Bellview, Linwood P. Whitney

Page 235: Center, article, *Houma Courier* July 27, 1944; bottom, M. Gene Burke, Terrebonne Parish Consolidated Government

Page 236: Top, first two, Angela F. Trahan; next three, Nicholls State University Archives

Page 237: All photos and wedding invitation R. R. Barrow Collection, Nicholls State University Archives; diploma, Angela F. Trahan

Pages 238-239: William Clifford Smith, PE, PLS

Page 240: All photos, R.R. Barrow Collection, Nicholls State University Archives; Garden Society, Rachel E. Cherry

Page 241: Aerial photo, R. R. Barrow Collection Nicholls State University Archives; fire photo, bed photo, Boehm birds, Rachel E. Cherry; birth certificate, Veranese E. Douglas

Pages 242-243: Regional Military Museum, Houma, and Luther F. Bragg, Jr.

Page 244: Article, *Houma Courier* April 26, 1945; photo Arabs, J. Louis Gibbens, Sr.

Page 245: All photos except Port Texaco, Regional Military Museum, Houma; Port Texaco, Ovide J. "Jock" Cenac Collection

Page 248: Regional Military Museum, Houma, and Luther F. Bragg, Jr.

Page 249: All photos, Regional Military Museum, Houma, and Luther F. Bragg, Jr.

Page 251: Regional Military Museum, Houma, and Luther F. Bragg, Jr.

Page 252: Background, Regional Military Museum, Houma; first article, *Terrebonne Press*, April 28, 1944; second article, *Houma Courier*, April 27, 1944; right center, C.J. Christ, *Houma Courier*; bottom, Regional Military Museum and Luther F. Bragg, Jr.

Page 253: Top article, Associated Press April 28, 1944, Regional Military Museum

Pages 254-255: All photos, Regional Military Museum, Houma, and Luther F. Bragg, Jr.

Page 256: Top, *Houma Courier* August 17, 1944; Grand Isle photo, *Houma Courier* July 15, 1944; Mess hall, Regional Military Museum and Luther F. Bragg, Jr.; far left center, *Houma Courier* July 14, 1944

Page 257: Top, *Houma Courier*, February 18, 1949; second and bottom, Regional Military Museum, Houma

Page 261: *Times Picayune* article, Laura A. Browning

Page 262: Lehmann painting, Daniel H. Davis; first two Weber paintings, Regional Military Museum, Houma; bottom painting, Stanley E. Yancey

Page 263: Providence-GSE Terral J. Martin, Jr., PLS

Page 264: Bottom, Regional Military Museum Houma, La.

Page 265: *Times Picayune* article, Laura A. Browning

Page 266: Top, Nicholls State University Archives; bottom, Litt Martin Collection,

PHOTO CREDITS | **495**

Nicholls State University Archives

Page 267: Top, Litt Martin Collection, Nicholls State University Archives; bottom, Nicholls State University Archives

Page 268: Nicholls State University Archives

Pages 270-271: Wallace R. Ellender III and Coralie H. Ellender

Page 272: All photos, Regional Military Museum, Houma, La.

Page 273: *Houma Courier* June 20, 1946

Page 274: Top, Dolly Domangue Duplantis; inset, *Houma Courier* April 10, 1926

Page 275: Top, Brian W. Larose; middle, Dolly Domangue Duplantis; bottom, R.R. Barrow Collection, Nicholls State University Archives

Page 276: Nicholls State University Archives

Page 277: All from Nicholls State University Archives

Page 278: Right top, R.R. Barrow Collection, Nicholls State University Archives; second right, Laura A. Browning; third right, R.R. Barrow Collection, Nicholls State University Archives; left top, *Houma Courier* February 5, 1887; second left, *Houma Courier* August 21, 1897; trespass notice, *Houma Courier* March 9, 1912; job injury, *Houma Courier* December 4, 1897; harvesting, *Houma Courier* November 14, 1903

Page 279: Top, Helen Wurzlow, *I Dug Up Houma Terrebonne*; all other photos the Rev. Michael Bergeron

Page 280: Both photos, Providence-GSE Terral J. Martin, Jr., PLS

Page 281: Top trespass notice, *Houma Courier* September 8, 1900; second notice, *Houma Courier* September 17, 1903; third notice, *Houma Courier* October 17, 1917

Page 282: Top, Thomas Blum Cobb, deceased; Viguerie photo, Helen Wurzlow, *I Dug Up Houma Terrebonne*; brand, Dr. Christopher E. Cenac, Sr., Livestock *Brands & Marks*; *Times Picayune* December 1927

Page 283: Right top, Gwen A. Pitre; left bottom, W.E. Butler, *Down Among the Sugar Cane*; one cent and 50 cents tokens, Janelle M. Moen

Page 284: Top, Office of State Lands, State of Louisiana; bottom photos, *Houma Courier* Magazine Edition 1906

Page 285: Right top, Judy A. Soniat, Terrebonne Parish Public Library; second down right, Dr. Herman E. Walker, Jr., and Lori R. Walker; next right, Richard Arcement

Page 286: Left, Dr. Christopher E. Cenac, Sr., Livestock *Brand & Marks*; others, Office of State Lands, State of Louisiana

Page 288: Office of State Lands, State of Louisiana

Page 290: Office of State Lands, State of Louisiana

Pages 292-293: Estate of George G. Rodrigue

Page 294: Top, Providence-GSE, Terral J. Martin, Jr., PLS

Page 295: Both photos, R.R. Barrow Collection, Nicholls State University Archives

Page 296: Terrebonne Parish Clerk of Court Office

Page 297: Top, Col. Joseph Tate Collection, Special Collections, Louisiana State University Library; right top, Donald W. Davis and Dr. Carl A. Brasseaux, Ph.D.

Page 298: Left top, *Houma Courier* April 11, 1946; trespass notice, *Houma Courier* March 14, 1903; railroad article, *New Orleans Crescent*, May 28, 1868

Page 299: Article, *Times Picayune*, 1898; photo, Helen Wurzlow *I Dug Up Houma Terrebonne*

Page 300: Top, Dr. Christopher E. Cenac, Sr., Livestock *Brands & Marks*; photo, Arlen B. Cenac, Sr. and Jacqueline Guidry Cenac

Page 301: Guidry photo, Arlen B. Cenac, Sr. and Jacqueline Guidry Cenac; brand, Dr. Christopher E. Cenac, Sr., Livestock *Brands & Marks*; St. Ann, Clara A. Redmond; all others, St. Ann Church 50th Anniversay (1908-1958) publication

Page 302: All photos, Laura A. Browning

Page 303: Top, St. Ann Church 50th Anniversary (1908-1958) publication; middle and bottom, Farmand J. Matherne, Jr.

Page 304: Top and center, Farmand J. Matherne, Jr.

Page 306: Photos from *Lures & Legends* by Brian Cheramie

Page 307: All photos from *Lures & Legends* by Brian Cheramie

Pages 308-309: All photos, Claude J. Bourg

Page 310: Ring, Dr. Craig M. Walker; middle Laura A. Browning; bottom, *The Rice Belt Journal* January 20, 1911

Page 311: Top, *Houma Courier* June 26, 1897; bottom, Wiliam T. and Lynn F. Nolan

Pages 312-313: Estate of George G. Rodrigue

Page 314: Brand, Dr. Christopher E. Cenac, Sr. Livestock *Brands & Marks*; photo, *Duplantis Legacy* by Lois Hebert Luke; map, Office of State Lands, State of Louisiana

Page 315: Top, Helen Wurzlow, *I Dug Up Houma Terrebonne*; bottom, *Terrebonne Magazine* December 1985, Douglas P. Patterson; middle right, Linda Caro

Page 316: Richard J. Bourgeois and Angela M. Cheramie

Page 317: Left and right, Richard J. Bourgeois and Angela M. Cheramie; bottom, Terrebonne Parish Consolidated Government

Page 318: Top left, Rudy R. Aucoin, deceased; top right and left center, Richard J. Bourgeois and Angela M. Cheramie; bottom left and right, Dr. Christopher E. Cenac, Sr., *LivestockBrands & Marks*

Page 319: Top right, Bayou History Center; below, The Historic New Orleans Collection

Page 320: Top, Providence-GSE Terral J. Martin, Jr., PLS; left center, *Houma Courier* October 8, 1972

Page 321: Top, Standing Site Survey, State of Louisiana 1981

Page 322: Top, Providence-GSE Terral J. Martin, Jr., PLS; left top, Office of State Lands, State of Louisiana; newspaper notice, *Houma Courier* November 7, 1903; Champagne, *The Historical Encyclopedia of Louisiana* Vol. II

Page 323: Right ad, *Houma Courier* October 24, 1976; map, Leryes J. Usie, *Terrebonne Parish, Louisiana: Our Bayou Parish*

Page 324: Mississippi Department of Archives and History

Page 325: Top, Dolly Domangue Duplantis; right center, *Houma Ceres* December 13, 1856

Page 326: Top, *Houma Courier* February 5, 1887; all brands, Dr. Christopher E. Cenac, Sr., Livestock *Brands & Marks*; Ellender children, Albert P. Ellender, DDS; oyster lugger photo, Patty A.Whitney, Bayou History Center

Page 327: Top left, Grant J. Dupre; brands, Dr. Christopher E. Cenac, Sr., Livestock *Brands & Marks*

Page 328: Lower right, Laura A. Browning

Page 329: Photo Claude J. Bourg; inset, *Houma Courier* August 5, 1945

Pages 330-331: Photo, Claude J. Bourg; Inset *Houma Courier* August 3, 1944

Page 331: Article, *Terrebonne Press* August 4, 1944

Page 332: All photos Claude J. Bourg

Page 333: Top, Claude J. Bourg; families photo, Laura A. Browning

Page 334: Both, Albert P. Naquin, Traditional Chief of the Isle de Jean Charles Band of Biloxi-Chitimacha-Choctaw

The Class of Marie Courrege, 1972, Oil on canvas, 36 x 24 inches, George G. Rodrigue

PHOTO CREDITS | **497**

Page 335: Left top, Laura A. Browning; center and next photo, the Rev. Roch Raphael Naquin; map, Office of State Lands, State of Louisiana

Page 336: All photos excluding center basket photo, Albert P. Naquin, Traditional Chief of the Isle de Jean Charles Band of Biloxi-Chitimacha-Choctaw

Page 337: All photos Albert P. Naquin, Traditional Chief of the Isle de Jean Charles Band of Biloxi-Chitimacha-Choctaw

Page 338: Newspaper photos, *Times Picayune* January 4, 1970, contributed by Russell W. Talbot

Page 339: Center photo of school, Helen Wurzlow, *I Dug Up Houma Terrebonne*; all other photos, Albert P. Naquin, Traditional Chief of the Isle de Jean Charles Band of Biloxi-Chitimacha-Choctaw

Page 340: All photos Albert P. Naquin, Traditional Chief of the Isle de Jean Charles Band of Biloxi-Chitimacha-Choctaw

Page 341: Photo and patch, Albert P. Naquin, Traditional Chief of the Isle de Jean Charles Band of Biloxi-Chitimacha-Choctaw

Page 342: Both photos Nicholls State University Archives

Page 343: Donald W. Davis

Page 344: All photos Nicholls State Universtiy Archives

Page 345: Nicholls State University Archives

Page 346-347: Estate of George G. Rodrigue

Page 348: George J. Jaubert

Page 349: Wallace R. Ellender III and Coralie H. Ellender

Page 350: Photos George J. Jaubert

Page 351: Deroche house, Standing Site Survey, State of Louisiana 1981

Page 352: Map, M. Gene Burke, Terrebonne Parish Consolidated Government; photos, Martha R. South

Page 353: Top, Bayou History Center; right center, Hill Memorial Library Special Collections, Louisiana State University Library; bottom right, Helen Wurzlow, *I Dug Up Houma Terrebonne*

Page 356: Left top, Col. Joseph Tate Collection, Hill Memorial Library, Louisiana State University Library; brand, Dr. Christopher E. Cenac, Sr., Livestock *Brands & Marks*; Bislands photo, Martha R. South; LeBlanc photo, *Houma Courier* Magazine Edition 1906; trespass notice, *Houma Courier* April 10, 1926

Page 357: Women photo, Patty A. Whitney, Bayou History Center; Bisland mill, Bayou History Center and Mrs. L.R. McGehee of Natchez, MS

Page 360-361: Aerial photo, Providence-GSE Terral J. Martin, Jr., PLS

Page 362: Left top, *The House that Sugarcane Built* by Donna McGee Onebane; left and right center, Helen Wurzlow, *I Dug Up Houma Terrebonne*

Page 363: Laura A. Browning

Page 364: Top left, R.R. Barrow Collection, Nicholls State University Archives; bottom left, *The New Orleans Bulletin* July 4, 1875

Page 365: Map, Grant J. Dupre

Page 366: Persac painting, The Historic New Orleans Collection; death notice, *Houma Courier* May 30, 1896

Page 367: Left, top ad, *Houma Courier* October 16, 1897; lower ad, *Houma Courier* July 17, 1897: oranges ad, *Houma Courier* November 29, 1913; top right, *Houma Courier* May21, 1892: tax sale, *Houma Courier* February 20, 1892: Viguerie photos, Helen Wurzlow, *I Dug Up Houma Terrebonne*; Mass article, *Houma Courier* January 23, 1897

Page 368: Top right, Helen Wurzlow, *I Dug Up Houma Terrebonne*; top left, Bazet Collection, Nicholls State University Archives

Page 369: Houses at right, Standing Site Survey, State of Louisiana 1981

Page 370: R. R. Barrow Collection Nicholls State UniversityArchives

Page 371: Top, Laura A. Browning; lower, Marshall Dunham Collection, Louisiana State University Library

Page 372: Top left, C. Mildred Smith, *A Southern Neo-Colonial Home*; second from left and second from right, *Houma Courier* Magazine Edition 1906; top right, Helen Wurzlow, *I Dug Up Houma Terrebonne*

Page 373: Right top and middle, R.R. Barrow Collection, Nicholls State University Archives; bottom left, Russell W. Talbot; bottom right, W.E. Butler, *Down Among the Sugar Cane*

Pages 374-375: Providence-GSE Terral J. Martin, Jr., PLS

Page 375: Inset, A.B. Gilmore, *The Louisiana Sugar Manual*, Nicholls State University Archives

Pages 376-377: The Historic New Orleans Collection

Page 378: Top left, W.E. Butler, *Down Among the Sugar Cane*; Chauvin photo, Russell W. Talbot; article, Laura A. Browning; shipping statement, Nicholls State University Archives; Viguerie photo, *Houma Courier* October 8, 1972; Sanders photo, Bazet Collection, Nicholls State University Archives; Guidry photo, Jacqueline G. Cenac; Ellender photo, Wallace R. Ellender III and Coralie H. Ellender; LeBlanc photo, *Houma Courier* Magazine Edition 1906

Page 379: Bottom, State Library of Louisiana

Pages 380-381: The Historic New Orleans Collection

Page 382: Janelle M. Moen

Page 383: Top right, Nicholls State University Archives; harvester photo and inset, Helen Wurzlow, *I Dug Up Houma Terrebonne*

Page 385: Right, Russell W. Talbot; bottom, Nicholls State University Archives

Page 386: Keneth L. Rembert

Page 387: Center left, middle right, middle left, and invoice, Sheri G. Bergeron; article, Laura A. Browning; right top, Helen Wurzlow, *I Dug Up Houma Terrebonne*

Page 388: Map, M. Gene Burke, Terrebonne Parish Consolidated Government; bottom left, *Houma Courier* June 11, 1910

Page 389: Grant J. Dupre and Providence-GSE Terral J. Martin Jr., PLS

Page 391: All photos Laura A. Browning

Page 392: Top right (ads) Sheri G. Bergeron; concert photo, Laura A. Browning

Page 393: Top left, bottom left, and Stoufflet photos, Laura A. Browning; Guidry's car photo, Sheri G. Bergeron; Redmond photos, Clara A. Redmond

Page 394: C. Mildred Smith, *A Southern Neo-Colonial Home*

Page 395: Right top, *Houma Courier* August 7, 1897; right center, Laura A. Browning; left and far left, C. Mildred Smith, *A Southern Neo-Colonial Home*

Page 396: Office of State Lands, State of Louisiana

Page 397: Top, C. Mildred Smith, *A Southern Neo-Colonial Home*; trespass notice, *Houma Courier* October 17, 1917; bottom, Bazet Collection, Nicholls State University Archives

Page 398: Both, Bazet Collection Nicholls State University Archives

Page 399: Lottie Gazzo photo, Bazet Collection Nicholls State University Archives; brand, Dr. Christopher E. Cenac, Sr., Livestock *Brands & Marks*; drilling notice, *Houma Courier* April 11, 1903

Pages 400-401: Office of State Lands, State of Louisiana; inset, Laura A. Browning

Page 401: Sheriff's Sale item, *Houma Courier* March 4, 1903

Page 403: Office of State Lands, State of Louisiana

Page 405: Bazet Collection, Nicholls State University Archives

Page 406: Tomb medallion, Laura A. Browning

Page 407: Bazet Collection Nicholls State University Archives

Page 408: Top, Sherwin Jonas Guidry, *Houma Courier* 1976; bottom left, Office of State Lands, State of Louisiana

Page 409: Top, Office of State Lands, State of Louisiana; middle right, Sherwin J. Guidry, *Houma Courier* c. 1976; brand, Dr. Christopher E. Cenac, Sr., Livestock *Brands & Marks*; left, Wallace R. Ellender III and Coralie H. Ellender

Page 410: Top left, William Clifford Smith, PE, PLS; bottom left, Dr. Craig M. Walker and Tina Hebert Walker; bottom right, Wallace R. Ellender III and Coralie Ellender

Page 411: Top left and bottom left, Allen J. Ellender Collection, Nicholls State University Archives; right center and right bottom, Dr. Albert J. Ellender, DDS

Page 412: Top left, Wallace R. Ellender III and Coralie H. Ellender; center left, Allen J. Ellender Collection Nicholls State University Archives; President Nixon photo, Thomas Blum Cobb, deceased

Page 413: Top, Wallace R. Ellender III and Coralie H. Ellender; brands, Dr. Christopher E. Cenac, Sr., Livestock *Brands & Marks*; background map, Laura A. Browning

Page 416: Brand, Dr. Christopher E. Cenac, Sr., Livestock *Brands & Marks*

Page 417: Keneth L. Rembert

Page 418: Keneth L. Rembert

Page 419: Right top and middle, Laura A. Browning; article, *Houma Courier* September 14, 1901 from Russell W. Talbot

Page 420: Newspaper short items, top two, *Houma Courier* April 30, 1892 and third from *The Thibodaux Sentinel* August 27, 1910; below, Russell W. Talbot

Page 421: Right, *Houma Courier* March 4, 1903; brands, Dr. Christopher E. Cenac, Sr., Livestock *Brands & Marks*; bottom, L. Philip Caillouet, Ph.D., FHIMSS

Page 422: A. Lirette photo, Helen Wurzlow, *I Dug Up Houma Terrebonne*; all others this page Laura A. Browning

Page 423: Article on right, *The Daily Picayune* May 1918

Page 425: Estate of George G. Rodrigue

Page 426: Top, Laura A. Browning

Page 428: All photos Laura A. browning

Page 429: Laura A. Browning

Page 431: Top right, Laura A. Browning; others from T.I. St. Martin Collection, Nicholls State University

Page 433: All documents and photos Sheri G. Bergeron

Page 434: Keneth L. Rembert

Page 435: Daigle photo, Helen Wurzlow, *I Dug Up Houma Terrebonne*; *Dixie* article, *Houma Courier* November 13, 1915; bottom left, F.G. Larose, Jr. Wright family photos

Page 436: Top, R.R. Barrow Collection, Nicholls State University Archives

Page 438: Top left, Standing Structure Survey 1981, State of Louisiana; top middle, Donald W. Davis; top right and bottom left, Laura A. Browning; brand, Dr. Christopher E. Cenac, Sr., *Livestock Brands & Marks*; bottom right, Standing Structure Survey 1981, State of Louisiana

Page 439: Map, Keneth L. Rembert

Page 441: Top, third from left, Laura A. Browning; second and third from right**,** Helen Wurzlow, *I Dug Up Houma Terrebonne*; brand, Dr. Christopher E. Cenac, Sr., *Livestock Brands & Marks*; trespass notice *Houma Courier* June 17, 1916; bottom left, Bazet Collection, Nicholls State University Archives

Page 442: Top, Helen Wurzlow, *I Dug Up Houma Terrebonne*

Page 443: Helen Wurzlow, *I Dug Up Houma Terrebonne*

Page 445: Top, Keneth L. Rembert; top right, Sherwin J. Guidry, *Houma Courier* May 19, 1977; bottom right, Laura A. Browning

Page 446: Keneth L. Rembert

Page 447: Aerial photo, Providence-GSE Terral J. Martin, Jr., PLS

Pages 448-449: Laura A. Browning

Pages 450-451: Laura A. Browning

Pages 452-453: Keneth L. Rembert

Page 454: Bazet Collection, Nicholls State University Archives

Page 455: Article and 1924 photo, Laura A. Browning; oyster house and camps and luggers, R.R. Barrow Collection, Nicholls State University Archives; Sanders camp 1918, boat races, and Sanders camp 1926, Bazet Collection, Nicholls State University Archives

Pages 456-457: Donald W. Davis

Page 458: Both photos, Michael T. Pellegrin

Page 459: Both photos, Michael T. Pellegrin

Pages 460-461: Brent J. Cenac and Christopher E. Cenac, Jr., M.D.; inset p. 460, Jason P. Theriot, Ph.D.

Page 466-467: Painting by Garth K. Swanson 2016

Page 425: Estate of George G. Rodrigue

Page 492: Estate of George G. Rodrigue

Page 497: Estate of George G. Rodrigue

Page 500: Estate of George G. Rodrigue

Doctor on the Bayou, *1982, Oil on canvas, 30 x 40 inches, George G. Rodrigue*

Just as Louisiana's unique history often chronicles examples of "hardscrabble" to "hallelujah," history sleuth Dr. Chris Cenac has intricately compiled a variety of narrative and pictorial representations to portray a truly rich historical evolution of Terrebonne Parish, Volume 1: Bayou Terrebonne—and this is only the first of four fact-filled volumes! A comprehensive resource like no other, this treasure will undoubtedly become the go-to Bible of the residents and friends of Terrebonne, as well as south Louisiana and beyond. The all-inclusive index alone will enrich the history enthusiast and educate the fortunate reader, making this work a collector's choice.

    Florent Hardy, Jr., Ph.D.
    Director of Archival Services
    Louisiana State Archives

Terrebonne Parish's preeminent historian, Dr. Chris Cenac, once again transports us back in time to discover a heretofore undocumented segment of local history. In this first volume of his new *Legacies* series, Cenac chronicles the pioneering families that settled along the banks of Bayou Terrebonne and illustrates the historical footprint of the great sugar plantations that once dotted the landscape and dominated the local economy. This book could not be timelier, as many of the once prosperous plantations have vanished into the past, taking with it another slice of the region's unique bayou culture.

    Jason P. Theriot, Ph.D.
    Jason P. Theriot Consulting, LLC
    American Energy, Imperiled Coast: Oil and Gas Development in Louisiana's Wetlands (LSU Press)

Mention Louisiana plantations and the mind immediately gravitates to the Mississippi River corridor with grand mansions shaded by rows of live oak trees. Nothing could be further from the truth; most plantation homes were modest and utilitarian. From *Hard Scrabble to Hallelujah – Legacies of Terrebonne Parish, Louisiana*, Dr. Cenac gives the reader an in-depth look at the history and people of bayou plantations within Terrebonne Parish. Stretching for 50 miles, Bayou Terrebonne was lined on both sides by plantations, towns and churches. The homes were only occasionally a statement of grandeur and prestige, but they were the exception. This engrossing book tells the rich history of these plantations, towns and religious institutions in which the people come alive on the page. Dr. Cenac has achieved a book that is both fascinating in its detail of plantation life and occupants, as well as making this reader wish for a time machine.

    Robert S. Brantley
    Author of *Henry Howard: Louisiana Architect*
    Princeton Architectural Press and The Historic New Orleans Collection
    Winner of The Henry-Russell Hitchcock Award, Victorian Society in America 2016

In *Hard Scrabble to Hallelujah – Legacies of Terrebonne Parish, Louisiana Vol I*, Dr. Chris Cenac has conceived a dynamic new model for the telling and teaching of local history. By elegantly weaving legal documents, stories and hundreds of images from Terrebonne Parish's past, he equally honors both people and place. Organized by family and geography, this will be a necessary reference for anyone studying Louisiana's social, agricultural, commercial, and architectural history. At the same time, his accessible style and presentation will serve as fine entertainment to the hobbyist—enter it at any chapter and you will not be able to put it down. In addition, each page is visually striking—composed like a panel extracted from a museum—and the reader anticipates the delights of the next section. This is exciting history–alive and well in Louisiana.

    Gregory Free, Architectural Historian, Austin, Texas

This first volume of Dr. Cenac's *Legacies of Terrebonne Parish* series provides readers with a well-documented and thoroughly researched volume on the two-hundred-year history of the parish along its namesake waterway. In this work, he has compiled photographs, maps, oral histories and other historical information, much of which has never been published, from numerous sources into a beautifully illustrated and encyclopedic resource. This work, in addition to Dr. Cenac's two previous books, helps to document parish history that is fast disappearing as the older generations who had first-hand knowledge of that history pass away. Readers will find detailed history of well-known plantations and communities, as well as many that are unknown, some whose names now exist only in street names or subdivisions. Scholars as well as current and future Terrebonneans will appreciate and cherish having a copy of this work in their personal library.

    Clifton P. Theriot, CA
    Archivist, Nicholls State University